Sound and Vibrations of Positive Displacement Compressors

Sound and Vibrations of Positive Displacement Compressors

Werner Soedel

CRC Press
Taylor & Francis Group
Boca Raton   London   New York

CRC Press is an imprint of the
Taylor & Francis Group, an informa business

CRC Press
Taylor & Francis Group
6000 Broken Sound Parkway NW, Suite 300
Boca Raton, FL 33487-2742

© 2007 by Taylor & Francis Group, LLC
CRC Press is an imprint of Taylor & Francis Group, an Informa business

No claim to original U.S. Government works
Printed in the United States of America on acid-free paper
10 9 8 7 6 5 4 3 2 1

International Standard Book Number-10: 0-8493-7049-3 (Hardcover)
International Standard Book Number-13: 978-0-8493-7049-6 (Hardcover)

This book contains information obtained from authentic and highly regarded sources. Reprinted material is quoted with permission, and sources are indicated. A wide variety of references are listed. Reasonable efforts have been made to publish reliable data and information, but the author and the publisher cannot assume responsibility for the validity of all materials or for the consequences of their use.

No part of this book may be reprinted, reproduced, transmitted, or utilized in any form by any electronic, mechanical, or other means, now known or hereafter invented, including photocopying, microfilming, and recording, or in any information storage or retrieval system, without written permission from the publishers.

For permission to photocopy or use material electronically from this work, please access www.copyright.com (http://www.copyright.com/) or contact the Copyright Clearance Center, Inc. (CCC) 222 Rosewood Drive, Danvers, MA 01923, 978-750-8400. CCC is a not-for-profit organization that provides licenses and registration for a variety of users. For organizations that have been granted a photocopy license by the CCC, a separate system of payment has been arranged.

Trademark Notice: Product or corporate names may be trademarks or registered trademarks, and are used only for identification and explanation without intent to infringe.

Library of Congress Cataloging-in-Publication Data

Soedel, Werner, 1936-
　Sound and vibrations of positive displacement compressors / by Werner Soedel.
　　　p. cm.
　Includes bibliographical references and index.
　ISBN 0-8493-7049-3 (alk. paper)
　1. Compressors. I. Title.

TJ990.S59 2006
621.5'1--dc22   2006044358

Visit the Taylor & Francis Web site at
http://www.taylorandfrancis.com

and the CRC Press Web site at
http://www.crcpress.com

Dedication

To my grandchildren: Amber, Nicholas, Ashleigh, April, Jackson, Broderic, Thomas, Carter, Kathleen, Ava, Chase, and the yet unborn

Preface

This book is based on my experiences as a researcher of sound and vibration problems in compressors, and on my experiences with developing and offering short courses in this area. I was also fortunate to have been invited over the years to serve as a consultant, not only by the refrigeration and air conditioning compressor industry, but also by air and gas compressor companies. I hope that I have contributed a few, practically realizable solutions. In the course of my work I have learned a lot from the practitioners of the art with respect to making compressors behave acceptably from a sound and vibration viewpoint.

My contributions as a consultant often consisted of interpretations of sound and vibration behavior measured in the laboratory, based on my theoretical knowledge, and of guiding the experimenters toward possible sound and vibration reduction solutions, again from a theoretical viewpoint. I have typically found, in most industrial compressor companies, highly developed sound and vibration measurement capabilities. My most effective efforts involved guiding the development engineer toward certain promising design changes based on mathematical models. Therefore, in this book, a number of relatively simple, or not so simple, mathematical models are presented, which are designed to help our understanding of the sound and vibration behavior of positive displacement compressors. The goal is to present a theoretical foundation for some of the noise and vibration phenomena wherever possible, and to extend the knowledge base in this area. Sound and vibration control measures will also be discussed.

An introductory knowledge of acoustics and vibrations is desirable when reading this book. Still, it is hoped that this publication will be suitable for study without the benefit of explanations given by a lecturer.

The choice of the way in which topics in sound and vibration of compressors are approached in this book is somewhat subjective, and someone else may very well prefer a different sequence or approach. It was decided not to simply present structural and acoustic finite-element methods. Rather, it was felt that the ideal approach would be to present relatively simple analytical approaches that would result in useful *design* information, preferably by easy-to-use formulas. This goal could not always be accomplished because of the difficulty of the subject, but what I hoped to accomplish was at least a compromise. Perhaps the theory presented is at times too opaque, but the end results illuminate certain ideas that must be kept in mind when controlling sound and vibration in compressors.

The book is somewhat slanted toward hermetically sealed refrigeration and air conditioning or heat pump compressors because of the large amount of research that has been done by this industry. This slant is reflected in the chapter on the compressor housing (the hermetic compressor shell) because air and nontoxic gas compressors are, as a rule, not hermetically sealed. But many topics, such as muffler design or the influence of valve impact on compressor casing vibrations, is of equal interest to air and gas compressor specialists.

Obviously, different types of compressors exhibit different sound and vibration effects, but there is enough commonality to justify a somewhat general approach to the book: most refrigerating and air conditioning compressors have a hermetically sealed housing, are mounted on isolation springs, produce gas pulsations, employ valves, experience casing vibrations, and so on, and similar sound- and vibration-producing features are also found in air and gas compressors. The introductory chapters do discuss certain differences between reciprocating, rotary vane, rolling piston, scroll, and screw compressors. This does not mean that there are not other designs deserving equal attention, but a somewhat subjective choice had to be made to limit the scope of the book. The choice was also influenced by my past research and practical experiences. Still, many of the mechanisms that are present in the compressor types discussed are also present in other types of compressors.

From the organizational viewpoint of this book, it would have made some sense to start with a chapter on noise radiation because it would immediately explain why so much importance is attached to vibration reduction when conducting noise control, but to understand sound radiation as presented requires an understanding of the way the compressor housing vibrates. Also, because the three-dimensional acoustic wave equations will be shown to be derivable from the equations of motion of three-dimensional elastic bodies (with application to the compressor casing), it was decided to discuss vibration mechanisms first.

The vibration problems are also important if one considers the classical noise control concept of source–path–receiver. The sources of sound (noise) in compressors, excluding the motor or the engine are, for example, the periodic compression process, valve vibrations, intake and exhaust gas pulsations, and kinematic shaking forces. The periodic compression, and valve vibrations and impacts, cause the compressor casing to vibrate and either radiate sound directly in case of air, gas or large refrigeration compressors, or to transmit the vibrations through support springs to a hermetic shell and/or the foundation. This transmission is the so-called mechanical path. The periodic intake and exhaust causes gas pulsations, which on the intake side in air compressors may radiate sound directly into the atmosphere, or in air conditioning compressors, excite the gas inside the hermetic shell into pulsations which in turn cause the hermetic shell to vibrate and radiate noise. In the process, gas pulsations in suction or discharge tubes cause these tubes to vibrate, which is then transmitted to the hermetic shell by the mechanical path, so the noise radiation is always the last item before the receiver.

Considering all of the above, it was decided to first present (in the introduction) an overview of the vibration- and sound-producing mechanisms many compressors have in common. Following this, a few commonly used compressor types are discussed.

The next four chapters deal with the free and forced vibrations of hermetically sealed compressor housing shells. The reason for this attention is that the housing shell is the ultimate sound radiator in refrigeration and air conditioning compressors. All sound- and vibration-producing mechanisms finally feed into the housing shell. Its vibration is, therefore, of utmost importance. Because shell mechanics are relatively complicated, shell vibration theory is reviewed in summary fashion, referring the reader to a text by the author on shell and plate vibrations (Soedel, 2004). It was felt that some of the effects observed by practicing compressor engineers can only be explained in this way. Such effects are mode splitting for nearly axisymmetric housing shells, stiffening to raise the natural frequencies, vibration localization involving the end caps, response to periodic forcing, dynamic absorbers, and so on.

The following two chapters present information about the vibrations of compressor casings, excited by impact or periodic forcing, and the vibration of other typical structural components such as suction and discharge tubes, and as part of a discussion of vibration isolation, surging in coil springs.

A large chapter is devoted to the design and vibrations of automatic plate or reed valves. The dynamic behavior of valves not only influences the thermodynamic efficiency of compressors, but also the sound and vibration output, and the life expectancy because valves are typically the parts that are most likely to fail. Therefore, typical procedures for valve design are presented and valve dynamics issues are discussed at length.

An equally extensive chapter is devoted to the analysis and design of suction and discharge mufflers, which serve to suppress gas pulsations. This includes Helmholtz approximations as well as one-dimensional acoustic wave theory.

Three-dimensional acoustics are applied to the gas resonance prediction in the interior of the housing shell. Simplified models of the sound radiation from the shell are given in order to underline the importance of minimizing the surface vibrations of the housing for the purpose of sound reduction. For example, the typical vibration amplitude reductions necessary to show noticeable improvement in perceived sound are discussed.

The book is not intended to be a measurement manual. Specialty books on this subject should be consulted. Still, it was deemed important to briefly discuss issues like repeatability of sound and vibration measurements, and certain measurements that allow identification of the dominant sound and vibration mechanisms.

The book is also not intended to present the sound and vibration behavior of power sources, such as electromotors or diesel engines, in case of certain air and gas compressors. Still, because in fractional-horsepower refrigeration and air conditioning compressors, electromotors are an integral part of the designs, surrounded by the same hermetic housing, a brief discussion of electromotors is included in the last chapter. The sound and vibration influence of the lubrication oil, which is typically collected at the bottom of housing shells, is also mentioned in the last chapter.

It should also be noted that this book is not intended to be a review of all the sound and vibration research that was ever published on compressors. References are, in general, given only when necessary. My work and the work of my students and colleagues is usually referenced, but not only because I am most familiar with it, but also because the Ray W. Herrick Laboratories, with whom I have been happily associated since 1963, have been and still are an international center for compressor research.

Every two years, since 1972, the Herrick Laboratories sponsor and organize an International Compressor Engineering Conference at Purdue University (from 1972 until 1984, the name was the Purdue Compressor Technology Conference). Typically, several sessions of each conference were devoted to papers on the sound and vibration of compressors. I have also drawn on this body of knowledge in writing this book. A large number of papers from these conferences are referenced, but the interested readers should also consult the proceedings of these conferences directly for additional information. They can be ordered from the Ray W. Herrick Laboratories directly.

Particular thanks must be extended to the Herrick family (especially Mr. Todd W. Herrick) of the Tecumseh Products Company, who made the compilation of this book possible by sponsoring the Herrick Professorship of Engineering, which I presently hold. Professor Raymond Cohen, also affectionately known as Professor Compressor to some, must be acknowledged as the person who put the Ray W. Herrick Laboratories on the map as a center for compressor research, ably assisted in this by my colleague, Professor James F. Hamilton, who has also made many contributions. Professor Cohen (the first Herrick Professor of Engineering) was also my mentor for many more years ago than I care to remember. His consistent and continuing good advice was and is very much appreciated. New and younger colleagues, who are presently active in compressor research, must also be mentioned: Professor D. Adams, Professor J. E. Braun, and Professor E. A. Groll.

Last but not least, the great help of D. K. Cackley (who typed the manuscript), and M. F. Soedel-Schaaf and A.S. Greiber-Soedel must be acknowledged. D. T. Soedel and H. J. Kim contributed to the chapter on compressor mufflers and S. M. Soedel and F. P. Soedel contributed to the vibration chapters. The research publications of many of my former graduate students are referenced in the text, and their contributions to my research are gratefully acknowledged.

Werner Soedel

Author

WERNER SOEDEL is Professor of Mechanical Engineering and the Herrick Professor of Engineering at the Ray W. Herrick Laboratories, School of Mechanical Engineering, Purdue University, West Lafayette, Indiana. The author of 240 papers in various journals and proceedings, he is a Fellow of the American Society of Mechanical Engineers and is or was a member of the Acoustical Society of America, the Society of Experimental Mechanics, and the American Academy of Mechanics, among others. Cited in *American Men of Science,* the *International Scholars Directory,* and other biographical works, he has been  a consultant to over 50 industrial companies and was the Americas Editor of the *Journal of Sound and Vibration* from 1989 until 2005. Professor Soedel received the equivalent of the B.S.M.E. degree (1957) from the Staatliche Ingenieurschule, Frankfurt, Germany and the General Motors Institute (Kettering University), (1961); and the M.S.M.E. (1965) and Ph.D. (1967) degrees in mechanical engineering from Purdue University.

Table of Contents

Chapter 1 Introduction 1
1.1 General 1
1.2 The Compression Process 1
1.3 Valves 2
1.4 Suction and Discharge Manifolds and Pipes 4
1.5 Casing 5
1.6 Hermetic Housing Shell 6
    1.6.1 Gas Pulsation Excitation 7
    1.6.2 Discharge Pipe Excitation 7
    1.6.3 Shell Resonances 7
    1.6.4 Excitation by Interior Gas Modes 7
    1.6.5 Excitation through Casing Isolation Springs 8
    1.6.6 Detuning Resonances and Damping 8
    1.6.7 Repeatability of Noise 8
1.7 Vibration Isolation Springs 9
1.8 Oil Sump 10

Chapter 2 Overview of Noise Source and Transmission Mechanisms
                  by Type of Compressor 11
2.1 Reciprocating Compressors 11
    2.1.1 Slider-Crank Mechanism 11
    2.1.2 Scotch Yoke Mechanism 13
    2.1.3 Reciprocating Swing Compressors 13
2.2 Rotary Vane Compressors 14
2.3 Rolling Piston Compressors 15
2.4 Scroll Compressors 16
2.5 Screw Compressors 17

Chapter 3 Natural Frequencies and Modes of Compressor Housings 19
3.1 Equations of Motion 19
    3.1.1 What Is a Shell? 19
    3.1.2 Reference Surface and Coordinates 20
    3.1.3 Strain–Stress Relationships 22
    3.1.4 Strain–Displacement Relationships 24
    3.1.5 Membrane Force and Bending Moment Resultants 26
    3.1.6 Energy Expressions 28

	3.1.7	Hamilton's Principle and General Equations of Motion	28
	3.1.8	How Do Shells Vibrate?	30
3.2	Boundary Conditions		30
3.3	Natural Frequencies and Modes		31
	3.3.1	The Compressor Housing Simplified as a Cylindrical Shell	32
3.4	Further Simplification		39
	3.4.1	Introducing Axial Curvature (Barreling)	41
	3.4.2	Spherical End Caps	43
3.5	Vibration Localization at End Caps		44

Chapter 4 Compressor Housings that Are Not Axisymmetric 47

4.1	Mode Splitting Caused by a Mass or Stiffness	47
4.2	Mode Splitting Caused by Ovalness	50
4.3	Example of Experimentally Obtained Housing Modes	51

Chapter 5 Modifications of Housing Natural Frequencies and Modes 55

5.1	Stiffening of Compressor Housing	55
5.2	The Influence of Residual Stresses	58

Chapter 6 Forced Vibration of Compressor Housing (Shell) 61

6.1	Modal Series Expansion Model	61
6.2	Steady-State Harmonic Response	63
6.3	Housing Dynamics in Terms of Modal Mass, Stiffness, Damping, and Forcing	65
6.4	Steady-State Response of Shells to Periodic Forcing	67
6.5	Remarks About Dissipative Damping	70
6.6	Dynamic Absorbers	72
6.7	Friction Damping	76

Chapter 7 Free and Forced Vibrations of Compressor Casings 81

7.1	The Three-Dimensional Equations of Motion for an Elastic Solid		81
	7.1.1	Strain–Stress Relationships	81
	7.1.2	Strain–Displacement Relationships	82
	7.1.3	Energy Expressions	83
	7.1.4	Hamilton's Principle and General Equations of Motion	83
	7.1.5	Boundary Conditions	84
	7.1.6	Example: Cartesian Coordinates	84
	7.1.7	One-Dimensional (Wave) Equation for Solids	86
7.2	Free and Forced Vibrations		87
7.3	Steady-State Harmonic Response		89
7.4	Response of the Casing to Impact		90
7.5	Steady-State Response of the Compressor Casing to Periodic Forcing		92

Chapter 8 Vibrations of Other Structural Components of a Compressor 95

8.1 Vibration and Force Transmission of Discharge or Suction Tubes 95
    8.1.1 Equation of Motion for a Straight Discharge Tube 95
    8.1.2 Natural Tube Frequencies Influenced by Pressure Changes 96
    8.1.3 Tube of Constant Curvature Vibrating
        in the Plane of Its Curvature 97
    8.1.4 Forces Transmitted into the Compressor
        Housing by a Vibrating Tube 98
    8.1.5 Effect of Mass Flow Rate on Tube Vibration 100
8.2 Vibration Isolation Considering Idealized Springs 101
    8.2.1 Review of the Standard Approach to Vibration Isolation 101
    8.2.2 Rotating Unbalance 104
    8.2.3 Reciprocating Compressor Unbalance 107
    8.2.4 Isolating the Flexural Vibrations of the Casing or Housing 110
8.3 Surging in Coil Springs Interfering with Vibration Isolation 112
    8.3.1 A Simplified Spring Surge Model 113
    8.3.2 The Effect of Surging on the Response 116
    8.3.3 Surging and Housing Resonances 119

Chapter 9 Sound and Vibration of Compressor Valves 123

9.1 Remarks About Types of Compressors and Their Valving 123
9.2 A Simplified Approach to Understanding Valve Design 125
    9.2.1 Thermodynamic Considerations 125
    9.2.2 Indicator Diagrams, Valve Timing, and Flow Velocity
        Estimates 126
    9.2.3 Required Valve Port Areas 129
    9.2.4 Allowable Valve Lift 131
    9.2.5 Advantage of a Valve Stop 131
    9.2.6 Estimating the Flow Force on the Valve and Selection
        of the Effective Valve Spring Rate 132
    9.2.7 Floating Valves or Spring Loaded Valves 133
    9.2.8 Reed Valve Shapes 134
    9.2.9 Suction and Discharge Volume Selection 136
    9.2.10 Reliability Considerations 136
    9.2.11 Valve Material Selection 137
    9.2.12 Multicylinder Compressors and Multistage Compressors 138
9.3 Useful Valve Calculations 139
    9.3.1 Effective Valve Flow Areas 139
        9.3.1.1 Orifices in Series 140
        9.3.1.2 Parallel Orifices 141
        9.3.1.3 Application Examples: Ring Valves 142
    9.3.2 Effective Valve Force Areas 143
    9.3.3 Measuring Effective Valve Flow and Force Areas 148
        9.3.3.1 Effective Valve Flow Areas 148
        9.3.3.2 Effective Valve Force Areas 150

9.4 Valve Dynamics 151
    9.4.1 Poppet Valves 151
    9.4.2 Flexible Reed or Plate Valves 162
        9.4.2.1 Equations of Motion 162
        9.4.2.2 Natural Frequencies and Modes 163
        9.4.2.3 Response of Beam or Platelike Valve Reeds
                to Forcing 165
        9.4.2.4 Approximate Flow Forces on Reed Valves 168
        9.4.2.5 Natural Frequencies and Modes by Experiment 172
    9.4.3 Pumping Oscillation 174
    9.4.4 Valve Stops and Damping 178
    9.4.5 Simulation of Valve Motions 180
9.5 Stresses in Valves 182
    9.5.1 Bending Stresses 182
    9.5.2 Impact Stresses 185

Chapter 10 Suction or Discharge System Gas Pulsations and Mufflers 195

10.1 Reducing Valve Noise 195
    10.1.1 Dissipative Mufflers 195
    10.1.2 Side-Branch Resonators 196
    10.1.3 Low Pass Filter Mufflers 196
    10.1.4 Impact Noise 197
    10.1.5 Noise Due to Turbulence 198
10.2 Reactive Mufflers Using the Helmholtz Simplification 198
    10.2.1 The Helmholtz Resonator 198
    10.2.2 The Helmholtz Resonator Approach Applied
            to Compressors 201
    10.2.3 Steady-State Harmonic Response to a Harmonic Volume
            Velocity Input and Design Criteria 203
    10.2.4 Discharge System with Two Resonators in Series 206
    10.2.5 Discharge System with Anechoic Pipe 208
    10.2.6 Resonator Plus Anechoic Pipe 209
    10.2.7 Discharge System for a Two-Cylinder Compressor 211
    10.2.8 Discharge System Model for More Than Two Cylinders 215
    10.2.9 Steady-State Harmonic Response of a Discharge System
            with Two Resonators in Series 219
    10.2.10 Low-Frequency Cutoff Formulas 222
    10.2.11 Oscillation Effects Caused by Cylinder Volume
            and Valve Passage Masses 223
    10.2.12 Discharge System with a Long Pipe Modeled
            in the Time Domain 224
10.3 The Continuous System Approach Applied to Tubelike
Compressor Suction and Discharge Manifolds 227
    10.3.1 The Wave Equation in One Dimension 227
    10.3.2 The Solution of the Undamped Wave Equation 231

	10.3.3	The Solution of the Damped Wave Equation	232
	10.3.4	Example: Open Pipe with Volume Velocity Input	234
	10.3.5	The Four Pole Concept	236
	10.3.6	Example: Open Pipe	238
	10.3.7	Global Four Poles from Local Element Four Poles: Tubes in Series	239
	10.3.8	Branched Tubes	242
	10.3.9	Anechoic Termination to a Muffler	246
	10.3.10	Termination Defined by an Impedance	249
	10.3.11	Multicylinder Compressors	250
	10.3.12	Gas Pulsations in Intercoolers	252
	10.3.13	Derivation of the Anechoic Termination Relationship	252
		10.3.13.1 Propagation of a Unit Volume Velocity Impulse in an Anechoic Pipe	252
		10.3.13.2 Response of Anechoic Pipe to General Volume Velocities at the Pipe Entrance	255
	10.3.14	Time Response of a Finite Gas Column	256
		10.3.14.1 Eigenvalues of the Gas Column	258
		10.3.14.2 Solution by Modal Expansion	259
10.4	Typical Behavior of Simple Compressor Suction or Discharge Mufflers		262
	10.4.1	Defining Transmission Loss	262
	10.4.2	Backpressure	263
	10.4.3	Reaction Muffler Element Arranged in Line	264
		10.4.3.1 Volume-Tailpipe Muffler	264
		10.4.3.2 Scaling	267
		10.4.3.3 Two-Volume Muffler with Inertia Tube and Tailpipe	269
		10.4.3.4 Triple-Volume Muffler	271
	10.4.4	Side-Branch Attenuators	271
		10.4.4.1 Helmholtz Resonator Side-Branch Attenuator	271
		10.4.4.2 Influence of Attachment Location	273
		10.4.4.3 Gas Column Side-Branch Attenuator	274
	10.4.5	Tube Penetrating Volumes	276
		10.4.5.1 Single-Volume Muffler with Penetrating Tailpipe	276
		10.4.5.2 Two-Volume Muffler with Penetrating Inertia Tube	278
	10.4.6	Influence of Temperature	279
	10.4.7	Muffler Synthesis Example	280
	10.4.8	Multiple Side-Branch Attenuators	282

Chapter 11 Multidimensional Compressor Sound 285

11.1	Three-Dimensional Acoustic Wave Equation	285
11.2	Sound Radiation from the Compressor Housing (or Casing)	288

		11.2.1	Monopole Source	288

        11.2.1 Monopole Source..........288
        11.2.2 Radiation from a Circular Cylindrical Housing..........290
        11.2.3 Practical Consideration..........294
11.3 Gas Pulsations in the Cavity between the Compressor Casing and Housing..........296
        11.3.1 Natural Frequencies and Modes of Gas in an Annular Cylinder..........296
        11.3.2 Natural Frequencies and Modes of Gas in a Circular Disk Volume..........299
        11.3.3 Natural Frequencies and Modes of Gas in a Volume Consisting of an Annular Cylinder and a Circular Disk..........303
        11.3.4 Rocking Vibration of Housing Shell..........304

Chapter 12 Remarks on Sound and Vibration Measurements and Source Identification..........305

12.1 Rooms for Measuring Sound..........305
12.2 Repeatability of Measurements..........307
12.3 Identification of Sound and Vibration Mechanisms..........309
        12.3.1 Exterior and Interior Measurements and Recommended Calculations..........309
        12.3.2 Modifications to the Compressor for Diagnostic Purposes..........310

Chapter 13 Miscellaneous Sound and Vibration Sources or Effects..........313

13.1 Electromotors..........313
13.2 Lubrication Oil..........314

References..........**317**

Author Index..........**335**

Subject Index..........**339**

Nomenclature

Note that some symbols are used more than once, in different chapters, following common usage. For example, the symbol T may mean kinetic energy in one chapter, or period of time in another.

ds	diagonal of an infinitesimal element
x, y, z	Cartesian coordinates
r, θ, z	cylindrical coordinates
ϕ, θ	spherical shell coordinates
$\alpha_1, \alpha_2, \alpha_3$	general curvilinear coordinates
A_1, A_2, A_3	Lamé parameters
R_1, R_2	radii of curvature
$\sigma_{11}, \sigma_{22}, \sigma_{33}$	normal stresses
$\sigma_{12}, \sigma_{13}, \sigma_{23}$	shear stresses
$\varepsilon_{11}, \varepsilon_{22}, \varepsilon_{33}$	normal strains
$\varepsilon_{12}, \varepsilon_{13}, \varepsilon_{23}$	shear stresses
μ	Poisson's ratio
E	Young's modulus
$(ds')_i$	deformed length
$(ds)_i$	undeformed length
$\varepsilon_{11}^o, \varepsilon_{22}^o, \varepsilon_{12}^o$	membrane strains
k_{11}, k_{22}, k_{12}	curvature changes
u_1, u_2, u_3	deflections of a point on the reference surface
β_1, β_2	slopes of deflection
θ_{12}	angle after shear deformation
N_{11}, N_{22}, N_{12}	membrane force resultants
M_{11}, M_{22}, M_{12}	bending moment resultants
K	membrane stiffness
D	bending stiffness
Q_{13}, Q_{23}	transverse shear resultants
h	thickness of housing shell
U	strain energy
T	kinetic energy

Symbol	Description
W_{nc}	nonconservative work
δ	variation
$\delta \bar{r}_i$	variation of displacement vector
ρ	mass density
t	time
q_1, q_2, q_3	forces per unit area
$N_{xx}, N_{\theta\theta}, N_{x\theta}$	membrane force resultants
$M_{xx}, M_{\theta\theta}, M_{x\theta}$	bending moment resultants
$Q_{x3}, Q_{\theta 3}$	transverse shear resultants
$\varepsilon^o_{xx}, \varepsilon^o_{\theta\theta}, \varepsilon^o_{x\theta}$	membrane strains
$k_{xx}, k_{\theta\theta}, k_{x\theta}$	curvature changes
u_x, u_θ, u_3	deflections
β_x, β_θ	slopes
q_x, q_θ, q_3	forces per unit area
L, a, h	dimensions of cylindrical shell
U_x, U_θ, U_3	natural mode components
m, n	integers
A, B, C	constants
$\omega, \omega_{mn}, \omega_k, \omega_n$	natural frequencies
ϕ	function
R, R_x, R_θ, a	radii
U_{3mn1}, U_{3mn2}	orthogonal mode sets
$\omega_{mn1}, \omega_{mn2}$	natural frequency pair
M	attached mass
M_s	housing shell mass
x^*, θ^*	location of attached mass
K	spring rate
$\varepsilon(\theta)$	periodic function
U_{1k}, U_{2k}, U_{3k}	natural modes
λ	damping factor
ζ_k	modal damping coefficient
η_k, η_{mn}, η	modal participation coefficient
$q_i^*(\alpha_1, \alpha_2)$	spatial distribution of loading
M_k, K_k, C_k	modal mass, stiffness and damping
T	period
Ω	operating speed
a_n, b_n	Fourier coefficients

η	hysteretic damping constant
$\alpha_{ij}, \beta_{ij}, \gamma_{ij}, \ldots$	receptances
$X_{A1}, X_{A2}, \ldots$	response amplitudes
$F_{A1}, F_{A2}, \ldots$	harmonic force amplitudes
x, y	deflections
P	normal force
μ_k	friction coefficient
E_d	energy per cycle
G	shear modulus
λ	bulk modulus
C_1	compression wave velocity
C_2	shear wave velocity
$B_k\{u_1, u_2, u_3\}$	boundary conditions
Λ_k	response magnitude
ϕ_k	phase lag
$\delta(t - t_1)$	Dirac delta function
M_i^*	distributed momentum change
v	velocity
m	impacting mass
p_s, p_d	suction and discharge pressure
I	area moment
A	cross-sectional area
W_c	displacement excitation amplitude
Q	shear force
m, ρ	mass per unit length of tube and liquid
F_o	force amplitude
F_T	transmitted force
$\tilde{F}_T$	amplitude of transmitted force
k, K	spring rates
c, C	damping constant
m, m_1, m_2	unbalanced mass
M	compressor mass
e	radius of unbalance
X	response amplitude
ζ	damping ratio
m'	mass per unit length of spring
M, K, C	mass, spring rate, and damping coefficient of spring

Symbol	Description
s, d	subscripts designating suction and discharge
T_s, T_d	suction and discharge temperatures
ρ_s, ρ_d	suction and discharge mass densities
V_i	volumes
A, A_d	cross-sectional suction and discharge port areas
λ	volumetric efficiency
Q_s, Q_d	suction and discharge volume velocities
Δp	pressure differential
M	Mach number
c	speed of sound
R	gas constant
k	adiabatic coefficient
k_s	contraction coefficient
A_e	effective area
C	circumference
h	valve lift height
t	valve thickness
v_s	setting velocity
K	flow coefficient
p_u, p_d	upstream and downstream pressure
$\dot{m}$	mass flow rate
$(KA)_e$	effective flow area
w	valve deflection
D_i	diameters
b	width
F	flow force on valve
A_i	areas on valve plate
A_v	measured effective flow area
B(w)	effective force area
w_s	stop height
$\in$	coefficient of restitution
$U(t - t_2)$	unit step function
a	slope of indicator diagram at time of valve closing
q'_3	force per unit length
A, B, C, D	constants
λ_k	roots of frequency equation
q	volume velocity

Q	volume velocity amplitude
Z	impedance
$\varepsilon_x, \varepsilon_y, \varepsilon_z$	normal strains
$\sigma_x, \sigma_y, \sigma_z$	normal stresses
$\varepsilon_{xy}, \varepsilon_{xz}, \varepsilon_{yz}$	shear strains
$\tau_{xy}, \tau_{xz}, \tau_{yz}$	shear stresses
N	crank speed
H	maximum valve displacement
h, H	subscripts designating valve plate and valve seat
f_h, g_h	stress waves in positive and negative directions
$v_h(o)$	impact velocity of valve plate
c_h, c_H	speeds of sound
$\sigma_{h\max}, \sigma_{H\max}$	peak impact stress amplitudes
ρ_h, ρ_H	mass densities
f_n	natural frequency in Hz
L	effective length of neck
A	cross-section of neck
V	volume
c	speed of sound
ξ	gas plug displacement
K_o	bulk modulus of gas
p_o	average pressure
p	acoustic pressure
D	equivalent viscous damping coefficient
ζ	damping factor
L_G	geometric length
f_{max}	highest frequency of interest
λ	wavelength
a	largest acoustic dimension
S, A	cross-sectional areas
r_1	equivalent viscous damping coefficient
v	effective kinematic viscosity
ζ	correction coefficient
k	wave number
k_1	modified wave number
c_1	modified speed of sound
A, B, C, D	four-pole parameters

$\dot{m}, \dot{M}$	mass flow rates
TL	transmission loss
f_c	low-frequency cutoff
d	distance
f_{em}	frequencies due to electromagnetic effects
R	number of rotor slots
P	number of poles
f_s	shaft speed
f	line frequency
f_a	aerodynamic frequency

1 Introduction

1.1 GENERAL

The term *positive displacement compressor* refers to a compressor where one or more volumes are decreased and increased kinematically. This means that the compression volumes are determined positively by a crank position or some other position input. This is in contrast to so-called *dynamic* compressors, such as centrifugal compressors, where the compression takes place by centrifugal and other gas dynamic forces, and not necessarily by volume shrinkage. Axial turbo compressors (as in aircraft engines) are another example of dynamic compressors.

The reason that this article concentrates on positive displacement compressors as a group is that they can be found practically anywhere. They are the hearts of refrigerators, air conditioners and heat pumps. They supply compressed air to machine tools, construction equipment such as jack hammers, medical devices such as dentist's drills and diving equipment. They pump natural gas through pipelines and compress oxygen and nitrogen. It is estimated that in the United States alone, there are more than 400 million compressors in existence as this is written. A typical household may have as many as six compressors: in a heat pump and central air conditioning system, one or two auxiliary window air conditioners, one or two refrigerators and freezers, and one shop air compressor. Industrial applications are numerous; sizes range from a fractional-horsepower compressor in a household refrigerator to thousand and more horsepower compressors to pump natural gas. As a group, they are not only one of the largest energy users, but also one of the largest noise polluters.

Most noise control research has been done on the smaller refrigeration, air conditioning, and heat pump compressors (see also Roys and Soedel, 1989) because the mass production makes research costs affordable, and because consumers who purchase such compressors are very aware of the noise they can make. This book reflects these facts, but the principles discussed can also be applied to air and gas compressors and can be extrapolated to compressors of all sizes.

1.2 THE COMPRESSION PROCESS

The fact that the gas is compressed from suction to discharge pressure conditions creates a time-varying, pressure-forcing function on the cylinder(s). This forcing creates a structural vibration of the compressor casing, which is transmitted through the vibration isolation spring, the discharge piping (in the case of a low-side design), and the suction piping (in the case of a high-side design), and the oil sump, and perhaps to a lesser extent, the gas filling the hermetic shell cavity. (Low-side and high-side designs refer to either the suction or discharge gas filling the hermetic shell cavity.)

What can be done about this? For a given volume rate to be compressed, it is best to use several smaller cylinders instead of one large one. Of course, the kinematics has to be such that the multicylinder compression processes are staggered. Instead of one compression per revolution, several compressions occur in the same time interval. Because of the smaller cylinder sizes, the compressor casing tends to be stiffer, and the staggering of the kinematics tends to smooth out the effective forcing function.

Another measure (feasible or not) is to use, given the same dimensions as boundary conditions, cast iron instead of, say, aluminum, to reduce the flexibility, increase the mass and also to take advantage of the higher internal damping.

While in smaller refrigeration compressors the compression process is close to isentropic because the relatively high speed (typically 3600 RPM) does not allow enough time for heat transfer to take place, there would be a theoretical benefit if one could approach isothermal compression by cooling because of the less steep ramping of the cylinder pressure–time curve during compression (and expansion). A less steep ramping will reduce the high frequency content of the excitation spectrum.

1.3 VALVES

In the foregoing discussion, ideal valves were assumed. Real valves often flutter, introducing ripples into the cylinder pressure–time diagrams during the suction or discharge portion. These pressure ripples will add to the frequency content of the casing excitation and should be avoided. Other effects of fluttering valves on noise will be discussed below, considering valves as another source of noise.

Valves produce noise because of the intermittent nature of the discharge or suction process, even if a vibrating valve reed or valve plate is not present. Therefore, so-called *gate valving*, as is used in screw, scroll, and certain rotary vane compressors, still produces noise. The definition of gate valving is that for a discharge valve, a port opens kinematically, which would otherwise be covered by, say, the rotor face of a screw compressor, allowing the valve to discharge. The port later closes at a predetermined interval. In time, we can view the gas volume flow as a periodic function that decomposes into Fourier components, which in turn produce sound by either direct radiation or by way of interaction of the discharge or suction process with a wave guide and its resonances.

Because gate valving is difficult to design when the discharge and suction pressures vary widely, and because certain designs, such as piston compressors, do not lend themselves easily to gate valving, most compressor designs use *automatic valves*, which can be floater valves, but are usually spring-loaded designs or flexible plate or reed valves where the spring and the port cover are combined. These valves open only on demand, for example, when the cylinder pressure is equal to or larger than the discharge pressure, in the case of a discharge valve. The durations of valve openings are therefore controlled by the pressure in the discharge or suction manifolds. This is a much more efficient design than the passive gate valve concept whenever discharge or suction operating pressures vary over relatively large ranges. Of course in some designs the gate valve opening could, in principle, be controlled by a servomechanism that electronically senses the difference between cylinder and manifold pressures, but this is an expensive solution. From a noise control viewpoint, the drawback is that spring-loaded valves tend to flutter, which (if it occurs) modulates the more steady flow that would be expected for a gate valve. This produces a superimposed tone on the general noise, which in the simplest sense is close to the frequency of the valve flutter, which as an approximation, occurs close to the dominant natural frequency of the spring-loaded valve or reed. Of course, if more than one mode is excited in a reedlike valve, several tones at the natural frequencies of the superimposed modes will be detectable.

Valve flutter is caused by two possible mechanisms. First, the fact that the valve opens relatively suddenly means that the valve will oscillate as a system that is forced by a unit step function. Given enough time and space, the flutter amplitude will be as much as twice the amplitude of the same valve if it could be opened very slowly. The other mechanism is the Bernoulli effect. A negative pressure may develop in the valve seat, which will tend to retard the opening of the valve until the cylinder pressures (again we are using the discharge valve as an example) has built up to such a high level that the valve is forced to open with a high velocity. This will make the valve overshoot the equilibrium position it would occupy if the flow had been steady. The effective spring will then force the valve to return to its seat way before the discharge is completed, assisted in the regions closer to the seat by the Bernoulli effect. The valve may actually close prematurely (before the cylinder pressure builds up sufficiently) and cause the flutter cycle to repeat.

In many designs, the two mechanisms will coexist.

It should also be noted that sound is not only produced by the intermittent, flutter-modulated flow through the valves, but also by valve impact. Such impact can occur during the opening phases, with the valve hitting a motion limiter, or during the closing phases when the valve hits the seat. A fluttering valve may have repeated seat impacts during the period it is open. The impacts will excite vibrations of the motion limiter or valve seat plate, and may actually transmit into the compressor casing, where it either radiates directly to the gas in the hermetic shell or transmits vibrations though the mounting springs or piping.

What can be done about valve-generated sound? First of all, flutter needs to be suppressed. While the effectiveness of practical damping applied to flutter suppression, while not zero, is relatively limited, it was found that by introducing a motion limiter, flutter can be either totally eliminated or at least partially suppressed. This limiter should be designed in such a way that the valve lies against it during its opening period and comes off only after the changing pressure differential across the valve forces it to close. Care has to be taken that the motion limit is not so low that it impedes the average flow and causes unacceptable energy losses.

To reduce impact noise, the material for the motion limiter should be soft. While the change in the momentum of the valve, and thus the impulse, will not be much affected as a whole, a soft motion limiter will increase the time of impact, and thus reduce the impact force with which the motion limiter, and thus the valve assembly and compressor casing, are excited. The same is true for the seat, except that valve seats are usually made of hardened steel for stress and wear-related reasons, and the seat area is subjected to a grinding operation for good, leak-free seating, making a softly cushioned seat design virtually impossible. However, where possible, the entire seat could theoretically be vibration isolated from the rest of the compressor casing. Unfortunately, this is a difficult (expensive) solution and should only be tried for cases where one is convinced that seat impact noise is important, and where no other measures are possible. Another possibility is to make the valve of soft material. There are designs where the disks of spring-loaded poppet valves are made of plastic, for example.

A motion limiter, as discussed above, will also reduce noise due to valve impact on the seat because it will prevent multiple impacts during the opening period. The ideal valve opens when it is supposed to, stays open, and closes (gently, if possible) when it is supposed to.

1.4 SUCTION AND DISCHARGE MANIFOLDS AND PIPES

The unavoidable periodically forced gas pulsations caused by the intermittent openings, regardless of whether there is valve flutter or not, should be suppressed as much as possible by gas pulsation mufflers. For example, the simplest discharge muffler design would be a flow-through Helmholtz resonator, with the volume of the manifold that communicates directly with the valve being the gas spring of the resonator. The neck of the Helmholtz resonator is a short passage or pipe that opens into a second volume—the *decoupling volume*—which in turn exits into the discharge pipe.

The simplest suction muffler design, for a low-side, hermetically sealed refrigeration compressor, is a manifold volume in front of the intake valve, and in front of it, a narrower passage or short pipe (the neck of the resonator), which communicates with the suction gas in the shell. This neck connects to the volume formed between the hermetic shell and the compressor casing.

In the case of air compressors, the principles on the discharge side are the same. On the suction side, where the air is taken in directly through a filter, the filter assembly could be combined with a flow-through Helmholtz resonator.

The flow-through Helmholtz resonator design acts like a low pass filter. Below the resonance frequency of the device, multiplied by $\sqrt{2}$, the flow-through Helmholtz resonator actually acts as an amplifier, but above it a gas pulsation transmission loss is produced.

The next step in muffler design would be to cascade several volumes and necks. Also, the necks could become long tubes for which the analytical simplification of viewing the gas in the neck as incompressible is no longer valid, and a wave approach has to be applied. In addition, side-branch Helmholtz resonators or tube resonators can be added which, at their resonance frequencies, drain pulsation energy from the system.

How do gas pulsations become noise? In open intake systems of air compressors, the gas pulsations radiate directly toward the receiver. In the case of an intake in a low-side, hermetically sealed compressor, the suction pulsation excites the interior, acoustic modes of the space between the hermetic shell and the compressor casing. The acoustic vibrations of the gas in this space excite the natural modes and frequencies of the hermetic shell or housing, which then radiates noise to the receiver. At resonances of the acoustic modes of the interior space, and at resonances of the housing (shell), the response is amplified.

A second path of noise transmission from the gas pulsations is, in the case of a low-side refrigeration compressor, the discharge pipe. Because discharge pipes in certain designs cannot be straight, but have to be bent to act as quasi-isolation springs that partially prevent a direct vibration excitation of the housing shell by the vibrating compressor casing, the pulsating change of momentum as the flow passes a bend will excite the discharge pipe into vibrations, which in turn excites housing (shell) vibrations. Another way the discharge pipe is excited is by the vibration of the casing. Because of its slenderness, a typical small heat pump compressor discharge pipe has a multitude of natural frequencies in the typical frequency range of concern, say 100 to 5000 Hz. Thus, discharge pipe resonances are almost unavoidable. Therefore, the shell vibration levels excited by discharge pipe resonances can be significant.

What can be done about vibrating discharge pipes (or vibrating suction pipes)? In refrigeration compressors, it often helps to introduce damping, in the form of external friction to the pipe, or to support the discharge pipe by soft suspensions that rest against the compressor casing, which will change the natural modes that are being excited. Of course eliminating, as much as possible, the gas pulsations by a muffler before they reach the flexible pipe is another possibility, as is a vibration reduction of the casing to which the pipe is connected.

To vibration isolate the pipes from the housing shell by flexible hoses is in principle effective, but practically not easily done because the refrigerant and oil environment makes certain isolation materials, such as polymer hose sections, unacceptable.

There is an additional potential noise source, which is a Helmholtz resonator effect involving the cylinder and the valve passage. Again, using the discharge valve as an example, it has to be realized that when the valve starts to open, there is a cylinder volume connected to the mass of the gas in the valve port (the vibrating valve reed or plate itself can be ignored as a first approximation). The gas in the cylinder volume is compressed slightly above the nominal discharge pressure, and once the valve opens, we have the effect of a champagne bottle being opened. An oscillation of the gas mass in the valve port on the air spring–like cylinder volume will produce an almost pure tone noise that is superimposed on the other types of noise. Because the cylinder volume changes during this event, the frequency of this tone changes. This effect was identified in reciprocating piston compressors, and was also found to be a dominant noise source in otherwise relatively low-noise designs of rotary vane compressors, as will be discussed later.

1.5 CASING

The casing of a compressor, which is defined as all of the compressor without the housing or hermetic shell for hermetic refrigeration and air conditioning compressors, is excited by the periodically varying compression and expansion of the cylinder pressures, possible piston slap, valve impact, and cylinder pressure modulations due to valve flutter; it is also excited by rotational imbalances and electromotor rotor and stator vibrations.

Because of the irregular shapes of casings, natural frequencies and modes cannot easily be estimated by relatively simple mathematical models, but must be determined by full blown, three-dimensional, finite element approaches. The flexible natural frequencies of the casing are usually relatively high (2000 Hz and higher, for typical small refrigeration compressors), but they can interact well with certain natural frequencies of the discharge or suction pipes that connect the casing mechanically to the housing. The other mechanical transmission path between the casing and housing, which is through the suspension spring, is also effective. The other two transmission paths, one through the refrigeration gas in the volume between the casing and the housing, and the other through the oil pool are probably less important, but have in a few cases been identified as significant.

1.6 HERMETIC HOUSING SHELL

In small to mid-sized refrigeration, air conditioning, and heat pump compressor designs, the compressor is hermetically sealed by a hermetic shell or housing to prevent refrigerant leakage. Figure 1.1 schematically illustrates a low-side arrangement (here for a reciprocating piston compressor), where the suction gas from the evaporator enters the cavity formed by the housing shell and the compressor casing. The suction gas then enters the suction manifold, flows through the suction valve into the cylinder volume, and is compressed and discharged through the discharge valve into the discharge manifold from where it flows to the condenser. For compressors

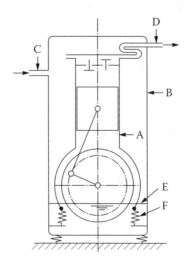

FIGURE 1.1 Schema of a low side, reciprocating refrigeration compressor system (A = compressor casing, B = compressor housing (hermetic shell), C = suction pipe, D = discharge pipe, E = oil sump, and F = support springs).

Introduction

that rely on forced lubrication, as in the rotary vane compressor, the housing cavity is filled with the discharge gas, while the suction gas enters through a pipe directly into the suction manifold. The higher-pressure discharge gas will force oil from the oil sump into the parts that have to be lubricated or sealed by oil.

Sound is either radiated directly from the vibrating housing surfaces, or vibrations are transmitted into the refrigerator or air conditioning structures or heat pump mounting platforms, which in turn radiate noise. Hermetic shells are basically not an air compressor design element, but they may be used for compressors of toxic gases. There are several excitation mechanisms that have to be considered.

1.6.1 Gas Pulsation Excitation

In a low-side refrigeration compressor, the suction gas fills the space between the compressor casing and the shell. Oscillations of the gas in this space are created by the intake manifold–port oscillations, which act as a volume source. The oscillation amplitudes of the gas are often amplified because of gas resonances in the gas that fills the space between the compressor shell and casing.

1.6.2 Discharge Pipe Excitation

As discussed above, for a low-side compressor, the discharge pipe is soldered to the housing shell, and its vibration due to the periodic change of momentum of the discharge gas flow is transmitted, typically as a moment excitation, to the shell, as is the vibration from the casing. Pipe resonances will amplify the input to the shell.

1.6.3 Shell Resonances

For refrigeration, air conditioning, and heat pump compressors, and certain compressors that pump toxic gases, the housing shell shapes approach approximately the form of a circular or oval cylindrical shell with top and bottom caps. At times, the cylindrical form is a barrel shape approaching, for special applications, a spherical shell. Whenever natural frequencies of the discharge pipe (or suction pipe for high-side compressors) coincide with natural frequencies of the shell, relatively high vibration amplitudes of the shell may result causing an increase of sound radiation at these frequencies. On the other hand, resonances involving the interaction of the natural frequencies of the gas inside the shell with the natural frequencies of the shell do not seem to be a particular problem, especially because the lowest natural frequencies of the interior gas are typically significantly lower than the lowest natural frequency of the housing shell, for typical small-horsepower compressors. In general, there is a mismatch of natural mode shapes at the interface between the interior gas and the inner surface of the shell. However, it has to be remembered that it does not necessarily take a resonance to create significant radiated noise from the shell.

1.6.4 Excitation by Interior Gas Modes

The shell can vibrate with significant amplitude at an off-resonance frequency if the excitation is sufficiently strong. This is the case involving *sloshing gas modes*. If the space between an approximately circular cylindrical housing shell and the compressor casing can be approximated as an annular cylinder, the fundamental sloshing mode is a volume velocity motion about the axis of the annulus that has two planes of zero motion at which the modal pressure amplitudes are at maximum. This type of mode is typically excited by the gas pulsations of the suction manifolding, and typically occurs at 200 to 500 Hz for fractional-horsepower types of hermetic compressors, where the lowest shell natural frequency may typically be 900 to 1200 Hz. This excitation may be strong enough to cause a definite positive bulge at 200 to 500 Hz in the sound spectra of these compressors. (A second bulge occurs typically at shell resonances in the 900 to 1200 Hz region.)

1.6.5 EXCITATION THROUGH CASING ISOLATION SPRINGS

Resonances of the housing can also be excited mechanically (besides through the suction and discharge pipe connections) through the vibration isolation springs that fix the compressor casing inside the housing. This will be discussed below.

1.6.6 DETUNING RESONANCES AND DAMPING

It does not seem feasible to space the natural frequencies of the housing by analytically predictive design in such a way that they are detuned from the excitation frequencies, because of a peculiarity of cylindrical shells that causes the natural frequencies to cluster in certain frequency bands. However, it is possible to shift (experimentally in the laboratory) the natural frequencies enough so that the worst resonances are avoided. Shifting resonances is accomplished by stiffening the housing, changing its effective modal masses, or changing its shape. Dynamic absorbers work in certain situations, but are not frequently used because of the added weight. To avoid resonances altogether is not feasible because the thickness of the housing shell would typically become unrealistic. A peculiarity of cylindrical housing shells is that natural frequencies are not proportional to shell thickness, except for bending-dominated modes. Membrane-dominated modes do not respond to thickness changes.

Another approach to controlling resonances is to introduce damping. In compressors that compress refrigerants, care has to be taken that the damping treatment does not appreciably change the heat transfer negatively, especially in high-side compressors where the hot discharge gas fills the housing interior and cooling it may rely on the housing as a part of the condenser system for an optimum coefficient of performance. Where this is not a problem, a viscoelastic sandwich treatment of the housing is expected to show beneficial results.

1.6.7 REPEATABILITY OF NOISE

The fact that there is sometimes a variation in noise output between batches of assembly line–produced compressors often has to do with tolerance variations and variations in manufacturing. This is illustrated here by the compressor housing.

The hermetic shell that surrounds smaller refrigeration, air conditioning and heat pump compressors is typically made of steel. Two more or less cylindrical shell

Introduction

halves are deep drawn from flat sheets, and after insertion of the compressor and motor, soldering the discharge or suction pipes to the respective shell halves, completing the electrical connections, and so on, they are welded together.

The welding process will introduce residual stresses into the finished assembly, which cannot be removed by annealing because of the temperature sensitivity of the interior equipment. Depending partially on the welding speed, for example, which may vary from batch to batch, the natural frequencies of the housing shells may be slightly different from batch to batch because of the slight differences in residual stresses. For typical slightly damped compressor shells, the response spectra show relatively narrow peaks. Slight changes due to residual stresses may make resonance peaks appear or disappear because of tuning or detuning of the shell's natural frequencies with one of the excitation harmonics or one of the natural frequencies of, say, the discharge pipe.

This all assumes, of course, that the noise measurement procedure is repeatable, and that the compressor to be measured has been warmed up properly, because another source of nonrepeatability is the variation in operating conditions, which will affect the speed of sound in the refrigerant gas and thus gas natural frequencies in the manifolds, piping, and interior gas spaces, as they interact with the natural frequencies of the housing.

1.7 VIBRATION ISOLATION SPRINGS

In certain compressor types, the kinematics are such that significant shaking forces occur, which require the compressor casing to be vibration isolated from the hermetic housing shell. Disregarding here the need to vibration isolate the discharge pipe (low-side compressor) so the compressor casing motion does not transmit through the discharge pipe into the housing shell, as discussed earlier, it must be realized that over the frequency range of interest, especially the 500-to 3000-Hz band in which, typically, the most easily excited shell modes are located for fractional-horsepower compressors, typical mounting springs do not act any longer as ideal massless springs, but have natural frequencies and modes of their own. If any of the spring frequencies are in resonance with any of the casing resonance frequencies, potentially high vibration levels may be transmitted into the compressor housing where they may interact with the natural frequencies of the housing shell and cause objectionable noise radiation.

Measures to control this mechanism are improved balancing, locating the isolation springs at certain node lines of the housing shell, or if feasible, locating the isolation springs on the compressor casing in regions where the casing vibration levels are at a minimum.

An interesting variation with respect to locating isolation springs on the shell depends on how the springs are mechanically attached to the shell. If the attachment is such that it results in an equivalent point force input to the shell, the node lines of the most objectionable shell mode are the best choice. On the other hand, if the spring is attached through, say, a cantilever bracket to the shell, the transmitted spring force is converted into a moment. In this case, the bracket should be located at an antinode.

1.8 OIL SUMP

Specifically in refrigeration, air conditioning, and heat pump compressors, lubrication oil with dissolved refrigerant often occupies the lower part of the space between the housing shell and the compressor casing. Thus, there is a transmission path from the vibrating compressor casing through the oil into the housing shell. In addition, any kind of churning of the oil, because of a crankshaft dipping into the oil, is an additional potential noise source.

On the other hand, it was observed that in instances when the refrigerant comes out of solution because of a pressure drop, it does so in the form of bubbles in the oil, up to the point of perhaps foaming. This will often actually have a beneficial effect as far as noise control is concerned, to the point where there was a design where the formation of bubbles in the oil was artificially created by bleeding refrigerant gas at discharge pressure into the oil of a low-side compressor.

The oil contribution, detrimental or beneficial, has not been investigated enough to draw definite conclusions about the oil transmission mechanism and its importance, but it is thought that one of the mechanisms is that the oil sump, with the oil containing bubbles, acts as a dynamic absorber on the housing shell.

2 Overview of Noise Source and Transmission Mechanisms by Type of Compressor

While the foregoing discussion applied to positive displacement compressors in a more or less general way, there are certain noise source and transmission mechanisms that are essentially unique to certain types of compressor kinematics and designs. A reciprocating compressor will behave differently than, say, a rotary vane compressor, depending in part on their unique designs.

While there are over 100 different kinematic compressor designs (see for example, Wankel (1963); Leemhuis and Soedel (1976); and Beard, Hall, and Soedel (1982), presently only a few seem to be widely used. They range from the reciprocating compressor, which continues to hold its own in spite of dating back into ancient China, to the rotating vane compressor, whose kinematics are also quite old, to the newer designs of the rolling piston compressor, the screw compressor, and the scroll compressor.

Only the unique noise-producing features of these five designs will be presented in the following. They all are, more or less, also subject to the noise mechanisms discussed so far.

2.1 RECIPROCATING COMPRESSORS

Reciprocating compressors typically require one suction and one discharge valve per cylinder, and are therefore subject to valve noise as discussed earlier. They require isolation springs, to support the casing inside the hermetic housing.

This category can be further subclassified by the way the reciprocating piston motion is created.

2.1.1 SLIDER-CRANK MECHANISM

The oldest compressor design from a historical viewpoint is still widely used. Figure 1.1 shows a schematic diagram of a one-cylinder compressor. A crankshaft, driven typically by an induction motor in small household refrigerators or air conditioners translates rotary motion into a reciprocating motion. Typically, gas is taken in through a suction valve and compressed gas is expelled through a discharge valve.

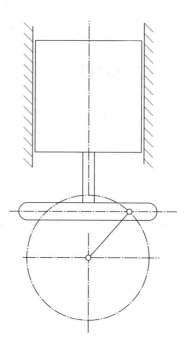

FIGURE 2.1 Kinematics of a scotch yoke, reciprocating piston compressor.

The crankshaft connecting rod interaction changes the direction of the normal force between the piston and the cylinder. Given enough clearance, this may result in what is called *piston slap*. Piston slap is more or less an impact mechanism, which can theoretically excite all the natural modes of the compressor casing, of which the cylinder is a part. Piston slap does not seem to be a prime noise producer in small bore compressors or well-lubricated compressors.

A second unique noise source originates from the shaking force caused by the slider-crank mechanism, which has an infinite number of harmonics of the rotational speed frequency. While the harmonics diminish in amplitude with increasing harmonic number, they are sufficiently present to excite not only the rigid body motion of the compressor casing, but also, and more importantly, certain natural modes of the casing. This vibration is then transmitted through the isolation springs or piping into the compressor housing where it further can interact with the natural housing modes.

To alleviate sound caused by shaking forces requires improvement of the dynamic balancing of the machine. Unfortunately, slider-crank mechanisms cannot be balanced perfectly in a practical way. The perfect balancing schemas, which were invented, for example, using auxiliary gearing, are costly and space consuming. An exception is a boxer arrangement for two cylinder compressors. In general, dividing a single cylinder compressor design into a multicylinder design, will reduce the overall shaking forces, if not perfectly, then at least approximately. Of course, even in a situation where perfect balancing cannot be achieved, one should at least optimize the balancing.

2.1.2 Scotch Yoke Mechanism

This is a variation of the previous design. A pin on the rotating crank slides in a slot that is rigidly attached to the piston, as illustrated schematically in Figure 2.1. This design seems to be used only in small, fractional-horsepower refrigeration or air conditioning compressors.

The many harmonics of the slider-crank mechanism are not present in this case. However, piston slap can occur. While the rotary component of the shaking force can be easily removed by balancing, the shaking force component of the reciprocating motion of the piston requires more complicated measures.

2.1.3 Reciprocating Swing Compressors

In this design, the reciprocating motion is created by an electrodynamic actuator, which is basically a magnet-coil combination. For example, see E. Pollak et al. (1978, 1979). The amplitude of motion is load dependent. It has found application in very small compressors and is not widely used, but because of recent interest, it is listed here. A schematic is shown in Figure 2.2.

There have been a number of design variations, which include the acoustic compressor where a gas column is oscillated in a reciprocating fashion by an

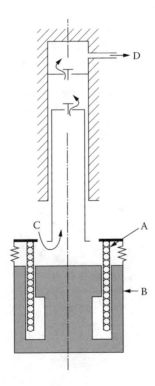

FIGURE 2.2 Example of a reciprocating swing compressor (A = coil, B = magnet, C = suction intake, D = discharge pipe).

electrodynamic actuator. There is basically only one force caused by unbalance at the frequency of the reciprocating motion. A boxer arrangement of two pistons moving against each other would balance this design.

2.2 ROTARY VANE COMPRESSORS

In this design, an eccentrically fixed rotating piston, carries multiple vanes located in radial or inclined slots. Centrifugal, gas, or spring forces press the vane tips against the circular or oval cylinder. Volumes between adjoining vanes decrease and increase, causing gas to be taken from the suction port, and compressed and exhaled through the discharge port. A schematic is shown in Figure 2.3.

A useful feature of this design from a noise control viewpoint is that this design does not require a suction valve other than an intake slot. Once any volume between two vanes passes the port slot, it is sealed by the trailing vane, and it starts compressing the gas from suction to discharge pressure. Once discharge pressure is reached, assuming that by then the volume is exposed to the discharge valve port, the discharge valve will open. The discharge valve can flutter, as discussed in the earlier section on compressor valves, and create noise by impact and generate gas pulsations in the discharge manifold, but on the suction side, while gas pulsations do occur, there is no fluttering or impacting valve.

Even more interesting, in the case of multivanes, say four, an automatic discharge valve is not necessary either, strictly speaking, because the compressed gas in the discharge system can only expand into the trailing volume, which cannot communicate with the intake. The compressor will operate without any valve, but the absence of a discharge valve will be punished by relatively large thermodynamic energy losses (the compressor will run hot). Also, the expansion of the already compressed discharge gas into the trailing volume will be done very suddenly, which

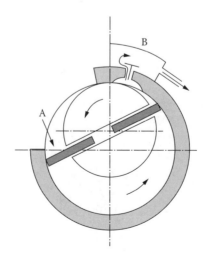

FIGURE 2.3 A two-bladed rotary vane compressor (A = suction port, B = discharge port).

will set up relatively large pressure oscillations not only in the discharge manifolding, but also in the trailing volume. Therefore, the absence of a discharge valve, while functionally possible, is not recommended from either an energy efficiency viewpoint or a noise control viewpoint. But the absence of a suction valve is a positive noise control advantage without energy loss punishment.

As it turns out, the presence of an automatic discharge valve still causes another noise mechanism to exist, in addition to valve flutter and impact. This mechanism is related to the previous discussion of explosive re-expansion of already compressed gas in the case of a missing discharge valve into the trailing volume. As it turns out, because the discharge valve cannot be designed to be flush with the interior cylinder wall, there has to be a port volume before the valve. For geometric reasons, this port volume may have to be of non-negligible size. Then the leading vane of the trailing volume passes the lip of the port volume, which is at a higher pressure than the pressure in the trailing volume, even if the valve should be closed. This port volume gas will expand explosively into the trailing volume and cause relatively large pressure oscillations in the trailing volume. These oscillations will excite the cylinder casing into vibrations that are large enough to make this mechanism an important, unique noise source in rotary vane compressors. See also, for example, Johnson and Hamilton (1972), and Yee and Soedel (1980, 1983, 1988).

Noise control measures are limited, other than to avoid an explosive re-expansion into the trailing volume, by using relief slots in the cylinder wall, which may not be desirable from a thermal efficiency viewpoint. Because the noise mechanism can be described in terms of a Helmholtz resonator with two volumes separated by a neck, which is the passage that opens as the leading vane passes the cutoff point to the discharge port, it is obvious that the oscillation frequencies can be changed by changing the port volume. This kind of detuning can be beneficial if it is determined that the original re-expansion oscillation frequencies are in resonance with other parts of the compressor system.

A positive noise control aspect of rotary vane compressors is that they can be balanced well, except for the small imbalances due to the vane masses.

At times, especially for two-vane compressors and when the difference between pressures in adjoining volumes between vanes is large, vane oscillations have been reported. The mass of a vane riding on the elastic oil film between vane and cylinder constitutes a simple oscillator, potentially excited by hydrodynamic forces. But this effect seems to occur infrequently in compressors when an oil film is not present, for example in air compressors.

2.3 ROLLING PISTON COMPRESSORS

In this design, a crank drives a piston that rolls against the inside of a circular cylindrical cylinder. Low- and high-side pressures are divided by a "stationary" vane that can move in a slot cut into the cylinder, and is pushed by spring or gas forces against the surface of the rolling piston. This is illustrated in Figure 2.4.

Again, in this design, an automatic suction valve is not needed. The suction volume fills with suction gas through a typically large intake slot, and only after the

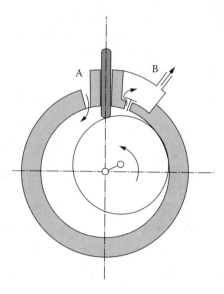

FIGURE 2.4 Rolling piston compressor (A = suction port, B = discharge port).

contact line between the rolling piston and the cylinder passes the trailing lip of the intake slot can compression begin.

The way the piston rolls off the interior of the cylinder does not cause noise problems similar to the re-expansion mechanism in rotary vane compressors. The single, stationary vane has been reported to oscillate at times on the oil film of the rolling piston, again usually when the differential pressure between discharge and suction is high.

A discharge valve is necessary. Perfect balancing of the rotor is of course possible; there is only a relatively small shaking force due to the vane moving in and out of its slot. For research published on rolling piston compressors, see Asami et al. (1982), Dreiman and Herrick (1998), Kumar et al. (1994), Seve et al. (2000), and others referenced later.

2.4 SCROLL COMPRESSORS

Scroll compressors compress gas in pockets formed by moving and stationary scroll elements that resemble spirals. Automatic valves are not needed; intake is directly from the suction gas supply into an open pocket. As the moving scroll orbits in a circular motion, the gas pockets diminish in size and the compressed gas is ejected when a particular pocket opens to the discharge manifold. Thus, the discharge valve is a gate valve. It opens at a kinematically prescribed position and is subject to over- or underpressure expansions at pressure ratios that do not conform to the design condition. To prevent backflow of already compressed gas into the pocket that starts to open to the discharge manifold, a check valve is used in some designs. The design is illustrated in Figure 2.5.

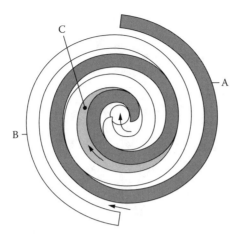

FIGURE 2.5 Schema of a scroll compressor (A = stationary scroll, B = orbiting scroll, C = gas pocket).

Noise is generated if the gas is overpressurized in the pocket that is about to open to the discharge manifold, it will discharge suddenly into the discharge manifold. On the other hand, if the kinematic timing is such that the gas has not yet reached discharge pressure when it opens to the discharge manifold, a sudden backflow into the pocket develops. (From a volumetric efficiency viewpoint, no loss is expected because the gas in the pocket is eventually pushed out anyway.)

The moving scroll can be balanced. This type of compressor can be mounted in the hermetic shell rigidly without springs and can result in a relatively compact design because of the relative absence of shaking forces. Of course, the general noise sources such as scroll and casing vibrations due to the pressure variations discussed earlier, and gas pulsations, still do potentially occur. For additional information, see, for example, Bukac (2004), Morimoto et al. (1996), Motegi and Nahashima (1996), and others referenced later.

2.5 SCREW COMPRESSORS

Screw compressors rely on volume changes between the lobes of male and female helical screws. (In the so-called *star screw compressor* design, there is only one helical screw that interacts with a star-shaped gear to form diminishing gas pockets.) No automatic valves are needed. Suction intake is direct into open pockets. The discharge valve is a kinematically fixed gate arrangement. At ideal design conditions, a particular gas pocket discharges gas at discharge pressure into the discharge manifold in a relatively smooth way, but if the compression ratios are such that the gas pressure in a pocket is higher than the discharge pressure as it opens to the discharge manifold, a sudden "explosive" discharge occurs, accentuating gas pulsations. If the pressure in the pocket is below discharge pressure when the pocket opens to the discharge manifold, then a sudden backflow will occur, again accentuating gas

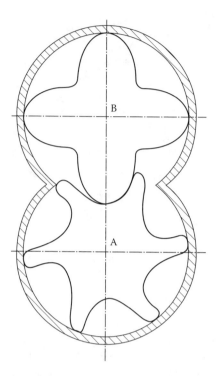

FIGURE 2.6 Cross-sectional view of a twin screw compressor (A = female screw rotor, B = male screw rotor).

pulsations. Gas pulsation–related noise is, therefore, lowest at the design condition for which the gate valve arrangement is designed. A typical design is illustrated in Figure 2.6.

An advantage of the screw compressor design is that there are multiple discharges per revolution, which is, in general, preferable to a single discharge per revolution from a gas pulsation viewpoint. On the other hand, the multiple discharges per revolution create a "pure" tone, which usually has to be suppressed by a discharge muffler.

Another potential mechanism of noise generation that is specific to screw compressors is chatter of the female helical screw on the driven, male screw. This is similar to gear chatter in spur gears, caused by kinematic imperfections, for example. Chatter seems to be least severe for high compression ratios because the gas pressure in the screw pockets prevents or diminishes kinematic separation.

Because balancing can be made perfect, no vibration isolation between casing and hermetic shell is needed; they become integral parts. For additional information, see Adams and Soedel (1992, 1994), Koai and Soedel (1990a, b), Shapiro (1992), and others.

3 Natural Frequencies and Modes of Compressor Housings

The reason that housing shell vibrations are discussed first is that for the very large field of hermetically sealed small and mid-sized refrigerating and air conditioning compressors, housing shell vibrations are the primary noise source as far as the customer is concerned. If all the interior noise and vibration sources could be prevented from being transmitting to the housing shell and attached piping, the system would be vibration-free and quiet. See, for example, Bush et al. (1992), Conrad and Soedel (1992), Gilliam and DiFlora (1992), Kelly and Knight (1992a), Soedel (1980 a), Tavakoli and Singh (1990), and others.

Sound is radiated from vibrating surfaces. The gas pulsations inside a hermetically sealed refrigeration compressor are converted into vibrations of the compressor housing shell and then radiated from this surface.

Therefore, it is important to develop some understanding of how a surface, be it a curved shell or a flat plate, vibrates, and what sources excite this vibration.

3.1 EQUATIONS OF MOTION

3.1.1 WHAT IS A SHELL?

A shell is a sheet structure that is curved, just as an arch is a beamlike structure that is curved. A shell that is flattened (mathematically a shell of zero curvature) becomes a plate, just as a flattened arch becomes a beam. A slice of a cylindrical shell in the circumferential direction is an arch or ring, just as a slice of a plate is a beam.

A shell may not be uniform. It does not have to be homogeneous or isotropic. (*Homogeneous* means that the material is the same everywhere in the shell, and *isotropic* means that the properties of the material are the same in all directions.)

A shell may be classified as thin or thick. For thin shells the stress distribution can be taken as linear through the thickness, similar to classical beam theory. A compressor housing is an example of a thin shell.

In the following, the general equations of motion of any thin shell or plate are derived by outline. *Thin* means that the thickness of the structure is small as compared to its surface dimensions. Interested readers should consult Soedel (2004) or a similar text for more details.

3.1.2 Reference Surface and Coordinates

In experimental work, locations on the shell are usually identified on the surface that is accessible to the observer or instrument. Locations may be defined by simply marking points or grids on that surface, without regard to any mathematical definition. Only if the experimental work is to confirm a theory or is to be supplemented by theory does conformation to mathematically formulatable coordinates become important.

In theoretical work, or where experimental work is to be used for theoretical considerations, more care has to be given to coordinate selection. Although any mathematically formulatable location description can be used, use of orthogonal coordinates provides a tremendous advantage. In orthogonal coordinate systems, any two-location lines drawn on the shell reference surface (or in an approximate sense, on the observed surface) cut each other at right angles. Use of orthogonal coordinates simplifies the equations of motion considerably.

The shell reference surface is that surface on which the coordinate location lines are defined. In homogeneous shells, the reference surface is usually taken as halfway between the outer and inner surface, and corresponds to the neutral surface, which is analogous to the neutral fiber of a beam, where zero bending stress occurs during pure bending. Selection of the neutral surface as the reference surface also has the advantage that the mathematical formulations of bending strains and membrane strains, in terms of local moment and force resultants, are uncoupled, which again simplifies the equations of motion.

Because it is time consuming to derive the equations of motion separately for cylindrical shells of oval or circular cross-section, for spherical shells, for circular or rectangular plates, or for stiffening rings or stringers, it has proven effective to derive the governing equations in curvilinear coordinates.

From the simplest conceptual viewpoint, we start with formulating the diagonal of an infinitesimal element. For a rectangular plate in Cartesian coordinates, it is

$$(ds)^2 = (dx)^2 + (dy)^2. \tag{3.1}$$

For a circular plate, it is

$$(ds)^2 = (dr)^2 + r^2(d\theta)^2. \tag{3.2}$$

For a circular cylindrical shell, in cylindrical coordinates, it is

$$(ds)^2 = (dx)^2 + a^2(d\theta)^2 \tag{3.3}$$

as shown in Figure 3.1. For a spherical shell, in spherical coordinates, it is

$$(ds)^2 = a^2(d\phi)^2 + a^2\sin^2\phi(d\theta)^2 \tag{3.4}$$

as shown in Figure 3.2.

Natural Frequencies and Modes of Compressor Housings

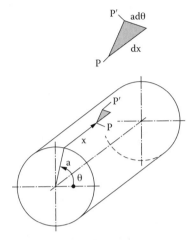

FIGURE 3.1 Cylindrical coordinates and the Lamé parameter triangle.

We note that all these expressions can be written as

$$(ds)^2 = A_1^2(d\alpha_1)^2 + A_2^2(d\alpha_2)^2. \tag{3.5}$$

This is the *fundamental form* and A_1 and A_2 are called the *Lamé parameters*, defined for the specific coordinate systems as given in Table 3.1, and α_1 and α_1 are the *curvilinear* coordinates.

Note again that the curvilinear coordinates describe the reference surface, often the *neutral* surface of the structure. The curvilinear coordinates must also be orthogonal.

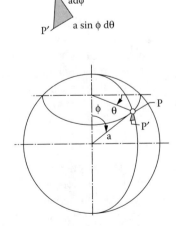

FIGURE 3.2 Spherical coordinates and the Lamé parameter triangle.

TABLE 3.1
Lamé Parameters

	α_1	α_2	A_1	A_2
Rectangular	x	y	1	1
Circular Plate	r	θ	1	r
Cylindrical	x	θ	1	a
Spherical Shell	φ	θ	a	a sin φ

At this point the reader may say: "I appreciate the idea of a reference surface and the desirability of this surface being the neutral surface. I am also willing to believe that there is great advantage in the use of orthogonal coordinates. But how do I go about defining orthogonal coordinates for the shell I am investigating?" Fortunately, it is known that on a curved surface, lines that define the direction of maximum curvature cross lines that define the direction of minimum curvature at right angles. If we use a curvature meter consisting of three points on a straight line with the center point attached to a dial indicator, we merely need to rotate this instrument until the indicator shows minimum elevation in order to find the local direction of minimum curvature. We can then move in this direction a small distance and establish a new direction of minimum curvature in the same way as before, and proceed to trace a line as we go. After the first line has been traced from boundary to boundary, or once around, we move a small distance parallel to it and repeat the process, until we have covered the shell with lines in the minimum curvature direction. We can do precisely the same for lines in the maximum curvature direction.

The third coordinate is normal to the reference surface and is used to define distance from the reference surface and the direction of normal displacement.

For certain classical geometries, the lines of minimum and maximum curvature direction are obvious; these include cylindrical and spherical surfaces, for which corresponding coordinate systems are well known.

For finite element analysis, one need not use orthogonal coordinates; a Cartesian coordinate system with an arbitrary origin location generally suffices for defining the location of the node points of the finite elements. However, in the formulation of the finite element properties, orthogonal coordinates are used.

3.1.3 Strain–Stress Relationships

In Figure 3.3, α_1 and α_2 are coordinate lines on the reference surface, usually halfway between the inner and outer surface of the shell, while α_3 is a coordinate normal to the reference surface. An infinitesimal cube is cut from the shell material and its possible normal and shear stresses are shown. The dimensions of the cube are $A_1(1+\alpha_3/R_1)d\alpha_1 \times A_2(1+\alpha_3/R_2) \times d\alpha_3$, where R_1 and R_2 are the radii of curvature of the shell (they are infinity for flat plates).

Natural Frequencies and Modes of Compressor Housings

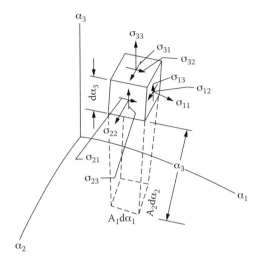

FIGURE 3.3 Stresses on an element.

The strains are related to stresses by

$$\epsilon_{11} = \frac{1}{E}[\sigma_{11} - \mu(\sigma_{22} + \sigma_{33})] \tag{3.6}$$

$$\epsilon_{22} = \frac{1}{E}[\sigma_{22} - \mu(\sigma_{11} + \sigma_{33})] \tag{3.7}$$

$$\epsilon_{33} = \frac{1}{E}[\sigma_{33} - \mu(\sigma_{11} + \sigma_{22})] \tag{3.8}$$

$$\epsilon_{12} = \frac{\sigma_{12}}{G} \tag{3.9}$$

$$\epsilon_{13} = \frac{\sigma_{13}}{G} \tag{3.10}$$

$$\epsilon_{23} = \frac{\sigma_{23}}{G} \tag{3.11}$$

where E and G are Young's modulus and shear modulus, respectively.

Because we are dealing with compressor housings that are typically thin shells, we may assume that normal stresses are negligibly small (at most equal to the normal pressure loading on the shell):

$$\sigma_{33} = 0. \tag{3.12}$$

Furthermore, for a thin structure, the transverse shear strain (not the integrated effect of transverse shear stress!) is negligible:

$$\epsilon_{13} = \epsilon_{23} = 0. \tag{3.13}$$

This reduces the equations we have to work with to

$$\epsilon_{11} = \frac{1}{E}(\sigma_{11} - \mu\sigma_{22}) \tag{3.14}$$

$$\epsilon_{22} = \frac{1}{E}(\sigma_{22} - \mu\sigma_{11}) \tag{3.15}$$

$$\epsilon_{12} = \frac{\sigma_{12}}{G}. \tag{3.16}$$

We also obtain a third equation,

$$\epsilon_{33} = -\frac{\mu}{E}(\sigma_{11} + \sigma_{22}), \tag{3.17}$$

which is of some interest in noise calculations because it is used to calculate the constriction of the shell thickness during vibration, which is an additional noise-generating mechanism besides the more important transverse vibration.

3.1.4 Strain–Displacement Relationships

It must be recognized that any point of a structure can vibrate in three directions, designated as u_1 in the α_1–direction, u_2 in the α_2–direction, and u_3 in α_3, the transverse direction, the displacements u_i being defined on the reference surface.

Employing the usual definitions of normal strain,

$$\epsilon_{ii} = \frac{(ds')_i - (ds)_i}{(ds)_i} \tag{3.18}$$

where $(ds')_i$ is the deformed length and $(ds)_i$ is the original, undeformed length, both in the i–direction, we obtain after some manipulation,

$$\epsilon_{11} = \epsilon_{11}^\circ + \alpha_3 k_{11} \tag{3.19}$$

$$\epsilon_{22} = \epsilon_{22}^\circ + \alpha_3 k_{22} \tag{3.20}$$

where the so-called membrane strains are (due to stretching of the reference surface)

$$\epsilon_{11}^o = \frac{1}{A_1}\frac{\partial u_1}{\partial \alpha_1} + \frac{u_2}{A_1 A_2}\frac{\partial A_1}{\partial \alpha_2} + \frac{u_3}{R_1} \tag{3.21}$$

$$\epsilon_{22}^o = \frac{1}{A_2}\frac{\partial u_2}{\partial \alpha_2} + \frac{u_1}{A_1 A_2}\frac{\partial A_2}{\partial \alpha_1} + \frac{u_3}{R_2}, \tag{3.22}$$

and where the changes of curvature due to bending of the housing are

$$k_{11} = \frac{1}{A_1}\frac{\partial \beta_1}{\partial \alpha_1} + \frac{\beta_2}{A_1 A_2}\frac{\partial A_1}{\partial \alpha_2} \tag{3.23}$$

$$k_{22} = \frac{1}{A_2}\frac{\partial \beta_2}{\partial \alpha_2} + \frac{\beta_1}{A_1 A_2}\frac{\partial A_2}{\partial \alpha_1} \tag{3.24}$$

and where

$$\beta_1 = \frac{u_1}{R_1} - \frac{1}{A_1}\frac{\partial u_3}{\partial \alpha_1} \tag{3.25}$$

$$\beta_2 = \frac{u_1}{R_2} - \frac{1}{A_2}\frac{\partial u_3}{\partial \alpha_2}. \tag{3.26}$$

The shear strain is the difference between the original 90° angle of the two faces forming a corner of a cubic infinitesimal element, and the angle θ_{12} of the corner after shear deformation has taken place:

$$\epsilon_{12} = \frac{\pi}{2} - \theta_{12}. \tag{3.27}$$

This results in

$$\epsilon_{12} = \epsilon_{12}^o + \alpha_3 k_{12} \tag{3.28}$$

where

$$\epsilon_{12}^o = \frac{A_2}{A_1}\frac{\partial}{\partial \alpha_1}\left(\frac{u_2}{A_2}\right) + \frac{A_1}{A_2}\frac{\partial}{\partial \alpha_2}\left(\frac{u_1}{A_1}\right) \tag{3.29}$$

$$k_{12} = \frac{A_2}{A_1}\frac{\partial}{\partial \alpha_1}\left(\frac{\beta_2}{A_2}\right) + \frac{A_1}{A_2}\frac{\partial}{\partial \alpha_2}\left(\frac{\beta_1}{A_1}\right) \tag{3.30}$$

3.1.5 MEMBRANE FORCE AND BENDING MOMENT RESULTANTS

To avoid extensive complexity of the equations, the next step is to integrate the stresses acting on each face of each cubic element over the thickness of the shell. For this purpose we imagine a sliver of material cut from the shell, as shown in Figure 3.4. From now on, we assume that the chosen reference surface is halfway between the inner and outer surface of the shell (or plate), the thickness being designated as h.

The force resultant per unit length of reference surface is, in the α_1-direction,

$$N_{11} = \int_{-h/2}^{h/2} \sigma_{11} d\alpha_3 = K\left(\epsilon_{11}^o + \mu\, \epsilon_{22}^o\right) \quad (3.31)$$

where K is

$$K = \frac{Eh}{1-\mu^2}. \quad (3.32)$$

Physically, K can be interpreted to be the in-plane stiffness, also called the *membrane stiffness*, and it is equivalent to the product EA in the equation of a vibrating rod, A being the cross-section.

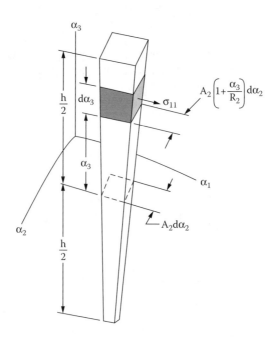

FIGURE 3.4 Integrating the effect of element stresses over the thickness of the shell.

Similarly, we obtain the force resultant in the α_2-direction,

$$N_{22} = K\left(\epsilon_{22}^o + \mu\,\epsilon_{11}^o\right), \tag{3.33}$$

and the in-plane shear force resultant

$$N_{12} = N_{21} = \frac{K(1-\mu)}{2}\epsilon_{12}^o. \tag{3.34}$$

To obtain the bending moment in the α_1-direction, we integrate

$$M_{11} = \int_{-h/2}^{h/2} \sigma_{11}\alpha_3 d\alpha_3 = D(k_{11} + \mu k_{22}), \tag{3.35}$$

where

$$D = \frac{Eh^3}{12(1-\mu^2)} \tag{3.36}$$

D is the so-called *bending stiffness*, equivalent to the product EI in the transverse vibration of beams, I being the area moment.

Similarly, we obtain the bending moment in the α_2-direction,

$$M_{22} = D(k_{22} + \mu k_{11}), \tag{3.37}$$

and the twisting moment in the tangential plane to the shell surface,

$$M_{12} = M_{21} = \frac{D(1-\mu)}{2}k_{12}. \tag{3.38}$$

While we are neglecting the transverse shear strains as small, we do not neglect the integrated effect of the transverse shear stresses. The transverse shear resultants are Q_{13} and Q_{23} at the faces of the sliver normal to the α_1 and α_2 directions, respectively, and are defined as

$$Q_{13} = \int_{-h/2}^{h/2} \sigma_{13}\,d\alpha_3 \tag{3.39}$$

$$Q_{23} = \int_{-h/2}^{h/2} \sigma_{23}\,d\alpha_3. \tag{3.40}$$

3.1.6 Energy Expressions

The equations of motion can be derived in at least two ways. We can draw a free body diagram of all the forces that act on our sliver of infinitesimal cross-section and use Newton's second law, or we can use an energy-based derivation such as Hamilton's principle. The former has the disadvantage of geometric complexity because of the curvature of the shell. Therefore, Hamilton's principle is typically preferred.

Also, the energy expressions are needed if we intend to develop a finite element approach.

The strain energy stored in the shell structure is the energy stored in one infinitesimal cubic element, integrated over the total volume of the structure:

$$U = \iiint_{\alpha_1 \alpha_2 \alpha_3} F \, dV \tag{3.41}$$

where

$$F = \frac{1}{2}(\sigma_{11} \epsilon_{11} + \sigma_{22} \epsilon_{22} + \sigma_{12} \epsilon_{12} + \sigma_{13} \epsilon_{13} + \sigma_{23} \epsilon_{23}) \tag{3.42}$$

$$dV = A_1 A_2 d\alpha_1 d\alpha_2 d\alpha_3. \tag{3.43}$$

In terms of strains, having integrated over α_3 from $-h/2$ to $+h/2$:

$$U = \frac{1}{2} \iint_{\alpha_1 \alpha_2} \left[K \left\{ \epsilon_{11}^{o2} + 2\mu \epsilon_{11}^o \epsilon_{22}^o + \epsilon_{22}^{o2} + \left(\frac{1-\mu}{2}\right) \epsilon_{12}^{o2} \right\} \right. \\ \left. + D \left\{ k_{11}^2 + 2\mu \, k_{11} k_{22} + k_{22}^2 + \left(\frac{1-\mu}{2}\right) k_{12}^2 \right\} \right] A_1 A_2 d\alpha_1 d\alpha_2. \tag{3.44}$$

The kinetic energy is, in terms of the three possible displacements at any point of the reference surface,

$$T = \frac{\rho h}{2} \iint_{\alpha_1 \alpha_2} \left(\dot{u}_1^2 + \dot{u}_2^2 + \dot{u}_3^2 \right) A_1 A_2 d\alpha_1 d\alpha_2. \tag{3.45}$$

3.1.7 Hamilton's Principle and General Equations of Motion

Hamilton (see Soedel, 2004) postulated that while there are usually several possible paths along which a dynamic system may move from one point to another in space and time, the path actually followed is the one that minimizes the time integral of the difference between kinetic and potential energies. In terms of the calculus of variations, it is usually stated as

$$\delta \int_{t_1}^{t_2} (T - U + W_{nc})\, dt = 0; \quad \delta \bar{r}_i = 0 \quad \text{at} \quad t = t_1 \text{ and } t_2. \qquad (3.46)$$

Newton's second law can be derived from this principle, or vice versa, the principle can be derived from Newton's second law.

The term W_{nc} designates any kind of energy input due to forcing, or with a negative sign, any kind of energy lost to the system, for example, due to damping.

The details of the mathematical procedure can be found in Soedel (2004). Here, it is sufficient to state the resulting equations. We obtain three coupled equations that describe the motion tangential to the shell reference surface,

$$-\frac{\partial(N_{11}A_2)}{\partial \alpha_1} - \frac{\partial(N_{21}A_1)}{\partial \alpha_2} - N_{12}\frac{\partial A_1}{\partial \alpha_2} + N_{22}\frac{\partial A_2}{\partial \alpha_1} - A_1 A_2 \frac{Q_{13}}{R_1} + A_1 A_2\, \rho \ddot{u}_1 = A_1 A_2 q_1 \qquad (3.47)$$

$$-\frac{\partial(N_{12}A_2)}{\partial \alpha_1} - \frac{\partial(N_{22}A_1)}{\partial \alpha_2} - N_{21}\frac{\partial A_2}{\partial \alpha_1} + N_{11}\frac{\partial A_1}{\partial \alpha_2} - A_1 A_2 \frac{Q_{23}}{R_2} + A_1 A_2\, \rho \ddot{u}_2 = A_1 A_2 q_2 \qquad (3.48)$$

and the motion in the transverse direction,

$$-\frac{\partial(Q_{13}A_2)}{\partial \alpha_1} - \frac{\partial(Q_{23}A_1)}{\partial \alpha_2} + A_1 A_2 \left(\frac{N_{11}}{R_1} + \frac{N_{22}}{R_2} \right) + A_1 A_2\, \rho \ddot{u}_3 = A_1 A_2 q_3, \qquad (3.49)$$

where q_1, q_2, and q_3 are the forcing functions.

In addition, we obtain two more equations that define for us the transverse shear resultants Q_{13} and Q_{23}, which were carried along as unknowns:

$$\frac{\partial(M_{11}A_2)}{\partial \alpha_1} + \frac{\partial(M_{21}A_1)}{\partial \alpha_2} + M_{12}\frac{\partial A_1}{\partial \alpha_2} - M_{22}\frac{\partial A_2}{\partial \alpha_1} - Q_{13}\, A_1 A_2 = 0 \qquad (3.50)$$

$$\frac{\partial(M_{12}A_2)}{\partial \alpha_1} + \frac{\partial(M_{22}A_1)}{\partial \alpha_2} + M_{21}\frac{\partial A_2}{\partial \alpha_1} - M_{11}\frac{\partial A_1}{\partial \alpha_2} - Q_{23}\, A_1 A_2 = 0. \qquad (3.51)$$

Including equivalent viscous damping or hysteresis damping, it is possible to write the equations in displacement form (by substituting the equations for force and moment resultants, and for the strains in terms of displacements).

3.1.8 How Do Shells Vibrate?

As we have seen, shell motions in general are governed by three coupled equations of motion, with vibration displacements defined in the three orthogonal directions. The fact that these equations are coupled implies, for example, that a transverse vibration has associated with it two in-plane vibration components. Consider the example of a circular cylindrical shell vibrating in a squash-in/squash-out natural mode, for which there occur four longitudinal lines of zero transverse motion on the shell surface. Although the transverse displacement is zero at these lines, the in-plane displacements are nonzero; zero in-plane displacements would require considerable stretching or compression (high membrane strains), which the shell is not likely to support because the nodal lines are not fixed boundaries that restrain in-plane motion. The reader may illustrate this behavior by fabricating two equal paper cylinders, squeezing one into an approximation of the squash-in/squash-out mode shape, and comparing it to the undisturbed cylinder. The in-plane motion will be clearly visible.

The coupling between the transverse and in-plane vibration of a shell is due to its curvature. This coupling occurs at even very small vibration amplitudes and should be of interest to the practical engineer who may have wondered about large transverse amplitudes of his shell in the absence of transverse load or edge-bending excitation. This coupling also implies that for strain gauge measurements, one should mount gauges on the inner and outer surface of the shell at the same reference location, in order to separate the strain into bending and membrane effects.

3.2 BOUNDARY CONDITIONS

The equations of motion require the specification of four boundary conditions on each edge (they are of the 8th order). It can be shown from Hamilton's principle (see Soedel, 2004), that we have to satisfy, on an α_2 edge:

- the in-plane deflection u_1 or a modified Kirchhoff shear resultant T_{21}
- the transverse deflection u_3 or a modified Kirchhoff shear resultant V_{23}
- the other in-plane deflection u_2 or the force resultant N_{22}
- and the slope β_2 or the bending moment M_{22}

Similar to beams, free edges make a shell less stiff than clamped edges, and other boundary conditions provide interim values of stiffness.

Just in passing, it is important for the experimenter who works with scaled models to understand the theoretical requirements and the role of boundary conditions. It is a common failing that shell models are scaled correctly but boundary conditions are not. We have similar choices along an α_1 edge. For details of the argument, see Soedel (2004).

3.3 NATURAL FREQUENCIES AND MODES

For an elastic system that is excited harmonically, there occur certain frequencies at which the vibration amplitudes become very large and at which the system vibrates in fixed patterns that are independent of the location of the excitation. These frequencies are called *natural frequencies*, and the patterns are called natural modes. There are as many of these modes as the system has degrees of freedom. The number of degrees of freedom is determined by the number of ways system masses or mass particles are able to move, that is, the minimum number of coordinates that are necessary to define these motions. In a continuous system, like a beam or a shell, there is an infinity of degrees of freedom, and thus an infinite number of natural frequencies and modes. Fortunately, only the first few are usually of technical importance.

Natural modes of beams have *node points*, at which the beam does not have any transverse vibration when it vibrates at the corresponding natural frequency. If the beam is, at the same time, vibrated in the longitudinal direction at the same frequency, then the node point is in the most general sense not vibration free, only free of transverse vibration. In general, the node points associated with vibrations in the longitudinal direction differ from node points associated with transverse vibration. Transverse vibration node points can be made visible by sprinkling sand or powder on the beam; the powder collects at these points.

An analogous situation occurs for plates or shells vibrating at a natural frequency in a natural mode, except that here there occur lines of zero normal displacements, and sand or powder will collect and describe these lines beautifully. This is easily done for plates or shallow shells, but for deep shells, the powder needs to have some adhesion, so that it does not fall off due to gravity alone. Making node lines visible on shells is actually so tricky that experimenters have turned to holography to obtain node line information.

It is of some historical interest (see historical discussion in Soedel (2004)) to note that node lines were first exhibited by the German engineer and physicist Chladni. When he demonstrated them to Emperor Napoleon Bonaparte of France, the emperor, a trained military engineer, was so fascinated by these lines that he donated a sum of money as a prize for the best explanation of the phenomenon. It was won by a young woman, Sophie Germaine, in competition with the leading minds of her time. Later on, a somewhat more rigorous explanation was given by Kirchhoff.

To us these lines are still very important because they enable us to visualize and sketch mode shapes at a glance, and to deduce from them some useful information. For example, from modal behavior, we can estimate locations of maximum curvature, and thus locations of maximum modal stress. This aids us in determining where strain gauges are to be placed. Because we know that dynamic forces acting at a node line do not excite the particular mode, and that dynamic moments acting at a node line and about that line excite the particular mode strongly, we can select force and application points to minimize modal response.

For more information on the natural modes of shells see Bucciarelli et al. (1992), Soedel (2004), Hsu and Soedel (1987), Kim and Soedel (1998), Lee and Kim (2000), and Soedel (1980a).

3.3.1 THE COMPRESSOR HOUSING SIMPLIFIED AS A CYLINDRICAL SHELL

While for accuracy of prediction the typical hermetic compressor shell has to be analyzed using finite elements, developed on the same theoretical basis as the foregoing energy expressions that led to the equations of motion, much can be learned from simpler models. For example, many hermetic compressor shells are approximately cylindrical. The cross-sections can be viewed as approximately circular (the influence of deviations from the circular cross-section will be discussed later). While the end caps will contribute their own dynamics (as will be discussed later), it is not a bad approximation to replace the end caps with simple support boundary conditions. A simple support suppresses transverse motion, but allows free rotation about the support.

The equations of motion are obtained by realizing that for cylindrical coordinates, $\alpha_1 = x$, $\alpha_2 = \theta$, $A_1 = 1$, $A_2 = a$, where a is the average radius of the shell (see also Figure 3.1). Substituting this into the general shell equations developed previously, they reduce to

$$\frac{\partial N_{xx}}{\partial x} + \frac{1}{a}\frac{\partial N_{\theta x}}{\partial \theta} + q_x = \rho h \frac{\partial^2 u_x}{\partial t^2} \quad (3.52)$$

$$\frac{\partial N_{x\theta}}{\partial x} + \frac{1}{a}\frac{\partial N_{\theta\theta}}{\partial \theta} + \frac{Q_{\theta 3}}{a} + q_\theta = \rho h \frac{\partial^2 u_\theta}{\partial t^2} \quad (3.53)$$

$$\frac{\partial Q_{x3}}{\partial x} + \frac{1}{a}\frac{\partial Q_{\theta 3}}{\partial \theta} - \frac{N_{\theta\theta}}{a} + q_3 = \rho h \frac{\partial^2 u_3}{\partial t^2} \quad (3.54)$$

where

$$Q_{x3} = \frac{\partial M_{xx}}{\partial x} + \frac{1}{a}\frac{\partial M_{\theta x}}{\partial \theta} \quad (3.55)$$

$$Q_{\theta 3} = \frac{\partial M_{x\theta}}{\partial x} + \frac{1}{a}\frac{\partial M_{\theta\theta}}{\partial \theta}. \quad (3.56)$$

The strain–displacement relations become:

$$\varepsilon_{xx}^o = \frac{\partial u_x}{\partial x} \tag{3.57}$$

$$\varepsilon_{\theta\theta}^o = \frac{1}{a}\frac{\partial u_\theta}{\partial \theta} + \frac{u_3}{a} \tag{3.58}$$

$$\varepsilon_{x\theta}^o = \frac{\partial u_\theta}{\partial x} + \frac{1}{a}\frac{\partial u_x}{\partial \theta} \tag{3.59}$$

$$k_{xx} = \frac{\partial \beta_x}{\partial x} \tag{3.60}$$

$$k_{\theta\theta} = \frac{1}{a}\frac{\partial \beta_\theta}{\partial \theta} \tag{3.61}$$

$$k_{x\theta} = \frac{\partial \beta_\theta}{\partial x} + \frac{1}{a}\frac{\partial \beta_x}{\partial \beta} \tag{3.62}$$

and β_1 and β_2 become:

$$\beta_x = -\frac{\partial u_3}{\partial x} \tag{3.63}$$

$$\beta_\theta = \frac{u_\theta}{a} - \frac{1}{a}\frac{\partial u_3}{\partial \theta}. \tag{3.64}$$

Setting $q_x = q_\theta = q_3 = 0$ for natural frequency and mode calculations, it turns out that for simple support boundary conditions with free motion in the axial direction,

$$u_3(o,\theta,t) = 0 \tag{3.65}$$

$$u_\theta(o,\theta,t) = 0 \tag{3.66}$$

$$M_{xx}(o,\theta,t) = 0 \tag{3.67}$$

$$N_{xx}(o,\theta,t) = 0 \tag{3.68}$$

and

$$u_3(L,\theta,t) = 0 \tag{3.69}$$

$$u_\theta(L,\theta,t) = 0 \tag{3.70}$$

$$M_{xx}(L,\theta,t) = 0 \tag{3.71}$$

$$N_{xx}(L,\theta,t) = 0. \tag{3.72}$$

This set of equations has an exact, closed form solution:

$$u_x(x,\theta,t) = U_x(x,\theta)e^{j\omega t} \tag{3.73}$$

$$u_\theta(x,\theta,t) = U_\theta(x,\theta)e^{j\omega t} \tag{3.74}$$

$$u_3(x,\theta,t) = U_3(x,\theta)e^{j\omega t} \tag{3.75}$$

where

$$U_x(x,\theta) = A \cos\frac{m\pi x}{L}\cos n(\theta-\phi) \tag{3.76}$$

$$U_\theta(x,\theta) = B \sin\frac{m\pi x}{L}\sin n(\theta-\phi) \tag{3.77}$$

$$U_3(x,\theta) = C \sin\frac{m\pi x}{L}\cos n(\theta-\phi) \tag{3.78}$$

and where $m = 1, 2, \cdots$ and $n = 0, 1, 2, \cdots$; ω represents the as yet unknown natural frequencies, and A, B, and C are as yet unknown constants. Substituting the solution terms in the equations of motion, we obtain

$$\begin{bmatrix} \rho h\omega^2 - k_{11} & k_{12} & k_{13} \\ k_{21} & \rho h\omega^2 - k_{22} & k_{23} \\ k_{31} & k_{32} & \rho h\omega^2 - k_{33} \end{bmatrix} \begin{Bmatrix} A \\ B \\ C \end{Bmatrix} = 0 \tag{3.79}$$

where

$$k_{11} = K\left[\left(\frac{m\pi}{L}\right)^2 + \frac{1-\mu}{2}\left(\frac{n}{a}\right)^2\right] \quad (3.80)$$

$$k_{12} = k_{21} = K\frac{1+\mu}{2}\frac{m\pi}{L}\frac{n}{a} \quad (3.81)$$

$$k_{13} = k_{31} = \frac{\mu K}{a}\frac{m\pi}{L} \quad (3.82)$$

$$k_{22} = \left(K + \frac{D}{a^2}\right)\left[\frac{1-\mu}{2}\left(\frac{m\pi}{L}\right)^2 + \left(\frac{n}{a}\right)^2\right] \quad (3.83)$$

$$k_{23} = k_{32} = -\frac{K}{a}\frac{n}{a} - \frac{D}{a}\frac{n}{a}\left[\left(\frac{m\pi}{L}\right)^2 + \left(\frac{n}{a}\right)^2\right] \quad (3.84)$$

$$k_{33} = D\left[\left(\frac{m\pi}{L}\right)^2 + \left(\frac{n}{a}\right)^2\right]^2 + \frac{K}{a^2}. \quad (3.85)$$

This matrix equation can only have a nonzero solution if the determinant of the matrix is zero. In expanded form, this gives:

$$\omega^6 + a_1\omega^4 + a_2\omega^2 + a_3 = 0 \quad (3.86)$$

where

$$a_1 = -\frac{1}{\rho h}(k_{11} + k_{22} + k_{33}) \quad (3.87)$$

$$a_2 = \frac{1}{(\rho h)^2}\left(k_{11}k_{33} + k_{22}k_{33} + k_{11}k_{22} - k_{23}^2 - k_{12}^2 - k_{13}^2\right) \quad (3.88)$$

$$a_3 = \frac{1}{(\rho h)^3}\left(k_{11}k_{23}^2 + k_{22}k_{13}^2 + k_{33}k_{12}^2 + 2k_{12}k_{23}k_{13} - k_{11}k_{22}k_{33}\right). \quad (3.89)$$

This cubic equation in ω^2 gives us three sets of natural frequencies, as illustrated in Figure 3.5, for a relatively small steel shell ($E = 20.6 \times 10^4$ N/mm², $\rho = 7.85 \times 10^{-9}$ N sec²/mm⁴, $\mu = 0.3$) of thickness $h = 2$ mm, radius $a = 100$ mm, and length $L = 200$ mm. This means that every mn combination, which determines the natural modes (there are $m-1$ interior nodal circles and $2n$ nodal lines), occurs at three different natural frequencies.

In general, for the lower set of natural frequencies, transverse motion dominates the modes. This is the set that is typically of most concern to the compressor-housing designer. The other two sets correspond to modes where tangential displacement dominates the motion. Conventional wisdom says that they are not important because they are not easily excited by gas pulsation pressures acting normal to the surface. But as knowledge is developed about the mechanisms of structurally transmitted noise, for instance, due to valve impact, a new importance may have to be attached to these higher-frequency branches.

The fact that the lower-frequency branch corresponds to modes where transverse motion dominates, and the higher branches correspond to modes where tangential motion dominates, can be illustrated by substituting the natural frequencies back into the matrix equation and solving for the ratios of the tangential mode component amplitudes to the transverse mode component amplitude:

$$\frac{A_i}{C_i} = -\frac{k_{13}\left(\rho h \omega_{imn}^2 - k_{22}\right) - k_{12}k_{23}}{\left(\rho h \omega_{imn}^2 - k_{11}\right)\left(\rho h \omega_{imn}^2 - k_{22}\right) - k_{12}^2} \tag{3.90}$$

$$\frac{B_i}{C_i} = -\frac{k_{23}\left(\rho h \omega_{imn}^2 - k_{11}\right) - k_{21}k_{13}}{\left(\rho h \omega_{imn}^2 - k_{11}\right)\left(\rho h \omega_{imn}^2 - k_{22}\right) - k_{12}^2}. \tag{3.91}$$

The expression for the natural modes becomes:

$$\begin{Bmatrix} U_x \\ U_\theta \\ U_3 \end{Bmatrix} = C_i \begin{Bmatrix} \dfrac{A_i}{C_i} \cos \dfrac{m\pi x}{L} \cos n(\theta - \phi) \\ \dfrac{B_i}{C_i} \sin \dfrac{m\pi x}{L} \sin n(\theta - \phi) \\ \sin \dfrac{m\pi x}{L} \cos n(\theta - \phi) \end{Bmatrix} \tag{3.92}$$

The ratios of the mode amplitudes A_i/C_i and B_i/C_i are illustrated in Figure 3.6 for the modes where transverse motion is dominant. It should be noted that for n = 2, the ratio B_i/C_i is −0.5, thus not completely negligible. This impacts modal mass calculations when evaluating the forced response of the compressor housing. The i = 2 and i = 3 sets of modes have motions that are primarily in directions tangential to the compressor housing shell and contribute less to sound radiation.

Shells exhibit some effects that plates do not. As we have seen, for a shell one can find identical surface node line arrangements that occur at three different natural frequencies. For any such arrangement, two of the three natural frequencies are usually higher than the lowest by an order of magnitude. However, what appear to be identical modes on the basis of the node line pattern prove to be in actuality three different modes with different ratios of transverse displacement to in-plane displacement. Because in a general shell transverse and in-plane displacements are not independent (as they are in a plate or beam), both transverse and in-plane displace-

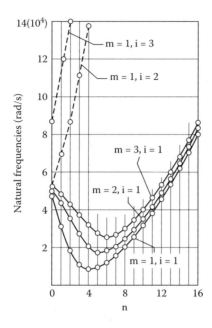

FIGURE 3.5 Natural frequencies of a typical compressor housing modeled as an equivalent circular cylindrical shell. For every (m, n) combination, there are three (i = 1, 2, 3) natural frequencies.

ments are part of a natural mode; different ratios, even with the same transverse pattern, constitute a different mode.

Another peculiarity of shells is that the mode with the simplest modal pattern generally is not associated with the lowest natural frequency (see Figure 3.5). The definition of the most simple pattern is quite clear for a beam. For instance, for a

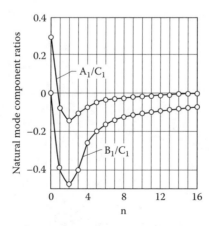

FIGURE 3.6 Mode component ratios for the m = 1 natural modes for which transverse displacement dominates (i = 1).

cantilevered beam, the simplest mode involves an up-and-down motion with no node point, the next simplest mode has one node point, the next two, and so on. For a plate or shell we may visualize by analogy that the simplest mode is associated with no node line, the next simplest with one node line and so on, realizing fully that node lines crisscross and the same number of node lines may occur twice.

Transversely dominant natural modes of an axisymmetric, cylindrical housing modeled as an equivalent circular cylindrical shell with simple end supports are presented in Figure 3.7, in terms their transverse displacements. The transverse node lines, the lines of zero transverse motion, are shown as dashed lines on the cylinders.

Depending on the relative dimensions, a cylindrical shell, for instance, may have its lowest natural frequency associated with a mode that has eight or ten node lines parallel to its axis, while the mode with the simplest nodal pattern is the so-called *breathing mode*, which has no node lines, and is associated with a very high natural frequency.

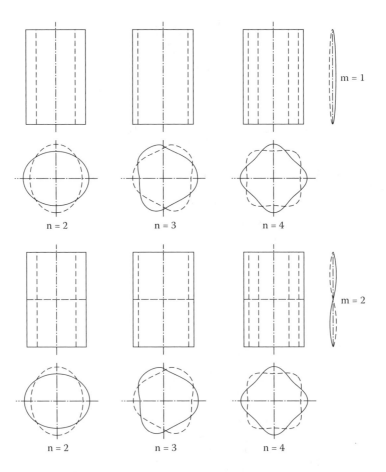

FIGURE 3.7 Illustration of shell natural modes for the $i = 1$ set of Figures 3.5 and 3.6 ($m = 1, 2$ and $n = 2, 3, 4$) in terms of the node lines of transverse displacement. Note that the compressor housing was approximated as an axisymmetric circular, cylindrical shell.

The explanation can be found by experimenting with a paper cylinder as before. Trying to create a breathing-type deflection is quite difficult; one in essence needs to force the paper cylinder over a circular cone (in which case it will almost certainly tear). The observer will agree that we have here a stretching (membrane) type of loading, rather than bending, and a lot of energy is required to create even the slightest displacement. On the other hand, to deform the paper cylinder into a squash-in/squash-out mode, which corresponds roughly to a cylindrical shell mode that has four node lines parallel to its axis, is quite easy. The deformation is mainly in bending, and very little energy is required to create appreciable deflection. Thus, if the reader remembers Raleigh's method for one-degree-of-freedom systems, he will appreciate that the case of the lowest ratio of potential to kinetic energy will be associated with the lowest natural frequency. For the paper cylinder, therefore, the lowest natural frequency corresponds to the squash-in/squash-out situation.

Things are only a little more complicated for a general shell. The potential strain energy is the sum of the membrane and bending strain energy contributions. Roughly speaking, the ratio of membrane strain energy to kinetic energy is high for modes with simple modal patterns and decreases toward zero as the number of node lines increases. The ratio of bending strain energy to kinetic energy is near zero for simple nodal patterns and increases as the pattern complexity increases. At some intermediate condition, where there is a minimum for the sum of these two energy ratios, there occurs the mode associated with the lowest frequency.

It is interesting to note that those natural frequencies that are controlled by membrane strain are approximately independent of shell thickness changes (just as the longitudinal vibrations of a beam are independent of its cross-sectional area), but that natural frequencies that are controlled by bending stiffness vary roughly in proportion to the shell thickness. This explains why increasing the thickness of the compressor housing, while in general a step in the right direction, often does not have the dramatically beneficial effect that is anticipated.

3.4 FURTHER SIMPLIFICATION

As stated before, the natural frequencies of the lower branch curves in Figure 3.5 are the ones of main interest in the noise control of compressors. Therefore, a theory that concentrates on this branch is of major interest. Please consult Soedel (2004) for details.

The first assumption is that contributions of in-plane deflections can be neglected in bending strain expressions (but not in the membrane strain expressions). In-plane loading is not permitted. Also, the transverse shear force terms $Q_{3,}/R_1$ and Q_{32}/R_2 are neglected. Then, a function ϕ is introduced that satisfies the first two equations. The third equation becomes:

$$D\nabla^4 u_3 + \nabla_K^2 \phi + \rho h \frac{\partial^2 u_3}{\partial t^2} = q_3 \qquad (3.93)$$

where

$$\nabla^2(\cdot) = \frac{1}{A_1 A_2} \left[\frac{\partial}{\partial \alpha_1} \left(\frac{A_2}{A_1} \frac{\partial(\cdot)}{\partial \alpha_1} \right) + \frac{\partial}{\partial \alpha_2} \left(\frac{A_1}{A_2} \frac{\partial(\cdot)}{\partial \alpha_2} \right) \right] \quad (3.94)$$

$$\nabla_K^2(\cdot) = \frac{1}{A_1 A_2} \left[\frac{\partial}{\partial \alpha_1} \left[\frac{1}{R_2} \frac{A_2}{A_1} \frac{\partial(\cdot)}{\partial \alpha_1} \right] + \frac{\partial}{\partial \alpha_2} \left[\frac{1}{R_1} \frac{A_1}{A_2} \frac{\partial(\cdot)}{\partial \alpha_2} \right] \right]. \quad (3.95)$$

We have in effect eliminated u_1 and u_2 but still have ϕ and u_3 to contend with. To obtain a second equation, we follow the standard procedure with Airy's stress function, namely to generate the compatibility equation. The way to do this is to take the six strain displacement relationships and eliminate from them the displacement by substitutions, additions, and subtraction. This gives

$$Eh\nabla_K^2 u_3 - \nabla^4 \phi = 0. \quad (3.96)$$

To obtain the natural frequencies and modes, we set $q_3 = 0$ and approach the solution with

$$u_3(\alpha_1, \alpha_2, t) = U_3(\alpha_1, \alpha_2) e^{j\omega t} \quad (3.97)$$

$$\phi(\alpha_1, \alpha_2, t) = \Phi(\alpha_1, \alpha_2) e^{j\omega t}. \quad (3.98)$$

Substituting this into the two equations of motion gives us

$$D\nabla^4 U_3 + \nabla_K^2 \Phi - \rho h \omega^2 U_3 = 0 \quad (3.99)$$

$$Eh\nabla_K^2 U_3 - \nabla^4 \Phi = 0. \quad (3.100)$$

These two equations may now be combined into a single equation:

$$D\nabla^8 U_3 + Eh\nabla_K^4 U_3 - \rho h \omega^2 \nabla^4 U_3 = 0. \quad (3.101)$$

To obtain the natural frequencies of a compressor housing modeled as a simply supported cylindrical shell of circular cross-section, we set $A_1 = 1$, $\alpha_1 = x$, $A_2 = a$, $\alpha_2 = \theta$.
This gives:

$$\nabla^4(\cdot) = \frac{1}{a^4} \frac{\partial^4(\cdot)}{\partial \theta^4} + \frac{\partial^4(\cdot)}{\partial x^4} + \frac{2}{a^2} \frac{\partial^4(\cdot)}{\partial x^2 \partial \theta^2} \quad (3.102)$$

$$\nabla_K^4(\cdot) = \frac{1}{a^2} \frac{\partial^4(\cdot)}{\partial x^4}. \quad (3.103)$$

Natural Frequencies and Modes of Compressor Housings

The mode shapes

$$U_{3mn}(x,\theta) = \sin \frac{m\pi x}{L} \cos n(\theta - \phi) \quad (3.104)$$

satisfy the primary single support conditions, and upon substitution into the eighth-order partial differential equation, result in a formula for the natural frequencies (of the lower branch):

$$\omega_{mn} = \frac{1}{a}\sqrt{\frac{\left(\frac{m\pi a}{L}\right)^4}{\left[\left(\frac{m\pi a}{L}\right)^2 + n^2\right]^2} + \frac{\left(\frac{h}{a}\right)^2}{12(1-\mu^2)}\left[\left(\frac{m\pi a}{L}\right)^2 + n^2\right]^2} \sqrt{\frac{E}{\rho}}. \quad (3.105)$$

The results of this equation agree well with the results of the full theory discussed earlier, and have the advantage that one clearly sees the relative influences of the various design parameters. The first term under the first square root is due to the membrane stiffness. As the number of axial node lines given by 2n increases, the influence of this term approaches zero. Its influence is strongest for the n = 0 mode. It is not a function of thickness h. The second term under the first square root is due to the bending stiffness. Its influence increases with the n-number and eventually becomes dominant. Its effect is proportional to thickness h. The second square root is the speed of sound of the material. It illustrates, for example, that if one makes the housing of aluminum, the natural frequencies would not necessarily shift dramatically because as E decreases, the mass density ρ decreases also.

At the lowest natural frequency for a particular m-number, which occurs at the n-number for which the first square root is a minimum, the influence of thickness is roughly proportional to $\sqrt{h}$. This illustrates that thickness changes do not shift the lowest natural frequencies appreciably.

3.4.1 INTRODUCING AXIAL CURVATURE (BARRELING)

A more effective way to increase the lowest natural frequency of the compressor housing is to convert the cylindrical shape into a barrel shape, as shown in Figure 3.8. It can be shown (see Soedel, 1973, 2004), using the Donnell-Mushtari-Vlasov theory, that

$$\omega_{Bmn}^2 = \omega_{Cmn}^2 + \frac{n^2\left[n^2\left(\frac{a}{R}\right)^2 + 2\left(\frac{a}{R}\right)\left(\frac{m\pi a}{L}\right)^2\right]}{a^2\left[\left(\frac{m\pi a}{L}\right)^2 + n^2\right]^2} \frac{E}{\rho} \quad (3.106)$$

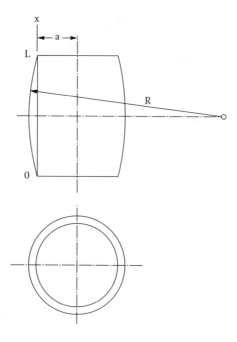

FIGURE 3.8 The cylindrical housing shell is converted to a barreled shell by introducing a radius of curvature R in the axial direction.

where

ω_{cmn} = the natural frequency of circular cylindrical shell, and
ω_{Bmn} = the natural frequency of barrel shell.

The equation is valid only if the axial curvature is not too pronounced, but it points out a trend. For the highest natural frequencies for a given volume and thickness of the shell, the Gaussian curvature, $1/(R_1 R_2)$, where R_1 and R_2 are the two orthogonal radii of curvature, should be maximized (a cylindrical shell has a Gaussian curvature of zero!). This leads us to the spherical shell. In a way, the spherical shell, for a given volume and thickness, is the optimal shape given the criteria that the lowest natural frequency should be as high as possible. This can also be shown by solving the shell equations of motion for the spherical shape.

Of course, we have to remember that shell shapes are dictated by other criteria besides high natural frequencies, for example, the shape of the compressor that the shell must hermetically enclose.

For example, by introducing a radius $R_x = 500$ mm in an axial direction for the cylindrical shell previously used, the lowest natural frequency was raised from about $0.9\,(10^4)$ rad/s to about $1.7\,(10^4)$ rad/s, without an appreciable increase in weight. This is shown in Figure 3.9.

At this point, the reader may ask: "Why is the lowest natural frequency so important?" It is important because below it, the shell will not be in resonance with

Natural Frequencies and Modes of Compressor Housings

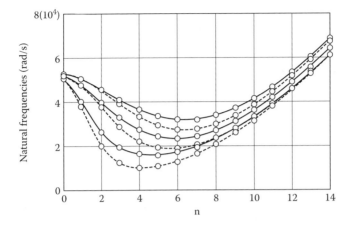

FIGURE 3.9 The natural frequencies of the technically most important i = 1 set of natural modes, for m = 1, 2, 3, are shown for the original cylindrical shape (--) and for the barreled shape of axial radius R = 500 mm (—). The lowest natural frequency has been raised approximately 60%.

the excitation mechanisms. At the lowest natural frequency, the first, usually strongest resonance may occur because the energy necessary to excite the lowest natural frequency mode, assuming equal damping, is lowest. Above it we can see from the illustrations that there are many potential resonances. Above the lowest natural frequency, it makes little sense to detune the excitation frequencies from the natural frequencies, because there are so many. Detuning one resonance may easily create another. Of course detuning may still be possible, but it will be more an experimental trial-and-error approach than a detuning that relies on theoretical calculations. Another difficulty involves the possible manufacturing variations, such as deep drawing and welding, which may differ enough from shell batch to shell batch so that some shells will be in resonance at certain excitation frequencies while others will not. This is one explanation why there are certain units that are noisier than others, from the same assembly line.

Just in passing, a negative Gaussian curvature lowers natural frequencies and is not recommended.

3.4.2 Spherical End Caps

The housing is terminated at both ends by *end caps*. Again, it is desirable to curve these end caps, as is already done in the industry. Using the same theory, it can be shown (Soedel, 1971, 2004) that the natural frequencies of spherically curved end caps, ω_{smn}, as shown in Figure 3.10, are higher than the natural frequencies of flat end caps, ω_{mn}, by

$$\omega_{smn}^2 = \omega_{mn}^2 + \frac{E}{\rho R^2} \tag{3.107}$$

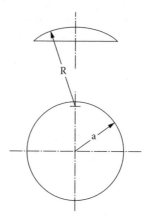

FIGURE 3.10 Converting a flat, circular housing shell end plate into a spherical cap of radius R raises cap-dominated natural frequencies.

where R is the radius of curvature. It can be shown that this relationship is true for end caps of any projection shape and boundary conditions. Therefore, end caps should always be curved or domed.

3.5 VIBRATION LOCALIZATION AT END CAPS

In previous sections, the compressor housing shell was approximated as an equivalent circular cylindrical shell with simply supported ends. While this approximation gives surprisingly valuable information about the behavior of real hermetic shells (Hsu and Soedel, 1987) and allows us to draw various important design conclusions, it cannot, of course, capture the vibration behavior of the top and bottom end caps and their influence. While in many hermetically sealed compressors the major radiated sound comes from the vibrating surface of the housing shell sidewall, there have been cases where it was found that at certain resonance frequencies the sidewall exhibits hardly any vibration, while the top cap vibrates at high amplitudes and thus becomes the major sound radiator.

In order to investigate this, various researchers have approximated the housing shell as a can consisting of two circular plates welded to the top and bottom of a circular cylindrical shell, for example, Faulkner (1969) and Tavakoli and Singh, (1990). The results obtained by Huang and Soedel (1993a, b) are discussed here; they demonstrated that it is possible that at certain natural frequencies one end cap, say the top one, can vibrate virtually alone, while the sidewall and the bottom end cap remain virtually vibration free. This effect is often referred to as *vibration localization*. One would intuitively expect frequent vibration localization where the cylindrical shell of the sidewall vibrates with little participation of the end caps (which is the basis of approximating the housing shell by an equivalent, simply supported circular cylindrical shell), or perhaps where both end caps exhibit the dominant vibration, but only one end cap vibrating is surprising.

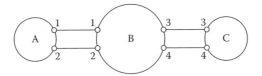

FIGURE 3.11 Receptance schema for joining two end plates B and C to cylinder A, to form an equivalent can model of a compressor housing. The plates are joined to the cylindrical shell by two coordinates each (deflections and slope).

The explanation is that slight stiffness variations between the top and bottom end caps cause this phenomenon. In the real compressor, this stiffness variation is provided by the different shapes of the two end caps; in the model discussed here, both end caps were assumed to be circular flat plates of identical radius, but with slight differences in thickness.

The analysis uses the receptance method, with each plate joined to the shell by two coordinates: deflection and slope transmitting forces and moments. Figure 3.11 shows the receptance schema, where A is the equivalent circular cylindrical shell, and B and C are the two end caps. The receptances used are line receptances, and their definitions can be found in Huang and Soedel (1993a, b).

Figure 3.12a shows the natural mode of the can for a case (shown in side profile only) where the top and bottom plates are identical and the vibration is localized in these two plates. Both have the same relative amplitude. Figure 3.12b shows the natural mode at virtually the identical natural frequency when the bottom plate thickness has increased by only 0.2%. Thus, given the natural thickness variations of deep drawn

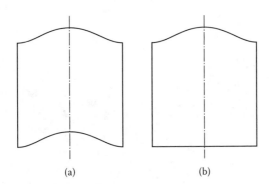

FIGURE 3.12 Examples of a natural mode of an equivalent can model of a compressor housing: (a) a mode where the vibration is localized in both end caps because they are identical; (b) only the top end cap participates in this natural mode because there is a 0.2% difference in the thicknesses of the two plates. This illustrates an often observed phenomenon.

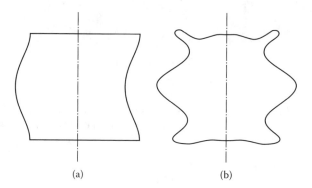

FIGURE 3.13 Two more examples of natural modes of an equivalent can model: (a) the vibration is localized in the circular cylindrical shell portion of the housing; (b) in this mode, there is no localization of vibrations: end plates and the cylindrical shell vibrate with more or less equal amplitudes. Note that the transverse vibrations are greatly exaggerated in all figures.

shells, and the typical geometric variations between top and bottom end caps, a vibration localization such as that in Figure 3.12a is not expected to occur in real compressors; the localization shown in Figure 3.12b is much more likely. Note also that there will be a natural mode at a nearby natural frequency where the localization is reversed: only the bottom plate vibrates and not the top plate.

For the sake of completeness, Figure 3.13a shows a natural mode example where vibration localization occurs in the circular cylindrical shell, and Figure 3.13b shows a case involving a natural mode where both the end plates and the cylindrical shell participate.

4 Compressor Housings that Are Not Axisymmetric

A perfectly axisymmetric housing is rare because the necessary electric connections, suction and discharge tube penetrations, and support arrangement prevent it from occurring, even if the shell itself is perfectly round. Still, it is useful to discuss this case because it shows the phenomenon of nonpreferential direction of axisymmetric modes. Taking the idealized case of a perfectly circular cylindrical shell, the experimenter notices that the wisdom he has absorbed in an ordinary introductory vibration course is no longer applicable. He has learned, from the example of a vibrating beam, most likely, that regardless of the location of the exciter, the mode shape is invariant. Yet in the case of the perfectly axisymmetric shell, he notices that the mode shape will orient itself such that one of its antinodes is always lined up with the exciter (see Figure 4.1). The physical reason is the axisymmetry because there should obviously be no preferential direction. In addition, what he may not realize is that he has recorded an incomplete modal set if he records only one mode at that particular frequency. For forced vibration considerations, he must record two orthogonal sets:

$$U_{3mn1}(x,\theta) = \sin(m\pi x/L) \cos n\theta \qquad (4.1)$$

$$U_{3mn2}(x,\theta) = \sin(m\pi x/L) \sin n\theta \qquad (4.2)$$

Chances are that the experimenter has only measured the first set. The node line pattern of both sets is the same, except that those of the second set are shifted by $\pi/2$. In my terminology, this case is a limiting case of node splitting because both sets occur at the same natural frequency. See Figure 4.2 for the basic mode set for $n = 2$.

4.1 MODE SPLITTING CAUSED BY A MASS OR STIFFNESS

Let us now take the same shell, but consider an attached small point mass that disturbs the axisymmetry. The result is that the modes now have a preferential direction. Their approximate theoretical expressions are still given by Equations 4.1 and 4.2, but their lineup is now such that in the set described by Equation 4.1 the antinode is located at the location of mass while the set described by Equation 4.2

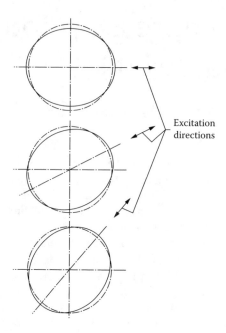

FIGURE 4.1 For an axisymmetric compressor housing, natural mode orientations in the circumferential direction change. The antinodes line up with the excitation direction.

will always have a node at the mass location (see Figure 4.3). That is, the origin of the θ coordinate is at the mass. The set described by Equation 4.2 occurs still at the natural frequencies of the shell without mass, namely

$$\omega_{mn1}^2 = \omega_{mn}^2. \tag{4.3}$$

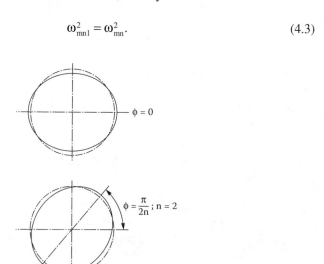

FIGURE 4.2 Base modes for n = 2.

Compressor Housings that Are Not Axisymmetric

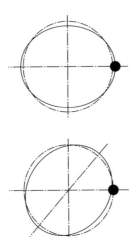

FIGURE 4.3 Split modes for n = 2 due to a point mass nonuniformity.

The set described by Equation 4.1 occurs now at a natural frequency (Soedel, 1980a):

$$\omega_{mn2}^2 = \omega_{mn}^2 \left(1 - \varepsilon_{mn}^2\right) \tag{4.4}$$

where, approximately,

$$\varepsilon_{mn}^2 = \frac{4M}{M_s} \sin^2(m\pi x^*/L) \tag{4.5}$$

and where M = the attached mass [Ns²/mm], M_s = the total shell mass [Ns²/mm], x* = the location of attached mass [mm], and L = the length of the shell [mm].

As it can be seen, for small antisymmetries, ε_{mn} is a small number and the two natural frequencies for modes of similar appearance are only separated by a small amount.

In the case of a stiffness added at a single location (Soedel, 1980a), the result is similar, except that we have now (see Figure 4.4):

$$\omega_{mn1}^2 = \omega_{mn}^2 \tag{4.6}$$

$$\omega_{mn2}^2 = \omega_{mn}^2 \left(1 + k_{mn}^2\right) \tag{4.7}$$

where

$$k_{mn}^2 = \frac{4K}{M\omega_{mn}^2} \sin^2(m\pi x^*/L) \tag{4.8}$$

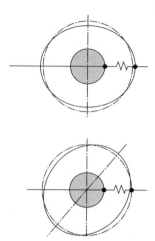

FIGURE 4.4 Split modes for n = 2 due to a point stiffness nonuniformity.

and where K = the attached stiffness [N/mm]. The only difference between the two cases is that the stiffness addition splits the modes so that one of the frequencies is larger than the original frequency, while the mass addition splits it so that one of the frequencies is smaller than the original one. It is easy to see that a combined addition of mass and stiffness will produce frequency pairs where one is lower and the other one is higher than the original frequency. If the added mass and stiffness have a natural frequency by themselves that coincides with one of the natural frequencies of the original shell, the split is roughly equal in both directions.

The effect of the nonaxisymmetric support arrangement and compressor suspension can be understood in the same way, except that variations of the discussed behavior may exist, which should perhaps be investigated in a similarly simple fashion.

4.2 MODE SPLITTING CAUSED BY OVALNESS

Interpreting an oval shell as a shell that is a deviation from an axisymmetric shell by defining a curvature in circumferential direction $1/R_\theta$ as a deviation from the equivalent curvature $1/a$ of a round shell:

$$\frac{1}{R_\theta^2} = \frac{1}{a^2}[1+\varepsilon(\theta)] \qquad (4.9)$$

where a = the radius of the equivalent circular cylindrical shell [mm] = $C/2\pi$, C = the circumference of the oval shell [mm], $\varepsilon(\theta)$ = a function periodic by 2π, R_θ = the actual radius of curvature [mm], one can show (Soedel, 1980a) that we again obtain two different sets of natural modes, which look similar when one counts the number

Compressor Housings that Are Not Axisymmetric

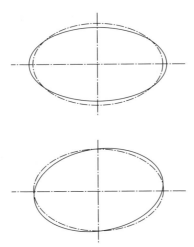

FIGURE 4.5 Split modes for n = 2 due to a shape deviation from axisymmetry (an oval housing).

of node lines. The natural frequencies are again split into two values (caused by the ovalness) and are approximately given by:

$$\omega_{mn1}^2 = \omega_{mn}^2 + A_{mn}\int_0^{2\pi} \varepsilon(\theta)\cos^2 n\theta\, d\theta \tag{4.10}$$

$$\omega_{mn2}^2 = \omega_{mn}^2 + A_{mn}\int_0^{2\pi} \varepsilon(\theta)\sin^2 n\theta\, d\theta \tag{4.11}$$

where

$$A_{mn} = \frac{E}{\rho a^2}\frac{(m\pi/L)^2}{[(n/a)^2 + (m\pi/L)^2]^2}. \tag{4.12}$$

The two basic mode sets are shown in Figure 4.5.

These simple models explain the often-observed phenomenon of mode splitting of compressor shells. They point out that it is not permissible for the experimenter to lump split modes together and not report them separately. One effect of mode splitting is that beating may occur under certain impact conditions. Finally, any kind of deviation from axisymmetry causes mode splitting.

4.3 EXAMPLE OF EXPERIMENTALLY OBTAINED HOUSING MODES

Figures 4.6a and 4.6b and Figures 4.7a and 4.7b show typical, measured natural modes of a compressor housing (Hsu and Soedel, 1987), in terms of the nodelines for zero transverse vibration. Each natural mode is shown in six views, for a complete

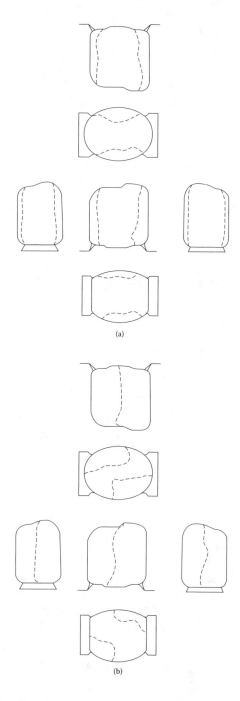

FIGURE 4.6 Typical split natural modes of the n = 2 type of a real, approximately oval compressor housing, shown in terms of the transverse node lines: (a) the approximately symmetric n = 2 mode having a natural frequency of 2490 Hz, and (b) the approximately asymmetric n = 2 mode having a natural frequency of 2642 Hz.

Compressor Housings that Are Not Axisymmetric

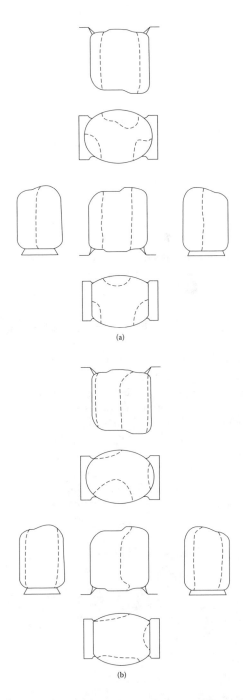

FIGURE 4.7 Typical split natural modes of the n = 3 type of a real, approximately oval compressor housing, shown in terms of transverse node lines: (a) the approximately symmetric n = 3 mode, having a natural frequency of 2200 Hz, and (b) the approximately asymmetric n = 3 mode having a natural frequency of 2450 Hz.

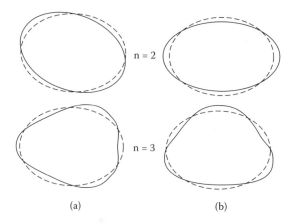

FIGURE 4.8 Aid to interpreting cylindrical sections of the natural modes of the real compressor housing examples of Figures 4.6 and 4.7.

presentation. If one concentrates on the cylindrical portion of the housing, one can identify the 2n axial node lines, which identify the mode number n. As expected, for each n-number, we find two sets—one symmetric and the other asymmetric. Figures 4.6a and 4.6b correspond to n = 2, and Figures 4.7a and 4.7b to n = 3. Their natural frequencies are pairs, the symmetric modes as defined in Figure 4.8b being of lower natural frequencies and the non-symmetric modes in Figure 4.8a being higher.

It is interesting to see that this housing, which is only cylindrical in a very approximate sense, conforms so well to the theoretically expected model forms for an equivalent cylinder. It has been the author's experience that it is possible to predict the lowest natural frequency of a real housing of approximately cylindrical shape with an approximate accuracy of ± 10% and has almost always found that the mode shapes agree with expectation. It is, therefore, believed that we understand the influence of deviations from axisymmetry fairly well.

5 Modifications of Housing Natural Frequencies and Modes

5.1 STIFFENING OF COMPRESSOR HOUSING

In cases where an excitation frequency (for general periodic forcing this includes any of the integer multiples of the fundamental frequency) coincides with one of the shell's natural frequencies, it may be desirable to shift the natural frequencies. This is a risky undertaking because avoidance of one resonance may create a resonance matching somewhere else in the frequency spectrum, except where all natural frequencies are shifted above a given excitation frequency. In any case, it is usually advantageous to stiffen rather than weaken a given shell to cause the desired shifts, because stiffening results in smaller vibration amplitudes in a general sense. Stiffening is generally accomplished by adding to a shell such supporting structures as beams (stringers), rings, shelves, and secondary shells.

As a rule of thumb, the natural frequency of the stiffening element by itself, in a natural mode that resembles the natural mode of vibration of the unstiffened shell, should be higher than the natural frequency of the unstiffened shell that is to be shifted upward. One may easily visualize that adding a piece of lead to our cylindrical shell will not increase the stiffness greatly, but will add quite a bit of mass, thus lowering the natural frequency of the combination. On the other hand, if we add a lot of stiffness and relatively little mass, the natural frequency of the combination will be increased.

Here, this basic concept of stiffeners is reviewed for a closed cylindrical shell (Figure 5.1) using the concept of receptances. A *receptance* is, in general, the ratio of harmonic response to harmonic forcing. The theory of point or line receptances is explained in Soedel (2004).

The natural frequencies of a cylindrical shell with a ring stiffener are given by Soedel (2004):

$$\alpha_{11} + \beta_{11} = 0 \qquad (5.1)$$

where α_{11} is the line receptance of the shell at the line of stiffener attachment and β_{11} is the line receptance of the stiffening ring. The line receptance of an equivalent, simply supported, cylindrical shell representing the closed compressor shell can be shown to be

$$\alpha_{11} = \frac{2}{\rho_s h_s L} \sum_{m=1}^{\infty} \frac{1}{\omega_{mn}^2 - \omega^2} \sin^2 \frac{m\pi x^*}{L} \qquad (5.2)$$

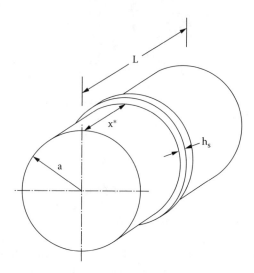

FIGURE 5.1 An equivalent circular cylindrical shell model of a compressor housing with a ring stiffener.

where L = the height of the shell, ρ_s = the mass density, h_s = the thickness of the shell, ω_{mn} are the natural frequencies of the shell before stiffening is applied, $n = 0, 1, 2, \ldots$ is the mode component number in the circumferential direction, $m = 0, 1, 2, \ldots$ is the mode component number in the axial direction, ω represents the as yet unknown new natural frequencies of the stiffened shell, and x^* is the location in the axial direction where the stiffening ring is applied.

The stiffening ring receptance β_{11} is (Soedel, 2004):

$$\beta_{11} = \frac{1}{\rho_s A (\omega_n^2 - \omega^2)} \qquad (5.3)$$

where ρ_s = the mass density of the ring material, A = the cross-sectional area, ω_n are the natural frequencies of the ring by itself before it is attached to the shell, and ω represents the as yet unknown natural frequencies of the stiffened shell. Therefore, the equation

$$\frac{2}{\rho h L} \sum_{m=1}^{\infty} \frac{1}{\omega_{mn}^2 - \omega^2} \sin^2\left(\frac{m\pi x^*}{L}\right) + \frac{1}{\rho_s A(\omega_n^2 - \omega^2)} = 0 \qquad (5.4)$$

must be solved for the values of ω that satisfy it.

To gain an intuitive insight, one can investigate the case where there is only a small stiffening effect such that ω is just a perturbation of the original ω_{mn}.

This allows one to write approximately, because only the one term in the series where ω is closest to ω_{mn} dominates,

$$\omega^2 = \omega_{mn}^2 \frac{1 + \left(\dfrac{2M_s}{M}\right)\left(\dfrac{\omega_n}{\omega_{mn}}\right)^2 \sin^2\left(m\pi \dfrac{x^*}{L}\right)}{1 + \left(\dfrac{2M_s}{M}\right)\sin^2\left(m\pi \dfrac{x^*}{L}\right)} \tag{5.5}$$

where M_s and M are the total masses of the ring stiffener and the shell, respectively:

$$M_s = 2\pi a A \rho_s \tag{5.6}$$

$$M = 2\pi L a h \rho. \tag{5.7}$$

This approximated solution shows immediately that the new shell natural frequency ω_k is increased or decreased according to the following law:

$$\begin{cases} \omega_k > \omega_{mn} & \text{if } \omega_n > \omega_{mn} \\ \omega_k < \omega_{mn} & \text{if } \omega_n < \omega_{mn} \end{cases} \tag{5.8}$$

as pointed out by Soedel (2004) and Kim and Soedel (1998). This means that the natural frequencies of the stiffening ring by itself must be higher than the natural frequencies of the unstiffened shell to have any beneficial effect. Based on my consulting experience, this basic condition of stiffening is often violated in practice (and to the surprise of the designers, the stiffening ring therefore has no beneficial effect).

For the reader to be able to make quick estimates, the natural frequencies of a stiffening ring are approximately (Soedel, 2004):

$$\omega_n = \frac{n(n^2 - 1)}{a^2 \sqrt{(n^2 + 1)}} \sqrt{\frac{EI}{\rho A}} \tag{5.9}$$

where a is the average radius, I is the area moment, A is the cross-sectional area, E is Young's modulus, and ρ is the mass density.

The equation to be used for the cylindrical housing (shell) frequencies ω_{mn} is the one derived earlier using the Donnell-Mushtari-Vlasov approximation.

The objective, for effective stiffening, is not only to make $\omega_n > \omega_{mn}$, but to make ω_n as large as feasible. Obviously we have to increase the area moment, and at the same time make the cross-sectional area as small as possible. Solid stiffeners should therefore be avoided, and T or U profiles should be welded to the shell.

Kim and Soedel (1998) discuss the influence of putting the stiffening ring on the shell under tension. A reasonable question is, of course, whether tension can be effectively introduced in a ring in a cost-effective and reliable way. It is not at all clear if this is possible. Also, practical results observed during my consulting activities were inconclusive. Uncontrolled experiments (let us try it and see if it works) using steel bands wrapped under tension around cylindrical housings sometimes worked and sometimes did not. What makes the issue even more complicated is that it was not clear if the ring bands under tension were slipping relative to the compressor housing or not. Slipping is, of course, a damping mechanism, and a sound pressure level reduction may be due to it and not to the detuning effect a stiffener normally provides.

5.2 THE INFLUENCE OF RESIDUAL STRESSES

We are concerned here with shells that are in a stressed state even when not vibrating (the corresponding static stresses are called *initial stresses*). The best-known cases are those of shells that are loaded by static pressure, either internally or externally. Although the general definition of these two terms is complicated for shells with complex shapes, the intent is clear for simple shells. A spherical shell that contains a gas under pressure is internally pressurized, while the same shell used as an underwater structure is externally pressurized.

The important point is that whenever the initial (residual) stresses are of a tensile nature (which usually occurs in internally pressurized shells), the shell experiences a stiffening effect, as for a slender beam that is under tension. This added stiffness is due to the same type of restoring forces as those that cause a stretched string or a stretched drumhead (membrane), which has negligible bending stiffness, to return to its equilibrium position. In general, initial tensile stresses cause all natural frequencies to increase. Conversely, initial compressive stresses, which usually occur in externally pressurized shells, make the shell less stiff and reduce its natural frequencies.

While pressurized shells are the most obvious example, we have to remember that compressor housing shells are fabricated by deep drawing and assembled using welding operations that tend to introduce initial stresses, whose presence and nature are rarely obvious. If we can heat the shell above the recrystallization point of the metal, we can remove these residual stresses; however, this approach is unrealistic for hermetic refrigeration compressors because they are welded with the compressor inside. It would not be feasible to anneal the assembly.

Nondestructive methods, like x-ray diffraction, are needed to investigate the initial stress state, in order to incorporate it in an analysis. Unfortunately, such methods are time consuming, and one may often do better by measuring the natural frequencies and modes under the influence of initial stresses as they are found, and by employing the measured modes for any further analysis.

As far as the compressor assembly is concerned, the presence of residual stresses offers another possible explanation why compressors from the same assembly line may differ in their noise spectra. Residual stresses are a function of welding speed, arc setting, and anything else that affects the welding process. The shifts in natural frequencies may be very small, caused by the slightly different residual stress distributions caused by welding variations, but resonances for slightly damped systems come in very sharp, so that one housing may produce a resonance and the other does not.

From an analytical viewpoint, only the most simple residual stress distributions allow solutions in terms of a simple formula. For example, it can be shown (Soedel, 2004) that if a cylindrical shell has a uniform residual stress that produces a uniform tension per unit length, T, in the axial direction,

$$\omega_{mnT}^2 = \omega_{mn}^2 + \frac{T}{\rho h}\left(\frac{m\pi}{L}\right)^2 \qquad (5.10)$$

where ω_{mnT} is the natural frequency of the shell with residual stress, and ω_{mn} is the natural frequency of the shell without residual stress. If we have compression, T would be negative and the natural frequencies decrease.

In an actual welded compressor housing, the distribution of residual stresses is, of course, much more complicated.

6 Forced Vibration of Compressor Housing (Shell)

6.1 MODAL SERIES EXPANSION MODEL

Up to this point, we have been concerned with the natural frequencies of the compressor housing. One of the reasons for this is that, eventually, the study of the forced response of the housing, either due to the gas pulsations inside or the vibration transmissions through the supports or other attachments. We can view the forced response of the housing as a superposition of the forced response of each of the natural modes.

The natural modes have an interesting and useful property called *orthogonality*. As for the sine and cosine functions of the Fourier series, orthogonality means that when two different normal mode functions (for instance, two sine functions, whose arguments differ by an integer multiple) are multiplied together and the result is integrated over the entire structural surface (which is analogous to the fundamental period of the sine functions), the result is zero. The boundary conditions are, of course, automatically satisfied because each natural mode satisfies all boundary conditions.

The usefulness of this lies in the fact that natural modes can be applied, just like sine or cosine functions in the case of the Fourier series, to express any deflection of the shell by an infinite series composed of natural modes. This series is called the *modal expansion series*. The factors multiplying each natural mode are called the modal participation factors and are determined in quite the same spirit as we determine Fourier coefficients, namely, from the forcing distribution and their time dependence. It turns out that in terms of the modal expansion, the shell response reduces to the solution of a set of nonhomogeneous one-degree-of-freedom oscillator equations—one equation for each mode that is to be considered. Thus, we may interpret the forced vibration of shells by considering the shell as composed of simple oscillators, where each oscillator consists of the shell restricted to vibrating in one of its natural modes. All these oscillators respond simultaneously, and the total shell vibration is simply the result of the addition (superposition) of all the individual vibrations.

Mathematically, this is expressed as:

$$u_i(\alpha_1,\alpha_2,t) = \sum_{k=1}^{\infty} \eta_k(t) U_{ik}(\alpha_1,\alpha_2) \qquad (6.1)$$

where $i = 1, 2, 3$, and U_{ik} are the natural mode components in the three principal directions. The modal participation factors η_k are unknown and have to be determined in the following.

In Soedel (2004), the equations of motion are of the form

$$L_i\{u_1, u_2, u_3\} - \lambda \dot{u}_i - \rho h \ddot{u}_i = -q_i \tag{6.2}$$

where λ is an equivalent viscous damping factor. The viscous damping term was introduced through the forcing term, replacing the original q_i by $q_i - \lambda \dot{u}_i$. Also note that the damping factor is assumed to be the same in all three principal directions. This is not necessarily true, but because damping values are notoriously difficult to determine theoretically (and thus have more qualitative than quantitative value), and because a uniform damping factor offers computational advantages, it was decided to adopt the uniform factor here. How this factor relates to the structural damping description, which uses a complex modulus, is discussed in Soedel (2004).

The operators L_i are defined, from before, as:

$$L_1\{u_1, u_2, u_3\} = \frac{1}{A_1 A_2}\left[\frac{\partial(N_{11} A_2)}{\partial \alpha_1} + \frac{\partial(N_{21} A_1)}{\partial \alpha_2} + N_{12}\frac{\partial A_1}{\partial \alpha_2} - N_{22}\frac{\partial A_2}{\partial \alpha_1} + A_1 A_2 \frac{Q_{13}}{R_1}\right] \tag{6.3}$$

$$L_2\{u_1, u_2, u_3\} = \frac{1}{A_1 A_2}\left[\frac{\partial(N_{12} A_2)}{\partial \alpha_1} + \frac{\partial(N_{22} A_1)}{\partial \alpha_2} + N_{21}\frac{\partial A_2}{\partial \alpha_1} - N_{11}\frac{\partial A_1}{\partial \alpha_2} + A_1 A_2 \frac{Q_{23}}{R_2}\right] \tag{6.4}$$

$$L_3\{u_1, u_2, u_3\} = \frac{1}{A_1 A_2}\left[\frac{\partial(Q_{13} A_2)}{\partial \alpha_1} + \frac{\partial(Q_{23} A_1)}{\partial \alpha_2} - A_1 A_2 \left(\frac{N_{11}}{R_1} + \frac{N_{22}}{R_2}\right)\right]. \tag{6.5}$$

We substitute the modal series solution in the equations of motion, and utilize the orthogonality relationship of natural modes (Soedel, 2004):

$$\iint_{\alpha_1 \alpha_2} (U_{1k} U_{1p} + U_{2k} U_{2p} + U_{3k} U_{3p}) A_1 A_2 d\alpha_1 d\alpha_2 = \begin{cases} N_k, & p = k \\ 0, & p \neq k \end{cases} \tag{6.6}$$

where

$$N_k = \iint_{\alpha_1 \alpha_2} \left(U_{1k}^2 + U_{2k}^2 + U_{3k}^2\right) A_1 A_2 d\alpha_1 d\alpha_2. \tag{6.7}$$

Forced Vibration of Compressor Housing (Shell)

This gives an ordinary differential equation from which the modal participation coefficients can be determined:

$$\ddot{\eta}_k + 2\zeta_k \omega_k \dot{\eta}_k + \omega_k^2 \eta_k = F_k(t) \tag{6.8}$$

where

$$F_k(t) = \frac{\int_{\alpha_2}\int_{\alpha_1}(q_1 U_{1k} + q_2 U_{2k} + q_3 U_{3k}) A_1 A_2 \, d\alpha_1 \, d\alpha_2}{\rho h N_k} \tag{6.9}$$

$$\zeta_k = \frac{\lambda}{2\rho h \omega_k}. \tag{6.10}$$

Note that ζ_k is called the *modal damping coefficient*. It is analogous to the damping coefficient in the simple oscillator problem.

For the case where $\zeta_k < 1$, which is typically the case for compressor housings, the general solution is

$$\eta_k(t) = e^{-\zeta_k \omega_k t}\left\{\eta_k(0)\cos\gamma_k t + [\eta_k(0)\zeta_k\omega_k + \dot{\eta}_k(0)]\frac{\sin\gamma_k t}{\gamma_k}\right\}$$
$$+ \frac{1}{\lambda_k}\int_0^t F_k(\tau) e^{-\zeta_k \omega_k (t-\tau)} \sin\gamma_k(t-\tau)d\tau \tag{6.11}$$

where

$$\gamma_k = \omega_k \sqrt{1-\zeta_k^2}. \tag{6.12}$$

6.2 STEADY-STATE HARMONIC RESPONSE

Because we are usually interested in the harmonic response of the housing (a periodic response is composed of a series of harmonic responses), in steady state, the first part of the solution, which is due to initial conditions and approaches zero as time increases, is not of importance.

The second part of the solution for an excitation that is harmonic in time,

$$q_i(\alpha_1, \alpha_2, t) = q_i^*(\alpha_1, \alpha_2) e^{j\omega t}, \tag{6.13}$$

simplifies to

$$\eta_k = \Lambda_k e^{j(\omega t - \phi_k)} \tag{6.14}$$

where the magnitude of the response is

$$\Lambda_k = \frac{F_k^*}{\omega_k^2 \sqrt{\left[1 - (\omega/\omega_k)^2\right]^2 + 4\zeta_k^2 (\omega/\omega_k)^2}} \tag{6.15}$$

and the phase lag is

$$\phi_k = \tan^{-1} \frac{2\zeta_k (\omega/\omega_k)}{1 - (\omega/\omega_k)^2} \tag{6.16}$$

and where

$$F_k^* = \frac{1}{\rho h N_k} \int_{\alpha_2} \int_{\alpha_1} (q_1^* U_{1k} + q_2^* U_{2k} + q_3^* U_{3k}) A_1 A_2 d\alpha_1 d\alpha_2. \tag{6.17}$$

As expected, a shell will behave in a manner similar to that of a collection of simple oscillators. Whenever the excitation frequency coincides with one of the natural frequencies, a peak in the response curve will occur.

If the excitation is periodic, as in most compressors, there are many harmonic components of excitation. Ideally, none of these excitation frequencies should coincide with any of the many natural frequencies of the housing. Accomplishing this is virtually impossible, but detuning certain resonances that have been identified as important is usually feasible.

We also see that a higher natural frequency, ω_k, will reduce the vibration response, assuming the amount of detuning stays the same. The detuning will, of course, not stay the same because as we change ω_k, we change the ratio ω_k/ω. But in an overall sense, it is probably best to make the compressor housing as stiff as possible (which increases ω_k). This is born out by reports of good noise performance of nearly spherical housings, which have higher ω_k than cylindrical housings.

A case can also be made for making the housing more massive, but it should not lower the natural frequencies ω_k. The controlling terms are such that the vibration response is, for transverse loading, proportional to:

$$u_i \propto \frac{q_3^*}{\rho h \omega_k^2}. \tag{6.18}$$

The term ρh is the mass per unit thickness. So if we add mass beyond ρh, we also need to add stiffness also so that the product $\rho h \omega_k^2$ is as large as possible.

6.3 HOUSING DYNAMICS IN TERMS OF MODAL MASS, STIFFNESS, DAMPING, AND FORCING

We may write the modal participation factor Equation 6.8 in the form

$$\rho h N_k \ddot{\eta}_k + \lambda N_k \dot{\eta}_k + \omega_k^2 \rho h N_k \eta_k = f_k \quad (6.19)$$

where N_k is defined by Equation 6.7, and f_k is

$$f_k = \int_{\alpha_2}\int_{\alpha_1} (q_1 U_{1k} + q_2 U_{2k} + q_3 U_{3k}) A_1 A_2 \, d\alpha_1 \, d\alpha_2. \quad (6.20)$$

See also Soedel (2004). Because Equation 6.19 is a one-degree-of-freedom oscillator equation, it makes some sense to view this equation in terms of modal mass, stiffness, damping, and forcing terms. *Modal mass* is defined as

$$M_k = \rho h N_k, \quad (6.21)$$

the modal stiffness (spring rate) is defined as

$$K_k = \omega_k^2 \rho h N_k = \omega_k^2 M_k, \quad (6.22)$$

and the modal forcing is defined by Equation 6.20.

Also identified is a modal damping constant

$$C_k = \lambda N_k. \quad (6.23)$$

The mode components U_{1k}, U_{2k}, U_{3k} can be defined as dimensionless ratios. Therefore, the unit of the modal mass is [kg], the unit of the modal stiffness is N/m, and the unit of the modal damping constant is Ns/m. The unit of the modal forcing term is [N]. Therefore, Equation 6.19 can be written as

$$M_k \ddot{\eta}_k + C_k \dot{\eta}_k + K_k \eta_k = f_k. \quad (6.24)$$

The interpretation of the modal mass is that of an adjusted or corrected mass of the structure for a particular mode, or an "equivalent" mass. The same is true for the modal stiffness. It is an "equivalent" spring rate. Therefore, for structures with a curvature, a correct description of

$$N_k = \int_{\alpha_2}\int_{\alpha_1} \left(U_{1k}^2 + U_{2k}^2 + U_{3k}^2 \right) A_1 A_2 \, d\alpha_1 \, d\alpha_2 \quad (6.25)$$

is important. An error is introduced if we ignore the contribution of the tangential mode components U_{2k} and U_{1k} when evaluating the transverse vibration response $u_3(\alpha_1,\alpha_2,t)$ for example, unless they are truly negligible, so that

$$U_{2k} \ll U_{3k} \quad \text{and} \quad U_{1k} \ll U_{3k}. \tag{6.26}$$

This is not the case for all natural modes of the housing shell. The error that may be caused if we ignore the U_{1k} and U_{2k} contributions can be analyzed for a particular mode k. If we define

$$\frac{\int\int_{\alpha_2 \alpha_1} \left(U_{1k}^2 + U_{2k}^2\right) A_1 A_2 \, d\alpha_1 \, d\alpha_2}{\int\int_{\alpha_2 \alpha_1} U_{3k}^2 A_1 A_2 \, d\alpha_1 \, d\alpha_2} = \xi_k, \tag{6.27}$$

we obtain

$$N_k = (1+\xi_k) \int\int_{\alpha_2 \alpha_1} U_{3k}^2 A_1 A_2 \, d\alpha_1 \, d\alpha_2. \tag{6.28}$$

Equation 6.19 becomes

$$\ddot{\eta}_k + \frac{\lambda}{\rho h}\dot{\eta}_k + \omega_k^2 \eta_k = \frac{f_k}{\rho h N_k} = \frac{f_k}{(1+\xi_k)\rho h \int\int_{\alpha_2 \alpha_1} U_{3k}^2 A_1 A_2 \, d\alpha_1 \, d\alpha_2}. \tag{6.29}$$

The calculated modal participation factor η_k, and thus the deflection u_3 will be inversely proportional to $(1+\xi_k)$:

$$\eta_k \propto \frac{1}{1+\xi_k}. \tag{6.30}$$

For example, if $\xi_k \neq 0$, but we take it as zero, our response will be larger by the ratio $(1+\xi_k)$ to (1). For example, if $\xi_k = 0.2$, our error for the kth mode participation will be 20%.

This type of error sometimes occurs, for example, when experimentally obtained natural modes of a housing shell are used in forced response calculations. Realistically, only the transverse mode components can reasonably be measured. The tangential mode components are often ignored. The forced response calculations will then be based on reduced modal masses, which means that they will be upper bounds.

Forced Vibration of Compressor Housing (Shell)

This does not necessarily mean that using only transverse mode components of a housing shell is wrong. It is still a valid approximation as long as one recognizes the modal mass error.

The concept of a modal mass is also of use when trying to define a resonance. One noise control measure is to add mass attachments to the housing shell. This means that the modal masses are increased and the natural frequencies are lowered. Another noise control measure is to stiffen the shell (which is practically difficult without also increasing the modal mass to some extent). To increase natural frequencies, more modal stiffness has to be added than modal mass.

It should be noted that in the foregoing discussions, the so-called *rigid body motion* of the housing is not considered because of the associated, typically low natural frequencies, although they participate in the vibration response, especially during starting and stopping of the compressor. For example, see Hamilton (1982) and Marriott (1998, 2000).

6.4 STEADY-STATE RESPONSE OF SHELLS TO PERIODIC FORCING

The steady-state periodic response of the housing shell to periodic forcing is a superposition of steady-state harmonic responses. The forcing is assumed here to be such that the spatial distribution does not change in time, but its amplitude is periodic in time. For compressor housings, this covers a large number of practical situations. It excludes cases where the spatial distribution itself changes periodically with time. The forcing function can be written as

$$q_1(\alpha_1, \alpha_2, t) = q_1^*(\alpha_1, \alpha_2) f(t) \tag{6.31}$$

where $f(t)$ is a function that is periodic in time. The period T of this periodic function is related to the frequency at which the function repeats, Ω, by

$$T = \frac{2\pi}{\Omega}. \tag{6.32}$$

In a compressor application, Ω is the steady state operating speed of a compressor in rad/s. for example, if the speed is 3600 RDM, then the fundamental frequency is $f = 1/T = 60$ Hz, or $\Omega = 377$ rad/s.

Expanding the function $f(t)$ in a Fourier series gives

$$f(t) = a_o + \sum_{n=1}^{\infty} (a_n \cos n\Omega t + b_n \sin n\Omega t) \tag{6.33}$$

where

$$a_o = \frac{1}{T}\int_0^T f(t)\,dt \qquad (6.34)$$

$$a_n = \frac{2}{T}\int_0^T f(t)\cos n\Omega t\,dt \qquad (6.35)$$

$$b_n = \frac{2}{T}\int_0^T f(t)\sin n\Omega t\,dt. \qquad (6.36)$$

The forcing function is, therefore

$$q_i(\alpha_1,\alpha_2,t) = q_i^*(\alpha_1,\alpha_2)\left[a_o + \sum_{n=1}^{\infty}(a_n\cos n\Omega t + b_n\sin n\Omega t)\right]. \qquad (6.37)$$

The coefficient a_o is a constant term and gives rise to a static deflection. It represents, if gas pulsations in the housing cavity excite the housing shell, for example, the influence of the average pressure. It is not of particular interest in noise and vibration control, but the static deflections caused by it can be calculated as (Soedel, 2004)

$$u_i(\alpha_1,\alpha_2) = \sum_{k=1}^{\infty}\frac{F_k^* a_o}{\omega_k^2}U_{ik}(\alpha_1,\alpha_2) \qquad (6.38)$$

where

$$F_k^* = \frac{1}{\rho h N_k}\int_{\alpha_2}\int_{\alpha_1}[q_1^*(\alpha_1,\alpha_2)U_{1k} + q_2^*(\alpha_1,\alpha_2)U_{2k} + q_3^*(\alpha_1,\alpha_2)U_{3k}]A_1 A_2\,d\alpha_1 d\alpha_2. \qquad (6.39)$$

The vibratory deflections due to each $a_n\cos n\Omega t$ term are

$$u_i^a(\alpha_1,\alpha_2,t) = \sum_{k=1}^{\infty}\eta_{kn}^a(t)U_{ik}(\alpha_1,\alpha_2), \qquad (6.40)$$

and where, assuming subcritical damping for all modes,

$$\eta_{kn}^a = \Lambda_{kn}^a\cos(n\Omega t - \phi_{kn}). \qquad (6.41)$$

Similarly, the solution to each $b_n \sin n\Omega t$ term is

$$u_i^b(\alpha_1,\alpha_2,t) = \sum_{k=1}^{\infty} \eta_{kn}^b(t) U_{ik}(\alpha_1,\alpha_2) \qquad (6.42)$$

where

$$\eta_{kn}^b = \Lambda_{kn}^b \sin(n\Omega t - \phi_{kn}) \qquad (6.43)$$

The superscripts a and b define the solutions to the $a_n \cos n\Omega t$ and $b_n \sin n\Omega t$ excitation terms.

The amplitudes and phase lags of the model participation factors are (Soedel, 2004) for the η_{kn}^a set,

$$\Lambda_{kn}^a = \frac{F_k^* a_n}{\omega_k^2 \sqrt{\left[1 - \left(\frac{n\Omega}{\omega_k}\right)^2\right]^2 + 4\zeta_k^2 \left(\frac{n\Omega}{\omega_k}\right)^2}} \qquad (6.44)$$

$$\phi_{kn} = \tan^{-1} \frac{2\zeta_k \left(\frac{n\Omega}{\omega_k}\right)}{1 - \left(\frac{n\Omega}{\omega_k}\right)^2} \qquad (6.45)$$

and for the η_{kn}^b set,

$$\Lambda_{kn}^b = \frac{F_k^* b_n}{\omega_k^2 \sqrt{\left[1 - \left(\frac{n\Omega}{\omega_k}\right)^2\right]^2 + 4\zeta_k^2 \left(\frac{n\Omega}{\omega_k}\right)^2}} \qquad (6.46)$$

with ϕ_{kn} of Equation 6.45 being the same for both sets.

The displacement response of the housing shell in the three directions are, therefore, given by

$$u_i(\alpha_1,\alpha_2,t) = \sum_{k=1}^{\infty} \frac{F_k^* a_0}{\omega_k^2} U_{ik}(\alpha_1,\alpha_2) + \sum_{k=1}^{\infty} \sum_{n=1}^{\infty} \left(\eta_{kn}^a(t) + \eta_{kn}^b(t)\right) U_{ik}(\alpha_1,\alpha_2) \qquad (6.47)$$

or in expanded form,

$$u_i(\alpha_1,\alpha_2,t) = \sum_{k=1}^{\infty} \frac{F_k^* a_o}{\omega_k^2} U_{ik}(\alpha_1,\alpha_2)$$

$$+ \sum_{k=1}^{\infty} \sum_{n=1}^{\infty} \frac{F_k^*[a_n \cos(n\Omega t - \phi_{kn}) + b_n \sin(n\Omega t - \phi_{kn})]U_{ik}(\alpha_1,\alpha_2)}{\omega_k^2 \sqrt{\left[1-\left(\frac{n\Omega}{\omega_k}\right)^2\right]^2 + 4\zeta_k^2\left(\frac{n\Omega}{\omega_k}\right)^2}}$$

(6.48)

Resonance occurs whenever

$$n\Omega = \omega_k \quad (6.49)$$

where $u = 1, 2, 3, \ldots, \infty$. Many types of compressors, especially, for example, reciprocating piston compressors, with a shaft speed of Ω in rad/sec, have higher harmonics present in the periodic, mechanical excitation, due to kinematics. Therefore, resonances in shell-like housings or other structural elements are often difficult to avoid. For example, a reciprocating piston compressor that rotates at approximately 3600 RPM creates a periodic forcing of the housing shell that may create potential resonances at 60, 120, 180, ..., 900, 960, 1020, 1080, 1140, 1200, 1260 Hz, ... etc. Hermetic housing shells of household refrigeration compressors have many natural frequencies clustered in various frequency bands, say, for example, 900 to 1200 Hz, and so on. Avoiding a resonance by trying to detune (space) the shell's natural frequencies from the excitation frequency harmonics at the design stage is virtually impossible because prediction tools are not that accurate. Any detuning (fine tuning) of the shell has to be done in the laboratory, if at all. For example, see Laursen (1990). A more practical approach seems to be the introduction of increased damping, and of course balancing, as Equation 6.48 shows.

6.5 REMARKS ABOUT DISSIPATIVE DAMPING

Dissipative damping is a very important tool for reducing vibrations. It is detrimental only for the off-resonance excitation frequencies that are, say, about halfway between the closest adjoining natural frequencies of the undamped shell; in this case, damping will increase the response amplitude. This is, however, usually not of concern, and in general we can expect damping to be quite beneficial. It reduces the resonance peaks in the response spectrum. The potentially detrimental effect of damping force transmission from the compressor casing to the housing, and from the housing to the foundation will be discussed later.

Application of a damping treatment of a viscoelastic nature to compressor housings in the refrigeration industry are not recommended when the coefficient of

Forced Vibration of Compressor Housing (Shell)

performance depends on the shell rejecting heat. The damping treatment tends to thermally insulate the shell. For those applications where it is feasible, application of damping material by spraying or gluing it to the shell surface is acceptable for thin layers, but will have only a very modest effect. Thick, open layers have a way of becoming ineffective because most of the straining action (and thus energy conversion into heat) will be confined to the layers close to the shell. It is therefore useful to increase the strain on the damping material by applying a backup layer of stiff plastic or metal foil. The design becomes a sandwich shell. Two shells, one inside the other, with a viscoelastic layer between them, are a good but expensive possibility, if heat transfer considerations will allow it.

Sometimes metal friction can be used to obtain dissipative compressor shell damping. Hermetic compressor shells are damped by pressing a steel strip inside the shell. Care must be taken to permit relative motion to occur. If the friction strip is pressed too tightly into the housing shell, no relative motion occurs and thus no energy is converted into heat, in which case one obtains only the questionable benefit of additional stiffness and mass.

In the foregoing modeling of the housing shell, a viscous damping formulation was used. It assumes that the distributed damping effect, which resists motion, is proportional to velocity. Another option would have been to use a material damping or hysteresis model.

The hysteresis model is based on the recognition that material damping, a so-called hysteresis looplike damping, can be described by replacing Young's modulus E by the complex modulus $E(1 + j\eta)$, where η is the hysteretic damping constant (see also Soedel, 2004).

As it turns out, the hysteresis model works fine for harmonic vibrations, but raises mathematical difficulties when other types of vibration responses are to be investigated, for example, the impulse (impact) response. Also, there are other types of damping, such as the energy of vibration, which is transferred through boundaries, friction damping, damping provided by an oil sump, and so forth. Therefore, we may just as well lump everything into an equivalent viscous damping model.

For example, it is shown in Soedel (2004) that an equivalent viscous damping coefficient can be formulated in terms of the hysteretic damping constant η:

$$\lambda = \rho h \omega_k \left(\frac{\omega_k}{\omega} \right) \eta \qquad (6.50)$$

where ω is the frequency of excitation, and ω_k is the natural frequency of a particular natural mode.

This results in a relationship for the equivalent modal damping coefficients of

$$\zeta_k = \frac{1}{2} \left(\frac{\omega_k}{\omega} \right) \eta. \qquad (6.51)$$

In conclusion, the viscous damping model is quite sufficient for our purposes.

6.6 DYNAMIC ABSORBERS

In the previous section, we discussed dissipative damping, where the energy is converted into heat. Dynamic damping, where the energy is shifted from the housing to an attached elastic system without necessarily much conversion into heat (where this would be objectionable), is another possibility. A device built on this principle is the dynamic absorber. This approach works well only if there is a clearly defined excitation frequency, at which one desires low vibration amplitude levels. A dynamic absorber is simply another dynamic system that has at least one natural frequency that coincides with the objectionable excitation frequency. In its simplest form, a dynamic absorber is a spring-mass oscillator. The amplitude of the shell at the attachment point of such a damper will become a minimum (the level depends on the amount of dissipative damping in the system), while the amplitude of the motion of the tuned dynamic absorber will be a maximum. But a dynamic absorber can also be a multi-degree-of-freedom spring-mass arrangement, a beam, or any other elastic system.

Besides the disadvantage that the dynamic absorber may take on a great amount of vibration energy and create noise on its own, we find a situation like that of Hercules battling the Hydra. For every head he chopped off the monster, two new ones grew in its place. By adding the dynamic damper, we not only shift the shell natural frequencies to new values, which may cause new troubles, but we also create at least one additional degree of freedom, which is likely to add a mode with a natural frequency in the range of interest. This mode needs to be taken into consideration in all further investigations.

But in certain special situations, dynamic absorbers have been effective tools of noise reduction.

To illustrate the dynamic absorber using a simple mathematical model, consider a housing to which the dynamic absorber is attached at one point in one direction only, and the housing is forced harmonically at one point only, as shown in Figure 6.1.

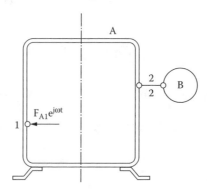

FIGURE 6.1 Harmonic force excitation of a housing shell A with an attached dynamic absorber B.

Forced Vibration of Compressor Housing (Shell)

The receptances (transfer functions) that have to be formulated for the housing alone are, in this case (Soedel, 2004):

$$\alpha_{11} = \frac{X_{A1}}{F_{A1}} \tag{6.52}$$

$$\alpha_{22} = \frac{X_{A2}}{F_{A2}} \tag{6.53}$$

$$\alpha_{21} = \alpha_{12} = \frac{X_{A2}}{F_{A1}} \tag{6.54}$$

where X_{A1}, X_{A2} are the response amplitudes of the harmonic transverse displacement of the shell at the point of forcing and at the dynamic absorber attachment point 2. F_{A1} is the forcing amplitude and F_{A2} is the amplitude of a harmonic force that we apply at location 2.

We also have to formulate, for the dynamic absorber alone, the receptance at point 2:

$$\beta_{22} = \frac{X_{B2}}{F_{B2}} \tag{6.55}$$

where F_{B2} is the amplitude of a harmonic force applied to the dynamic absorber at location 2 and X_{B2} is the harmonic response amplitude.

These receptances are obtained, theoretically or by measurement, before the dynamic absorber is attached to the housing.

Next, when we attach the dynamic absorber, the response amplitudes of the housing are modified because both F_{A1} and also F_{A2} (the force amplitude that the dynamic absorber exerts on the housing) contribute. They are now (Soedel, 2004):

$$X_{A1} = \alpha_{11} F_{A1} + \alpha_{12} F_{A2}$$

$$X_{A2} = \alpha_{21} F_{A1} + \alpha_{22} F_{A2}.$$

It also must be that

$$F_{A2} = -F_{B2},$$

which means that at the dynamic absorber junction, the two forces add up to zero. F_{B2} is given by

$$F_{B2} = \frac{X_{B2}}{\beta_{22}}$$

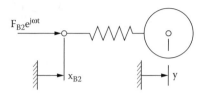

FIGURE 6.2 One-degree-of-freedom dynamic absorber.

where $X_{B2} = X_{A2}$ because A and B are attached at 2.

Solving these equations for the amplitude X_{A2} of the harmonic displacement at the dynamic absorber attachment point gives

$$X_{A2} = \frac{\beta_{22}\alpha_{21}}{\alpha_{22} + \beta_{22}} F_{A1}$$

or, in time ($x_{A2} = X_{A2} e^{j\omega t}$):

$$x_{A2} = \frac{\beta_{22}\alpha_{21}}{\alpha_{22} + \beta_{22}} F_{A1} e^{j\omega t}.$$

This equation is quite general, and is independent of the specific geometry of the housing and the number of degrees of freedom of the dynamic absorber. It tells us that if we can tune the dynamic absorber to a particular, objectionable frequency such that

$$\beta_{22} = 0, \tag{6.56}$$

there will be a zero vibration response (for zero damping) of the housing at the point where the dynamic absorber is attached.

Let us take as an example the simplest conceivable dynamic absorber, which is a spring-mass system as shown in Figure 6.2. The equations of motion of this base excited system are

$$M\ddot{y} + Ky = Kx_{B2} \quad \text{and} \tag{6.57}$$

$$(y - x_{B2})K + F_{B2} e^{j\omega t} = 0. \tag{6.58}$$

In steady-state harmonic motion at excitation frequency ω, we have

$$y = Y e^{j\omega t} \quad \text{and} \tag{6.59}$$

$$x_{B2} = X_{B2} e^{j\omega t}. \tag{6.60}$$

Forced Vibration of Compressor Housing (Shell)

Substitution gives

$$\beta_{22} = \frac{X_{B2}}{F_{B2}} = -\frac{1}{M\omega^2}\left[1 - \left(\frac{\omega}{\omega_n}\right)^2\right] \tag{6.61}$$

where

$$\omega_n = \sqrt{\frac{K}{M}}. \tag{6.62}$$

Examining Equation 6.61, we see that $\beta_{22} = 0$ when

$$\omega = \omega_n = \sqrt{\frac{K}{M}}. \tag{6.63}$$

At this excitation frequency, an undamped absorber will force the housing response to be zero at the attachment location.

This does not mean that the housing response is zero at other locations (Soedel, 2004). One will have to investigate, on a case-by-case basis, whether the effect averaged over all locations is beneficial enough to justify the expense of this design. However, the fact that a dynamic absorber can be designed such that it prevents a response at the attachment point is of interest.

It must be pointed out that if damping is present, the response will not quite go to zero at the attachment point (Soedel, 2004). But the response will still be diminished, and on the plus side, damping will reduce response peaks at the resonances.

Note also that multi-degree-of-freedom dynamic absorbers have the ability to reduce the responses at more than one excitation frequency, depending on how many tuning (natural) frequencies the dynamic absorber possesses. But again, one should not forget that more potential resonances are also created.

Finally, a point needs to be made about the size (mass) of the dynamic absorber. While in a completely undamped system, a one-degree-of-freedom system absorber of any mass (therefore also a very small mass) will create a zero response at its attachment point when $\beta_{22} = 0$, the excitation frequency at which this happens is very sensitive to the tuning frequency. The slightest deviation may create large responses. Also, the slightest amount of damping will fill in the response curve at the tuning frequency and the response will be nonzero. It can be shown that as the mass of our one-degree-of-freedom dynamic absorber is increased, these sensitivities diminish. Therefore, the absorber mass has to have a minimum value (the tuning frequency is still that given by Equation 6.63, however) on the order of, say, 10% of the housing mass. In other words, one should not design a dynamic absorber the size of a peanut attached to a watch spring, but one must boldly start with a mass of the size of a fist when dealing, for example, with a household heat pump compressor housing. Of course this size requirement makes dynamic absorbers less attractive.

6.7 FRICTION DAMPING

As discussed in the foregoing, a way of damping a housing shell that is sometimes effective is to press a ringlike metal strip into the shell, or to spot weld one or more cantilever-type springs to the inside or outside of the shell (the inside is preferable because the strip is hidden from view) in such a way that the tips are pressed against the shell with more or less constant force. The relative motion between the friction metal tip and the shell surface removes energy from the vibrating shell.

Let us investigate a simple model of this (Figure 6.3). Let us assume that the friction force location relative to the nonvibrating shell does not vary. In reality, because the strip will have to be located on the shell surface itself, this will not be quite true, but the approximation is good enough for our purposes here.

The tangential motion of the contact point B on the outside surface of the shell, where the point force P is applied, is in the circumferential direction (see also Section 3.1.4 and Soedel, 2004):

$$U_\theta(x,\theta,\alpha_3,t) = u_\theta(x,\theta,0,t) + \alpha_3 \beta_\theta(x,\theta,0,t) \tag{6.64}$$

where $u_\theta(x,\theta,0,t)$ is the tangential displacement at point A on the reference surface of the housing shell (at $\alpha_3 = 0$); β_θ is the slope created by bending; x, θ are the locations of the friction contact on the shell; and $\alpha_3 = \frac{h}{2}$ on the outer surface of the shell and $\alpha_3 = -\frac{h}{2}$ on the inner surface of the shell, depending on whether the friction is applied to the outside or inside. As derived in Soedel (2004), the slope β_θ is related to u_θ and u_3 (the latter being the transverse deflection of the housing shell) by (using the cylindrical shell approximation)

$$\beta_\theta = \frac{u_\theta}{a} - \frac{1}{a}\frac{\partial u_3}{\partial \theta} \tag{6.65}$$

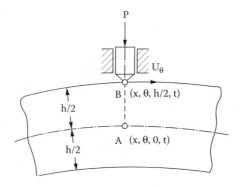

FIGURE 6.3 A very simplified model of friction damping: a constant force P is in contact with the housing shell surface at point B. The motion of the shell at this point is U_θ relative to the force application point.

Forced Vibration of Compressor Housing (Shell)

where, using the *inextensional approximation* (again, see Soedel, 2004), the tangential deflection at the reference surface, u_θ, is related to the transverse deflection, u_3, by

$$\frac{\partial u_\theta}{\partial \theta} = -u_3. \tag{6.66}$$

Therefore, Equation 6.65 becomes

$$\beta_\theta = \frac{1}{a}\left(u_\theta + \frac{\partial^2 u_\theta}{\partial \theta^2}\right) \tag{6.67}$$

and Equation 6.64 becomes

$$U_\theta = u_\theta\left(1 + \frac{\alpha_3}{a}\right) + \frac{\alpha_3}{a}\frac{\partial^2 u_\theta}{\partial \theta^2}. \tag{6.68}$$

When the housing shell (for convenience approximated as a cylinder) vibrates at a natural frequency defined by the mode numbers (m, n), it can be shown (Soedel, 2004) that

$$u_\theta = U_{\theta mn} \sin \omega t \tag{6.69}$$

where

$$U_{\theta mn} = B_{mn} \sin\frac{m\pi x}{L} \sin n(\theta - \phi) \tag{6.70}$$

and where B_{mn} is the vibration amplitude of the tangential, natural mode component, and ϕ is an arbitrary phase angle that describes the fact that in a perfectly axisymmetric compressor shell, there is no preferred direction of the natural modes as far as the θ–coordinate is concerned. The value of ω is ω_{mn}, the natural frequency of the mode in question in rad/s.

Equation 6.68 becomes, therefore (for $\alpha_3/a \ll 1$)

$$U_\theta = B_{mn}\left(1 - \frac{\alpha_3 n^2}{a}\right)\sin\frac{m\pi x}{L}\sin n(\theta - \phi)\sin \omega t \tag{6.71}$$

or

$$U_\theta = \left(1 - \frac{\alpha_3 n^2}{a}\right)U_{\theta mn}\sin \omega t . \tag{6.72}$$

The motion $U_\theta(x,\theta,\alpha_3,t)$ is opposed by the friction force $f = P\mu_k$, where μ_k is the dynamic friction coefficient. Therefore, the energy differential dE_d, which is removed from the vibrating housing shell, is

$$dE_d = P\mu_k \left(\frac{dU_\theta}{dt}\right) dt \qquad (6.73)$$

or, after integration over one cycle of oscillation, keeping in mind that the friction force $P\mu_k$ always opposes motion, we obtain

$$E_d = 4 P\mu_k \left| B_{mn}\left(1 - \frac{\alpha_3 n^2}{a}\right) \sin\frac{m\pi x}{L} \sin n(\theta - \phi) \right|. \qquad (6.74)$$

Obviously, the amount of energy per cycle, E_d, that is removed is proportional to the contact force P, the dynamic friction coefficient μ_k, the vibration amplitude of the mode of interest, B_{mn}, and dependent on the location α_3 (is the friction force applied on the outside or inside?), the effective cylinder radius, a, and most important for our purpose here, the location x,θ of load application. To maximize E_d, we must maximize $\sin\frac{m\pi x}{L} \sin n(\theta - \phi)$. This means that the friction application must be at an antinode of the tangential mode component in the $\theta-$ direction. If it is at a node line, $E_d = 0$. At an antinode, $|\sin\frac{m\pi x}{L} \sin n(\theta - \phi)| = 1$ and we remove, per oscillation, an amount of energy equal to

$$E_d = 4 P\mu_k \left| B_{mn}\left(1 - \frac{\alpha_3 n^2}{a}\right) \right|. \qquad (6.75)$$

Note also that if the friction force is created by the tip of a cantilever-like strip (beam), it must be clamped (by spot welding, for example), at a node in the $\theta-$ direction. While the nonpreferential direction effect may make it difficult to do this, it applies only to a cylinder. In a real compressor housing, say, of oval cross-section, the modes are usually fixed (have a preferential direction). Note that for circular cylindrical shells, antinodes of the tangential mode components in the $\theta-$ direction are typically at the nodes of the transverse mode components.

It may be somewhat amusing to retell a story I heard many years ago (I have forgotten from whom). A two-pronged friction metal strip was properly designed with a calculated curvature that was less than the shell curvature. When spot welded to the inside of the shell, presumably to a node point in the $\theta-$ direction (the tips were at the antinodes) by clamping it at the same time, a bending moment was created, and thus forces at the tips of the strip pressed against the housing shell at antinodes, as calculated. Prototype measurements showed good results, and for a while good results were also obtained in the production compressors coming from the assembly line. But after a few weeks, the compressors had become as noisy as before. Engineers tried unsuccessfully for a long time to figure out why. Finally they found that the operator whose task it was to spot weld the strips to the housing shell had decided that it would speed up his job (and save the company money) if he prebent each strip to match the housing curvature exactly, making clamping during spot welding unnecessary. Obviously, the bending moment mechanism, which created the friction force, had been eliminated. This illustrates that workers must be viewed as important team members, and that one must explain to them why certain measures are important, especially in sound and vibration control.

7 Free and Forced Vibrations of Compressor Casings

7.1 THE THREE-DIMENSIONAL EQUATIONS OF MOTION FOR AN ELASTIC SOLID

When it comes to the vibrations of the thick, irregularly shaped compressor casing (which includes the cylinders, the crank case, often the motor housing, and so on), the equations of motion for shells or plates are only useful in the most approximate sense. Rather, the compressor casing has to be viewed as a three-dimensional solid. It is, therefore, worthwhile to examine the derivation of the governing equations, even while in a practical sense, finite element approaches will most likely have to be used because of the complexity of the geometry. However, these equations (or their equivalent energy expressions) are the foundation of the finite element formulation, and we will use these equations to derive the three-dimensional acoustic wave equation by reduction. See also Ramani et al. (1994a, b).

The foundation of any derivation is that we require the solution of an orthogonal coordinate system. Because either Cartesian or cylindrical coordinates will most likely be used in practice, it is best to simulate the equations in terms of orthogonal curvilinear coordinates, as was done for the shell or plate structures.

The first difference (in comparison to the shell equation development) is that the fundamental form has another branch (see Soedel, 2004):

$$(ds)^2 = A_1^2(d\alpha_1)^2 + A_2^2(d\alpha_2)^2 + A_3^2(d\alpha_3)^2. \tag{7.1}$$

7.1.1 STRAIN–STRESS RELATIONSHIPS

It is no longer possible to neglect certain stresses or strains, the full equations from three-dimensional elasticity must be used:

$$\varepsilon_{11} = \frac{1}{E}[\sigma_{11} - \mu(\sigma_{22} + \sigma_{33})] \tag{7.2}$$

$$\varepsilon_{22} = \frac{1}{E}[\sigma_{22} - \mu(\sigma_{11} + \sigma_{33})] \tag{7.3}$$

$$\varepsilon_{33} = \frac{1}{E}[\sigma_{33} - \mu(\sigma_{11} + \sigma_{22})] \tag{7.4}$$

$$\varepsilon_{12} = \frac{\sigma_{12}}{G} = \varepsilon_{21} \tag{7.5}$$

$$\varepsilon_{13} = \frac{\sigma_{13}}{G} = \varepsilon_{31} \tag{7.6}$$

$$\varepsilon_{23} = \frac{\sigma_{23}}{G} = \varepsilon_{32} \tag{7.7}$$

In a general sense, none of the stresses and strains can be neglected (but when applying it to certain problems concerning the vibration of the compressor casing, we will be able to simplify).

7.1.2 Strain–Displacement Relationships

Employing the standard definitions for normal and shear strains, recognizing that any point in the structure can have deflections u_1, u_2, u_3 in the three orthogonal coordinate directions, the strain–displacement relationships become:

$$\varepsilon_{11} = \frac{1}{A_1}\left(\frac{\partial A_1}{\partial \alpha_1}\frac{u_1}{A_1} + \frac{\partial A_1}{\partial \alpha_2}\frac{u_2}{A_2} + \frac{\partial A_1}{\partial \alpha_3}\frac{u_3}{A_3}\right) + \frac{\partial}{\partial \alpha_1}\left(\frac{u_1}{A_1}\right) \tag{7.8}$$

$$\varepsilon_{22} = \frac{1}{A_2}\left(\frac{\partial A_2}{\partial \alpha_1}\frac{u_1}{A_1} + \frac{\partial A_2}{\partial \alpha_2}\frac{u_2}{A_2} + \frac{\partial A_2}{\partial \alpha_3}\frac{u_3}{A_3}\right) + \frac{\partial}{\partial \alpha_2}\left(\frac{u_2}{A_2}\right) \tag{7.9}$$

$$\varepsilon_{33} = \frac{1}{A_3}\left(\frac{\partial A_3}{\partial \alpha_1}\frac{u_1}{A_1} + \frac{\partial A_3}{\partial \alpha_2}\frac{u_2}{A_2} + \frac{\partial A_3}{\partial \alpha_3}\frac{u_3}{A_3}\right) + \frac{\partial}{\partial \alpha_3}\left(\frac{u_3}{A_3}\right) \tag{7.10}$$

$$\varepsilon_{12} = \frac{A_1}{A_2}\frac{\partial}{\partial \alpha_2}\left(\frac{u_1}{A_1}\right) + \frac{A_2}{A_1}\frac{\partial}{\partial \alpha_1}\left(\frac{u_2}{A_2}\right) \tag{7.11}$$

$$\varepsilon_{13} = \frac{A_1}{A_3}\frac{\partial}{\partial \alpha_3}\left(\frac{u_1}{A_1}\right) + \frac{A_3}{A_1}\frac{\partial}{\partial \alpha_1}\left(\frac{u_3}{A_3}\right) \tag{7.12}$$

$$\varepsilon_{12} = \frac{A_2}{A_3}\frac{\partial}{\partial \alpha_3}\left(\frac{u_2}{A_2}\right) + \frac{A_3}{A_2}\frac{\partial}{\partial \alpha_2}\left(\frac{u_3}{A_3}\right) \tag{7.13}$$

The details of the derivation can be found in Soedel (2004).

7.1.3 Energy Expressions

Because none of the stresses or strains can be neglected, the strain energy is

$$U = \frac{1}{2}\int_{\alpha_1}\int_{\alpha_2}\int_{\alpha_3}(\sigma_{11}\varepsilon_{11} + \sigma_{22}\varepsilon_{22} + \sigma_{33}\varepsilon_{33} + \sigma_{12}\varepsilon_{12} + \sigma_{13}\varepsilon_{13} + \sigma_{23}\varepsilon_{23})$$
$$\times A_1 A_2 A_3 d\alpha_1 d\alpha_2 d\alpha_3 \tag{7.14}$$

and the kinetic energy is

$$T = \frac{1}{2}\int_{\alpha_1}\int_{\alpha_2}\int_{\alpha_3}\rho\left(\dot{u}_1^2 + \dot{u}_2^2 + \dot{u}_3^2\right)A_1 A_2 A_3 d\alpha_1 d\alpha_2 d\alpha_3. \tag{7.15}$$

7.1.4 Hamilton's Principle and General Equations of Motion

As before for the housing (shell), Hamilton's principle $\delta\int_{t_1}^{t_2}(T - U + W_{nc})\,dt = 0$, $\delta\bar{r}_i = 0$ at $t = t_1$ and t_2 is applied. The details of the mathematical operation can be found in Soedel (2004). We obtain three equations of motion, as one would expect:

$$\frac{1}{A_1^2 A_2 A_3}\left[\frac{\partial}{\partial\alpha_1}(\sigma_{11}A_1 A_2 A_3) + \frac{\partial}{\partial\alpha_2}(\sigma_{12}A_1^2 A_3) + \frac{\partial}{\partial\alpha_3}(\sigma_{23}A_2^2 A_1)\right]$$
$$-\left(\frac{\sigma_{11}}{A_1^2}\frac{\partial A_1}{\partial\alpha_1} + \frac{\sigma_{22}}{A_1 A_2}\frac{\partial A_2}{\partial\alpha_1} + \frac{\sigma_{33}}{A_1 A_3}\frac{\partial A_3}{\partial\alpha_1}\right) - \rho\ddot{u}_1 = -q_1 \tag{7.16}$$

$$\frac{1}{A_2^2 A_1 A_3}\left[\frac{\partial}{\partial\alpha_2}(\sigma_{22}A_1 A_2 A_3) + \frac{\partial}{\partial\alpha_1}(\sigma_{12}A_2^2 A_3) + \frac{\partial}{\partial\alpha_3}(\sigma_{23}A_2^2 A_1)\right]$$
$$-\left(\frac{\sigma_{11}}{A_1 A_2}\frac{\partial A_1}{\partial\alpha_2} + \frac{\sigma_{22}}{A_2^2}\frac{\partial A_2}{\partial\alpha_2} + \frac{\sigma_{33}}{A_2 A_3}\frac{\partial A_3}{\partial\alpha_2}\right) - \rho\ddot{u}_2 = -q_2 \tag{7.17}$$

$$\frac{1}{A_3^2 A_1 A_2}\left[\frac{\partial}{\partial\alpha_3}(\sigma_{33}A_1 A_2 A_3) + \frac{\partial}{\partial\alpha_1}(\sigma_{13}A_3^2 A_2) + \frac{\partial}{\partial\alpha_2}(\sigma_{23}A_3^2 A_1)\right]$$
$$-\left(\frac{\sigma_{11}}{A_1 A_3}\frac{\partial A_1}{\partial\alpha_3} + \frac{\sigma_{22}}{A_2 A_3}\frac{\partial A_2}{\partial\alpha_3} + \frac{\sigma_{33}}{A_3^2}\frac{\partial A_3}{\partial\alpha_3}\right) - \rho\ddot{u}_3 = -q_3 \tag{7.18}$$

7.1.5 Boundary Conditions

The equations of motion require the specifications of three boundary conditions on each surface. For example, on a typical $\alpha_1 - \alpha_2$ surface:

1. either normal stress σ_{33} or normal displacement u_3 have to be specified
2. either the shear stress σ_{31} or the displacement u_1 in the α_1–direction have to be specified
3. either the shear stress σ_{32} or the displacement u_2 in the α_2–direction have to be specified

On a $\alpha_1 - \alpha_2$ surface of a compressor housing, which is normal to the α_3 –direction, we specify, for example, if there is no load on it (a free surface), that $\sigma_{33} = 0$, $\sigma_{31} = 0$, and $\sigma_{32} = 0$.

7.1.6 Example: Cartesian Coordinates

A widely used orthogonal coordinate system is the Cartesian coordinate system. To convert the foregoing equations, we realize that the diagonal of an infinitesimal element is

$$(ds)^2 = (dx)^2 + (dy)^2 + (dz)^2. \tag{7.19}$$

We compare this to the fundamental form for curvilinear coordinates,

$$(ds)^2 = A_1^2(d\alpha_1)^2 + A_2^2(d\alpha_2)^2 + A_3^2(d\alpha_3)^2 \tag{7.20}$$

and find $A_1 = A_2 = A_3 = 1$, $\alpha_1 = x$, $\alpha_2 = y$, $\alpha_3 = z$. The equations of motion become, therefore,

$$\frac{\partial \sigma_{xx}}{\partial x} + \frac{\partial \sigma_{xy}}{\partial y} + \frac{\partial \sigma_{xz}}{\partial z} - \rho \ddot{u}_x = -q_x \tag{7.21}$$

$$\frac{\partial \sigma_{yy}}{\partial y} + \frac{\partial \sigma_{xy}}{\partial x} + \frac{\partial \sigma_{yz}}{\partial z} - \rho \ddot{u}_y = -q_y \tag{7.22}$$

$$\frac{\partial \sigma_{zz}}{\partial z} + \frac{\partial \sigma_{xz}}{\partial x} + \frac{\partial \sigma_{yz}}{\partial y} - \rho \ddot{u}_z = -q_z \tag{7.23}$$

and the strain–displacement relationships become

$$\varepsilon_{xx} = \frac{\partial u_x}{\partial x} \tag{7.24}$$

$$\varepsilon_{yy} = \frac{\partial u_y}{\partial x} \tag{7.25}$$

$$\varepsilon_{zz} = \frac{\partial u_z}{\partial z} \tag{7.26}$$

$$\varepsilon_{xy} = \frac{\partial u_x}{\partial y} + \frac{\partial u_y}{\partial x} \tag{7.27}$$

$$\varepsilon_{xz} = \frac{\partial u_x}{\partial z} + \frac{\partial u_z}{\partial x} \tag{7.28}$$

$$\varepsilon_{yz} = \frac{\partial u_y}{\partial z} + \frac{\partial u_z}{\partial y}. \tag{7.29}$$

Because of the relative simplicity of these expressions, it is customary to substitute and write the equations of motion in displacement form:

$$G\left(\frac{\partial^2 u_x}{\partial x^2} + \frac{\partial^2 u_x}{\partial y^2} + \frac{\partial^2 u_x}{\partial z^2}\right) + (\lambda + G)\left(\frac{\partial^2 u_x}{\partial x^2} + \frac{\partial^2 u_y}{\partial y \partial x} + \frac{\partial^2 u_z}{\partial z \partial x}\right) - \rho \ddot{u}_x = -q_x \tag{7.30}$$

$$G\left(\frac{\partial^2 u_y}{\partial x^2} + \frac{\partial^2 u_y}{\partial y^2} + \frac{\partial^2 u_y}{\partial z^2}\right) + (\lambda + G)\left(\frac{\partial^2 u_x}{\partial x \partial y} + \frac{\partial^2 u_y}{\partial y^2} + \frac{\partial^2 u_z}{\partial z \partial y}\right) - \rho \ddot{u}_y = -q_y \tag{7.31}$$

$$G\left(\frac{\partial^2 u_z}{\partial x^2} + \frac{\partial^2 u_z}{\partial y^2} + \frac{\partial^2 u_z}{\partial z^2}\right) + (\lambda + G)\left(\frac{\partial^2 u_x}{\partial x \partial z} + \frac{\partial^2 u_y}{\partial y \partial z} + \frac{\partial^2 u_z}{\partial z^2}\right) - \rho \ddot{u}_z = -q_z \tag{7.32}$$

where G is the shear modulus,

$$G = \frac{E}{2(1+\mu)} \tag{7.33}$$

and λ is the bulk modulus,

$$\lambda = \frac{\mu E}{(1+\mu)(1-2\mu)}. \tag{7.34}$$

7.1.7 One-Dimensional (Wave) Equation for Solids

Setting $q_x = q_y = q_z = 0$ in the equation of motion and allowing u_x, u_y and u_z to be only functions of x (an assumption that looks only at oscillations in the x-direction) results in, from the first of the three equations,

$$G\frac{\partial^2 u_x}{\partial x^2} + (\lambda + G)\frac{\partial^2 u_x}{\partial x^2} = \rho \ddot{u}_x \tag{7.35}$$

or

$$\frac{\partial^2 u_x}{\partial x^2} - \frac{1}{C_1^2}\ddot{u}_x = 0 \tag{7.36}$$

where

$$C_1^2 = \frac{\lambda + 2G}{\rho} = \frac{E(1-\mu)}{\rho(1+\mu)(1-2\mu)}. \tag{7.37}$$

C_1 is the speed of sound of a compression wave. From the second equation we obtain

$$G\frac{\partial^2 u_y}{\partial x^2} = \rho \ddot{u}_y \tag{7.38}$$

or

$$\frac{\partial^2 u_y}{\partial x^2} - \frac{1}{C_2^2}\ddot{u}_y = 0 \tag{7.39}$$

where C_2 is the shear velocity,

$$C_2^2 = \frac{G}{\rho}. \tag{7.40}$$

Similarly, the shear wave equation for the $z - x$ plane is given by

$$\frac{\partial^2 u_z}{\partial x^2} - \frac{1}{C_2^2}\ddot{u}_z = 0. \tag{7.41}$$

7.2 FREE AND FORCED VIBRATIONS

For a three-dimensional elastic system, where the simplifications to a thin shell do not apply, the equations of motion can be written, just as for the shell, as

$$L_i\{u_1, u_2, u_3\} - \lambda \dot{u}_i - \rho \ddot{u}_i = -q_i. \tag{7.42}$$

Note that q_i is in (N/m³), λ is in Ns/m⁴, and ρ is in kg/m³.

First, we obtain the natural frequencies and modes by solving the free vibration equation

$$L_i\{u_1, u_2, u_3\} - \rho \ddot{u}_i = 0 \tag{7.43}$$

with boundary conditions,

$$B_k\{u_1, u_2, u_3\} = 0. \tag{7.44}$$

At any natural frequency,

$$u_i(\alpha_1, \alpha_2, t) = U_i(\alpha_1, \alpha_2)e^{j\omega t}. \tag{7.45}$$

This gives

$$L_i\{U_1, U_2, U_3\} + \rho h \omega^2 U_i = 0. \tag{7.46}$$

The boundary conditions become

$$B_K\{U_1, U_2, U_3\} = 0. \tag{7.47}$$

The natural modes and frequencies are obtained from these equations. Figure 7.1 illustrates what a typical natural mode of the casing might look like. See also, for example, Ramani et al. (1994a, b). Natural modes can become very complicated and numerical approaches have to be used to obtain them.

It is shown in Soedel (2004) that the natural modes are orthogonal and satisfy

$$\int\int\int_{\alpha_3\alpha_2\alpha_1}(U_{1k}U_{1p} + U_{2k}U_{2p} + U_{3k}U_{3p})A_1A_2A_3 d\alpha_1 d\alpha_2 d\alpha_3 = \begin{cases} N_k, & \text{if } p = k \\ 0, & \text{if } p \neq k \end{cases} \tag{7.48}$$

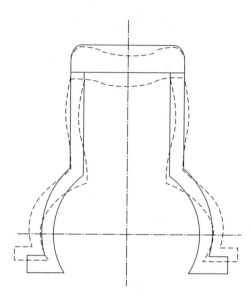

FIGURE 7.1 A typical natural mode of a compressor casing, in cross-sectional view. Note that there are nonnegligible thickness variations: simplifications used in plate or shell models do not apply.

where

$$N_k = \iiint_{\alpha_3 \, \alpha_2 \, \alpha_1} \left(U_{1k}^2 + U_{2k}^2 + U_{3k}^2 \right) A_1 A_2 A_3 d\alpha_1 d\alpha_2 d\alpha_3. \tag{7.49}$$

To solve the forced vibration case, we approach the solution with a modal series:

$$u_i(\alpha_1, \alpha_2, \alpha_3, t) = \sum_{k=1}^{\infty} \eta_k(t) U_{ik}(\alpha_1, \alpha_2, \alpha_3). \tag{7.50}$$

Substituting this into the equation of motion with forcing and equivalent viscous damping, and evoking the orthogonality condition, gives an equation from which we can obtain the modal expansion coefficient η_k:

$$\ddot{\eta}_k + 2\zeta_k \omega_k \dot{\eta}_k + \omega_k^2 \eta_k = F_k(t) \tag{7.51}$$

where

$$F_k(t) = \frac{1}{\rho N_k} \iiint_{\alpha_3 \, \alpha_2 \, \alpha_1} (q_1 U_{1k} + q_2 U_{2k} + q_3 U_{3k}) A_1 A_2 A_3 d\alpha_1 d\alpha_2 d\alpha_3 \tag{7.52}$$

$$\zeta_k = \frac{\lambda}{2\rho \omega_k}. \tag{7.53}$$

Free and Forced Vibrations of Compressor Casings   89

The analogy to the forced vibration of shells is obvious. Here, we have a triple integral and the units of λ, ρ, and q_i are different. That is all.

For subcritical damping, which is typical for compressor casings, $\zeta_k < 1$, and the general solution is

$$\eta_k(t) = e^{-\zeta_k \omega_k t} \left\{ \eta_k(0) \cos \gamma_k t + [\eta_k(0)\zeta_k \omega_k + \dot{\eta}_k(0)] \frac{\sin \gamma_k t}{\gamma_k} \right\} \quad (7.54)$$
$$+ \frac{1}{\gamma_k} \int_0^t F_k(\tau) e^{-\zeta_k \omega_k (t-\tau)} \sin \gamma_k (t-\tau) d\tau$$

where

$$\gamma_k = \omega_k \sqrt{1 - \zeta_k^2}. \quad (7.55)$$

Again, as for the compressor housing, the transient solution due to initial conditions is of lesser interest and is often simply set to zero.

7.3 STEADY-STATE HARMONIC RESPONSE

The compressor casing is excited periodically (which can be viewed as a sum of harmonic excitations) by the cylinder pressure, and by unbalances due to kinematics, and so forth. Periodic valve impacts can also be viewed as a series of harmonic excitations. Therefore, it is of special interest to study the harmonic response due to forcing:

$$q_i(\alpha_1, \alpha_2, \alpha_3, t) = q_i^*(\alpha_1, \alpha_2, \alpha_3) e^{j\omega t} \quad (7.56)$$

The solution is

$$u_i = \sum_{k=1}^{\infty} \eta_k(t) U_i(\alpha_1, \alpha_2, \alpha_3) \quad (7.57)$$

where

$$\eta_k = \Lambda_k e^{j(\omega t - \phi_k)} \quad (7.58)$$

and where the magnitude of the response is

$$\Lambda_k = \frac{F_k^*}{\omega_k^2 \sqrt{\left[1 - (\omega/\omega_k)^2\right]^2 + 4\zeta_k^2 (\omega/\omega_k)^2}}. \quad (7.59)$$

The phase lag is

$$\phi_k = \tan^{-1} \frac{2\zeta_k(\omega/\omega_k)}{1-(\omega/\omega_k)^2} \qquad (7.60)$$

and the function F_k^* is

$$F_k^* = \frac{1}{\rho N_k} \int_{\alpha_3}\int_{\alpha_2}\int_{\alpha_1} \left(q_1^* U_{1k} + q_2^* U_{2k} + q_3^* U_{3k}\right) A_1 A_2 A_3 \, d\alpha_1 d\alpha_2 d\alpha_3. \qquad (7.61)$$

As for the housing, we see that resonances occur whenever one of the excitation frequencies coincides with a natural frequency of the casing. Given these resonance possibilities, in general we can state that in the vicinity of the kth resonance, the response amplitudes are proportional to the forcing q_i and inversely proportional to the mass density ρ and the square of the kth natural frequency:

$$u_i \propto \frac{q_i}{\rho \omega_k^2} \qquad (7.62)$$

where ρ is the mass per unit volume. This proportionality explains why cast iron is better than, say, aluminum from a noise viewpoint, assuming approximately the same resonance distribution. Of course, grey cast iron also provides more damping than aluminum, which lowers the resonance peaks.

7.4 RESPONSE OF THE CASING TO IMPACT

Another vibration, and thus sound source, is the periodic impact produced by mechanical valves or by piston slap. If we view it as a sequence of single events, the forcing can, in principle, be expressed as

$$q_i(\alpha_1, \alpha_2, \alpha_3, t) = M_i^*(\alpha_1, \alpha_2, \alpha_3)\, \delta(t-t_1) \qquad (7.63)$$

where $M_i^*(\alpha_1, \alpha_2, \alpha_3)$ is a distributed momentum change per unit volume, with units of Ns/m^3. This approach assumes that the compressor casing response to an impact has died out before the next impact occurs. An alternative is to view periodic impact as a forcing function that can be formulated as a steady-state periodic forced response problem.

$\delta(t-t_1)$ is the Dirac delta function, which defines the occurrence of impact at time $t = t_1$. Its definitions are (for a discussion see Soedel, 2004):

$$\delta(t-t_1) = 0 \quad \text{if } t \neq t_1 \qquad (7.64)$$

$$\int_{t=-\infty}^{t=\infty} \delta(t-t_1)\, dt = 1 \qquad (7.65)$$

Free and Forced Vibrations of Compressor Casings 91

The unit of $\delta(t-t_1)$ is second^{-1}.
The value of F_k^* is

$$F_k^* = \frac{1}{\rho N_k} \int_{\alpha_3}\int_{\alpha_2}\int_{\alpha_1} \left(M_1^* U_{1k} + M_2^* U_{2k} + M_3^* U_{3k}\right) A_1 A_2 A_3 \, d\alpha_1 d\alpha_2 d\alpha_3. \quad (7.66)$$

Substituting this in the general solution gives, for subcritical damping and zero initial conditions,

$$\eta_k(t) = \frac{F_k^*}{\gamma_k} e^{-\zeta_k \omega_k (t-t_1)} \sin\gamma_k(t-t_1), \quad (7.67)$$

and the solution is:

$$u_i = \sum_{k=1}^{\infty} \eta_k(t)\, U_i(\alpha_1, \alpha_2, \alpha_3). \quad (7.68)$$

The result shows that in principle all natural modes of the compressor casing are excited by impact. This is analogous to ringing a bell. On the other hand, there is no resonance if the vibration diminishes to zero before the next impact occurs. However, because of the periodic nature of valve impact, and the relatively low damping of typical casings, it can be shown that resonances are occurring.

The resulting vibration amplitudes of the kth mode are proportional to the momentum of impact and inversely proportional to the mass density of the casing and to the kth natural frequency:

$$u_i \propto \frac{M_i^*}{\rho \omega_k}. \quad (7.69)$$

The momentum term is proportional to valve impact velocity and to valve mass:

$$M_i^* \propto mv. \quad (7.70)$$

Therefore, we see that effective ways to lower noise due to impact are to decrease the valve impact velocity, decrease the mass of the valves, make the compressor casing of high-density material with high internal damping, and to make the compressor casing as stiff as possible so that the natural frequencies are high. Similar arguments apply to piston slap.

7.5 STEADY-STATE RESPONSE OF THE COMPRESSOR CASING TO PERIODIC FORCING

If the response of the compressor casing does not die out between, for example, successive valve impacts, one needs to approach the response as a periodic phenomenon. The excitation by the cylinder pressure (in terms of a pressure-time diagram) is also periodic. Again, as in the chapter on the steady-state response of the compressor housing to periodic forcing, we assume that the spatial distribution of the forcing does not change with time, only that the temporal distribution changes. This allows us to express the forcing functions as

$$q_i(\alpha_1, \alpha_2, \alpha_3, t) = q_i^*(\alpha_1, \alpha_2, \alpha_3) f(t) \tag{7.71}$$

where f(t) is a function that is periodic in time. The period T of this periodic function is related to the frequency at which the function repeats, Ω, by

$$T = \frac{2\pi}{\Omega}. \tag{7.72}$$

In a compressor application, Ω is the steady-state operating speed of a compressor in rad/s. For example, if the speed is 3600 RDM, then the fundamental frequency is $f = 1/T = 60$ Hz, or $\Omega = 377$ rad/s.

As before, expanding the function f(t) in a Fourier series gives

$$f(t) = a_o + \sum_{n=1}^{\infty}(a_n \cos n\Omega t + b_n \sin n\Omega t) \tag{7.73}$$

where

$$a_o = \frac{1}{T}\int_o^T f(t)\,dt \tag{7.74}$$

$$a_n = \frac{2}{T}\int_o^T f(t)\cos n\Omega t\,dt \tag{7.75}$$

$$b_n = \frac{2}{T}\int_o^T f(t)\sin n\Omega t\,dt. \tag{7.76}$$

The forcing function is, therefore,

$$q_i(\alpha_1, \alpha_2, \alpha_3, t) = q_i^*(\alpha_1, \alpha_2, \alpha_3)\left[a_o + \sum_{n=1}^{\infty}(a_n \cos n\Omega t + b_n \sin n\Omega t)\right]. \tag{7.77}$$

Free and Forced Vibrations of Compressor Casings

The coefficient a_o is a constant term and is, as for the housing, of no significance. It is typically zero and is not of particular interest in sound and vibration control, which centers on the casing.

The vibratory deflections due to each $a_n \cos n\Omega t$ term are

$$u_i^a(\alpha_1, \alpha_2, \alpha_3, t) = \sum_{k=1}^{\infty} \eta_{kn}^a(t) U_{ik}(\alpha_1, \alpha_2, \alpha_3), \qquad (7.78)$$

and where, assuming subcritical damping for all modes,

$$\eta_{kn}^a = \Lambda_{kn}^a \cos(n\Omega t - \phi_{kn}). \qquad (7.79)$$

Similarly, the solution to each $b_n \sin n\Omega t$ term is

$$u_i^b(\alpha_1, \alpha_2, \alpha_3, t) = \sum_{k=1}^{\infty} \eta_{kn}^b(t) U_{ik}(\alpha_1, \alpha_2, \alpha_3) \qquad (7.80)$$

where

$$\eta_{kn}^b = \Lambda_{kn}^b \sin(n\Omega t - \phi_{kn}). \qquad (7.81)$$

The superscripts a and b define the solutions to the $a_n \cos n\Omega t$ and $b_n \sin n\Omega t$ excitation terms.

The amplitudes and phase lags of the model participation factors are (see Soedel, 2004) for the η_{kn}^a set,

$$\Lambda_{kn}^a = \frac{F_k^* a_n}{\omega_k^2 \sqrt{\left[1 - \left(\frac{n\Omega}{\omega_k}\right)^2\right]^2 + 4\zeta_k^2 \left(\frac{n\Omega}{\omega_k}\right)^2}} \qquad (7.82)$$

$$\phi_{kn} = \tan^{-1} \frac{2\zeta_k \left(\frac{n\Omega}{\omega_k}\right)}{1 - \left(\frac{n\Omega}{\omega_k}\right)^2} \qquad (7.83)$$

where

$$F_k^* = \frac{1}{\rho h N_k} \int_{\alpha_2} \int_{\alpha_1} [q_1^*(\alpha_1,\alpha_2,\alpha_3) U_{1k} + q_2^*(\alpha_1,\alpha_2,\alpha_3) U_{2k} + q_3^*(\alpha_1,\alpha_2,\alpha_3) U_{3k}]$$

$$\times A_1 A_2 A_3 d\alpha_1 d\alpha_2 d\alpha_3$$

$$(7.84)$$

and for the η_{kn}^b set,

$$\Lambda_{kn}^b = \frac{F_k^* b_n}{\omega_k^2 \sqrt{\left[1 - \left(\frac{n\Omega}{\omega_k}\right)^2\right]^2 + 4\zeta_k^2 \left(\frac{n\Omega}{\omega_k}\right)^2}} \tag{7.85}$$

with ϕ_{kn} of Equation 7.83 being the same for both sets.

The displacement response of the compressor casing in the three directions are, therefore, given by

$$u_i(\alpha_1, \alpha_2, \alpha_3, t) = \sum_{k=1}^{\infty} \frac{F_k^* a_o}{\omega_k^2} U_{ik}(\alpha_1, \alpha_2, \alpha_3) + \sum_{k=1}^{\infty} \sum_{n=1}^{\infty} \left(\eta_{kn}^a(t) + \eta_{kn}^b(t)\right) U_{ik}(\alpha_1, \alpha_2, \alpha_3) \tag{7.86}$$

or, in expanded form,

$$u_i(\alpha_1, \alpha_2, \alpha_3, t) = \sum_{k=1}^{\infty} \frac{F_k^* a_o}{\omega_k^2} U_{ik}(\alpha_1, \alpha_2, \alpha_3)$$

$$+ \sum_{k=1}^{\infty} \sum_{n=1}^{\infty} \frac{F_k^* [a_n \cos(n\Omega t - \phi_{kn}) + b_n \sin(n\Omega t - \phi_{kn})] U_{ik}(\alpha_1, \alpha_2, \alpha_3)}{\omega_k^2 \sqrt{\left[1 - \left(\frac{n\Omega}{\omega_k}\right)^2\right]^2 + 4\zeta_k^2 \left(\frac{n\Omega}{\omega_k}\right)^2}} \tag{7.87}$$

Similar to the compressor housing, resonances of the compressor casing occur whenever

$$n\Omega = \omega_k \tag{7.88}$$

where n = 1, 2, 3, ⋯ and k = 1, 2, 3, ⋯. Again, many types of compressors (especially for example reciprocating piston compressors) that have a shaft speed of Ω in rad/sec, have higher harmonics present in the periodic, mechanical excitation, due to the kinematics and due to the cylinder pressure. The periodic valve impact also produces, obviously, higher harmonics. Therefore, resonances in compressor casings are often difficult to avoid. Similar to the housing, avoiding a resonance by trying to detune (space) the casing's natural frequencies from the excitation frequency harmonics at the design stage is virtually impossible because prediction tools are not accurate enough. However, if the compressor casing is well isolated from the compressor housing, avoiding casing resonances is often less important than avoiding housing resonances. On the other hand, if casing resonances are identified as noise sources, damping treatment of the casing is even more difficult than an effective damping treatment of the housing, and detuning may have to be attempted. What comes to mind is that because grey cast iron has higher internal damping than, say, aluminum, the former is preferable for damping out resonance peaks.

8 Vibrations of Other Structural Components of a Compressor

8.1 VIBRATION AND FORCE TRANSMISSION OF DISCHARGE OR SUCTION TUBES

On an idealized model of a straight, simply supported discharge tube (Figure 8.1), fully realizing that real discharge tubes are curved, it is demonstrated how the discharge and suction pressures influence the resonance frequencies and thus the force transmission from the compressor to the compressor shell.

It is frequently observed that the sound pressure, radiated from the shell of a refrigeration compressor, changes when operating conditions change. One of the possible mechanisms is that when the speed of sound of the gas in the shell changes with a condition change, gas resonances in the cavity formed between the casing and the housing change (see, for example, Johnson and Hamilton, 1972 and 1988).

There is another possible mechanism. As the discharge-suction pressure difference changes, discharge or suction tubes change their resonance frequencies and thus their vibration transmission into the shell.

From a physical viewpoint, the reader is invited to study a tubelike toy balloon and observe how after blowing up the initially limp balloon, it becomes a stiff structure capable of beamlike vibrations. The balloon beam increases its stiffness with an increasing pressure differential, and thus increases its resonance frequency. The effect can also be observed on a Borden tube.

8.1.1 EQUATION OF MOTION FOR A STRAIGHT DISCHARGE TUBE

Let us examine the idealized case of a straight discharge tube in a low-side refrigeration compressor that is free to expand in an axial direction. Following any standard approach of derivation, such as Hamilton's principle, one obtains the following equation of motion (Soedel, 1980a):

$$EI\frac{\partial^4 w}{\partial x^4} + [\rho v^2 - (P_d - P_s)A]\frac{\partial^2 w}{\partial x^2} + 2\rho v\frac{\partial^2 w}{\partial x \partial t} + (m+\rho)\frac{\partial^2 w}{\partial t^2} = 0 \qquad (8.1)$$

where w = the transverse deflection, ρ = the mass of gas per unit length, P_d = the discharge pressure, P_s = the suction pressure, A = the internal area of tube,

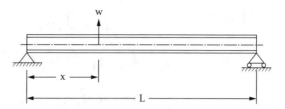

FIGURE 8.1 Model of a straightened discharge tube.

v = the average gas velocity, E = Young's modulus, I = the area moment, and m = the mass of the tube per unit length.

The equation can be physically understood by arguing that all terms represent forces per unit length of the tube, which either tend to restore the tube to its equilibrium position or displace it.

8.1.2 Natural Tube Frequencies Influenced by Pressure Changes

For the case of a simple supported tube of length L (which is another idealization that does not conform to reality, but serves to explain the mechanism), we obtain as a solution, the natural frequencies f_n [Hz] with $n = 1, 2, \ldots$:

$$f_n^2 = \frac{\pi^2 n^4}{4L^4} \frac{EI}{m} \left[1 + \frac{(P_d - P_s) AL^2}{EI \pi^2 n^2} \right] \quad (8.2)$$

where m = the mass per unit length, and where the influence of average gas velocity was neglected. We observe a stiffening effect of the tube with an increased pressure differential. While this stiffening effect shifts the natural frequencies for a typical refrigeration or air conditioning compressor only a few Hertz, it is sufficient to cause a noticeable change in vibration transmission because it may shift the tube in or out of a resonance if there is little damping (the resonances come in very sharp and a few Hertz make all the difference). To minimize this effect, it is best to introduce damping in the form of a friction wire or by other means, such as allowing the tube to support itself against the casing through a plastic attachment if the resonances cannot be avoided. In the case of a suction tube, P_s and P_d are reversed and a decrease in natural frequency results from an increase in the pressure differential. It also should be noted that the typically long and slender discharge tubes (low-side compressors) or suction tubes (high-side compressors) have natural frequencies that are relatively closely spaced (see also Wang et al., 2004). If the fundamental natural frequency of such an idealized straight tube is, say, 20 Hz, the natural frequencies (without the pressure influence) would be spaced 20, 80, 180, 320, 500 Hz … .

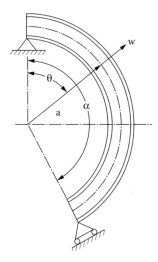

FIGURE 8.2 Model of a discharge tube that has constant, planar curvature.

8.1.3 Tube of Constant Curvature Vibrating in the Plane of Its Curvature

To illustrate that the presence of curvature does not lead to any different conclusion, the case of a tube that is a ring segment free to expand circumferentially (as shown in Figure 8.2) is investigated next (see also Soedel, 1980a). Again, by Hamilton's principle, one obtains, after velocity effects are neglected and the standard simplifications are made,

$$\frac{EI}{a^4}\left\{\frac{\partial^4 w}{\partial \theta^4} + \left[2 - (P_d - P_s)\frac{Aa^2}{EI}\right]\frac{\partial^2 w}{\partial \theta^2} + w\right\} + m\frac{\partial^2 w}{\partial t^2} = 0 \qquad (8.3)$$

where a is the radius of the curved tube.

For the case of a simple supported tube of arc length $L = a\alpha$ ($\theta = 0$ to $\theta = \alpha$) we obtain

$$f_n^2 = \frac{\pi^2 n^4}{4L^4}\left(1 - \frac{L^2}{\pi^2 n^2 a^2}\right)^2\left(\frac{EI}{m}\right)\left[1 + \frac{(P_d - P_s)AL^2}{EI\pi^2 n^2\left(1 - \frac{L^2}{\pi^2 n^2 a^2}\right)^2}\right]. \qquad (8.4)$$

We conclude that the presence of the curvature 1/a does not change the basic effect observed for the straight tube. Similar conclusions could be drawn had we investigated the out-of-plane motion of a curved tube.

FIGURE 8.3 Model of a discharge tube with a compressor casing motion as input. Of interest is the force transmission into the housing shell, modeled as a simple support on the right end.

8.1.4 Forces Transmitted into the Compressor Housing by a Vibrating Tube

To illustrate the force transmission mechanism into the shell, let us solve the idealized situation shown in Figure 8.3. The compressor motion is represented by w(o,t), while the housing motion is modeled as negligibly small, which allows the calculation of a force Q(L,t), which is transmitted into the housing. The boundary conditions are:

$$w(o,t) = W_c e^{j\omega t} \tag{8.5}$$

$$\frac{\partial^2 w}{\partial x^2}(o,t) = 0 \tag{8.6}$$

$$w(L,t) = 0 \tag{8.7}$$

$$\frac{\partial^2 w}{\partial x^2}(L,t) = 0 \tag{8.8}$$

where W_c is the amplitude of the compressor motion and ω is the particular compressor harmonic that is to be investigated. Making the substitution

$$w(x,t) = u(x,t) + W_c \left(1 - \frac{x}{L}\right) e^{j\omega t} \tag{8.9}$$

gives (Soedel, 1980a)

$$EI\frac{\partial^4 u}{\partial x^4} - (P_d - P_s)A\frac{\partial^2 u}{\partial x^2} + m\frac{\partial^2 u}{\partial t^2} + \lambda\frac{\partial u}{\partial t} = m\omega^2 W_c \left(1 - \frac{x}{L}\right) e^{j\omega t}. \tag{8.10}$$

Vibrations of Other Structural Components of a Compressor

An equivalent viscous damping coefficient λ in Ns/mm² has been introduced. The boundary conditions are now homogeneous:

$$u(0, t) = 0 \tag{8.11}$$

$$\frac{\partial^2 u}{\partial x^2}(0, t) = 0 \tag{8.12}$$

$$u(L, t) = 0 \tag{8.13}$$

$$\frac{\partial^2 u}{\partial x^2}(L, t) = 0. \tag{8.14}$$

The solution is

$$u(x, t) = \sum_{n=1}^{\infty} \eta_n \sin \frac{n\pi x}{L} \tag{8.15}$$

where

$$\eta_n = \Lambda_n e^{j(\omega t - \phi_n)} \tag{8.16}$$

$$\Lambda_n = \frac{F_n^*}{\omega_n^2 \sqrt{\left[1 - \left(\frac{\omega}{\omega_n}\right)^2\right]^2 + 4\zeta_n^2 \left(\frac{\omega}{\omega_n}\right)^2}} \tag{8.17}$$

$$\phi_n = \tan^{-1}\left[\frac{2\zeta_n\left(\frac{\omega}{\omega_n}\right)}{\left(1 - \frac{\omega}{\omega_n}\right)^2}\right] \tag{8.18}$$

$$F_n^* = \frac{2\omega^2 W_c}{n\pi}(1 - 2\cos n\pi) \tag{8.19}$$

and where

$$\zeta_n = \frac{\lambda}{2m\omega_n} \tag{8.20}$$

$$\omega_n = 2\pi f_n \tag{8.21}$$

and where f_n was obtained earlier.

The force transmitted into the shell is

$$Q(L,t) = -EI \frac{\partial^3 u}{\partial x^3}(L,t) \tag{8.22}$$

or

$$Q(L,t) = EI \left(\frac{\pi}{L}\right)^3 \sum_{n=1}^{\infty} \eta_n n^3 \cos n\pi. \tag{8.23}$$

If there is a single tube resonance in the vicinity of a compressor harmonic ω, this reduces approximately to

$$Q(L,t) = \frac{2EIW_c \pi^2}{L^3} \left(\frac{\omega}{\omega_n}\right)^2 \frac{n^2 \cos n\pi (1 - 2\cos n\pi) e^{j(\omega t - \phi_n)}}{\sqrt{\left[1 - \left(\frac{\omega}{\omega_n}\right)^2\right]^2 + 4\zeta_n^2 \left(\frac{\omega}{\omega_n}\right)^2}}. \tag{8.24}$$

The unit of the force is [N].

The slow convergence of Equation 8.23 should be noted because of the n^2 term in the numerator. Convergence of modal expansion solutions is not necessarily always assured and needs to be checked. Still, even without convergence proof, the result allows us to draw certain conclusions.

The result obviously illustrates that the force transmitted into the compressor housing is proportional to the compressor casing amplitude W_c. W_c should be reduced by good balancing of the compressor kinematics and by avoiding compressor casing resonances, as much as is feasible. Vibration isolating the tube from the compressor casing or the compressor housing would be a good measure if it is practical. Note that a slope excitation at the casing and of the tube will lead to the same conclusions.

This result also illustrates clearly the importance of a small deviation in ω_n, which may occur when the suction and discharge pressures change. As the tube damping is increased, this importance diminishes because the resonance peaks are less sharp.

8.1.5 Effect of Mass Flow Rate on Tube Vibration

Neglecting all effects but the flow velocity effect in the equation of motion, for the straight tube, for example, we obtain (Soedel, 1980a)

$$EI \frac{\partial^4 w}{\partial x^4} + \rho v^2 \frac{\partial^2 w}{\partial x^2} + (m+\rho) \frac{\partial^2 w}{\partial t^2} = 0. \tag{8.25}$$

Following the standard approach we obtain, for a simply supported tube,

$$f_n^2 = \frac{\pi^2 n^4}{4L^4}\left(\frac{EI}{m+\rho}\right)\left[1 - \frac{\rho v^2 L^2}{EI\pi^2 n^2}\right]. \tag{8.26}$$

We see that the natural frequency of the tube decreases as the flow velocity increases. This effect was found to be important for tubes conveying liquids, but is thought to be of much lesser importance than the pressure effect in typical refrigeration compressor applications.

However, what is important is the fact that this result is related to the mechanism with which gas pulsations can excite the vibration of curved tubes. Oscillating mass flow through a curved tube tends to locally straighten the tube at the frequency of oscillation.

8.2 VIBRATION ISOLATION CONSIDERING IDEALIZED SPRINGS

Compressor casings are usually vibration isolated from the housing by support springs (the fact that the casing is also connected to the housing by suction or discharge tubes was the topic of the last section). Similarly, compressor housings are vibration isolated from the support structure.

The common approach when examining isolation springs is to consider the springs to be massless. This provides valuable insights as the following will show. How these insights are affected by the spring masses will be discussed in Section 8.3. See also, for example, Reynolds (1981) and Hamilton (1982, 1988).

8.2.1 Review of the Standard Approach to Vibration Isolation

While in principle the compressor casing has six degrees of freedom of motion (motions in three orthogonal directions and three rotating motions), we will consider only one motion degree of freedom here. The standard problem, as treated in many texts on mechanical vibrations, is a single mass representing the casing, and a single isolation spring. A viscous damper is also considered, as shown in Figure 8.4.

We are interested in the force that is transmitted through the isolation. It will be our goal to minimize this force. Drawing a free body and applying Newton's second law in the x–direction, the equation of motion is

$$m\ddot{x} + c\dot{x} + kx = F_o \sin \omega t. \tag{8.27}$$

The steady-state solution is of the form

$$x = X \sin(\omega t - \phi). \tag{8.28}$$

The force transmitted from the compressor casing into the housing will be

$$F_T = kx + c\dot{x}. \tag{8.29}$$

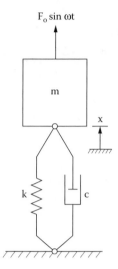

FIGURE 8.4 One-degree-of-freedom model of a vibration isolated compressor casing.

Substituting Equation 8.28 into Equation 8.29 gives

$$F_T = X[k \sin(\omega t - \phi) + c\omega \cos(\omega t - \phi)]. \tag{8.30}$$

Writing Equation 8.30 in the form

$$F_T = A \sin(\omega t - \phi + \beta) \tag{8.31}$$

we get, by standard procedure,

$$A = X\sqrt{k^2 + (c\omega)^2} \tag{8.32}$$

$$\beta = \tan^{-1} \frac{c\omega}{k} = \tan^{-1} 2\zeta \frac{\omega}{\omega_n} \tag{8.33}$$

where $\zeta = c/(2m\omega_n)$ and where $\omega_n = \sqrt{k/m}$. The values of X and ϕ are known from the steady-state solution of Equation 8.27 and are

$$X = \frac{F_o}{\sqrt{(k - m\omega^2)^2 + (c\omega)^2}}, \quad \phi = \tan^{-1} \frac{2\zeta \dfrac{\omega}{\omega_n}}{1 - \left(\dfrac{\omega}{\omega_n}\right)^2}. \tag{8.34}$$

Vibrations of Other Structural Components of a Compressor

Also, if we define $-\alpha = -\phi + \beta$, we obtain

$$-\alpha = -\tan^{-1} \frac{2\zeta \frac{\omega}{\omega_n}}{1 - \left(\frac{\omega}{\omega_n}\right)^2} + \tan^{-1} 2\zeta \frac{\omega}{\omega_n}. \tag{8.35}$$

Equations 8.32, 8.34, and 8.35 into Equation 8.31 gives

$$F_T = \frac{F_o \sqrt{k^2 + (c\omega)^2}}{\sqrt{(k - m\omega^2)^2 + (c\omega)^2}} \sin(\omega t - \alpha) \tag{8.36}$$

or

$$F_T = \tilde{F}_T \sin(\omega t - \alpha) \tag{8.37}$$

where $\tilde{F}_T$ can also be written

$$\tilde{F}_T = \frac{F_o \sqrt{1 + \left(2\zeta \frac{\omega}{\omega_n}\right)^2}}{\sqrt{\left[1 - \left(\frac{\omega}{\omega_n}\right)^2\right]^2 + \left(2\zeta \frac{\omega}{\omega_n}\right)^2}}. \tag{8.38}$$

Thus, the ratio of the transmitted force amplitude $\tilde{F}_T$, to the excitation force amplitude F_o, is

$$\frac{\tilde{F}_T}{F_o} = \frac{\sqrt{1 + \left(2\zeta \frac{\omega}{\omega_n}\right)^2}}{\sqrt{\left[1 - \left(\frac{\omega}{\omega_n}\right)^2\right]^2 + \left(2\zeta \frac{\omega}{\omega_n}\right)^2}}. \tag{8.39}$$

The ratio F_T/F_o is the force transmission ratio, and is sketched in Figure 8.5.
Vibration isolation takes place whenever

$$\frac{\tilde{F}_T}{F_o} < 1. \tag{8.40}$$

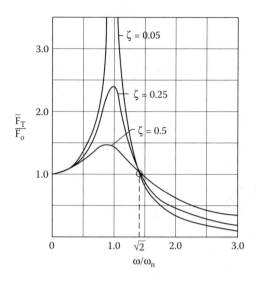

FIGURE 8.5 Force transmission ratio as a function of the ratio of excitation frequency to natural frequency, for various values of the damping ratio. The effective isolation region is to the right of $\omega/\omega_n = \sqrt{2}$.

From Equation 8.39, we conclude that to achieve vibration isolation, we must design springs such that

$$\omega_n \ll \frac{\omega}{\sqrt{2}} \tag{8.41}$$

or

$$k \ll \frac{m\omega^2}{2}. \tag{8.42}$$

After the spring is properly designed, we should choose low damping, or better, zero damping. A compromise may become necessary if the allowed rigid body motion amplitude X of the compressor casing is limited.

8.2.2 Rotating Unbalance

In this somewhat more advanced model, an unbalance consisting of an arm e and a mass m, rotating with constant angular velocity ω, is considered. Again, we allow only one degree of freedom of motion, as shown in Figure 8.6, to keep the model simple. The mass of the casing is M, which includes the unbalanced mass m.

At time t, the unbalanced mass is at location y, measured from the static equilibrium position when $\omega t = 0$. The center mass is at location x. Thus,

$$y = x + e \sin \omega t. \tag{8.43}$$

Vibrations of Other Structural Components of a Compressor

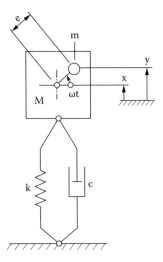

FIGURE 8.6 One-degree-of-freedom model of vibration isolation when the casing is excited by a rotating unbalanced mass.

Drawing free body diagrams in a vertical direction and applying Newton's second law gives

$$M\ddot{x} + C\dot{x} + Kx = me\omega^2 \sin\omega t. \tag{8.44}$$

The steady-state response is then

$$x_p = X\sin(\omega t - \phi) \tag{8.45}$$

where

$$X = \frac{\frac{m}{M}e\left(\frac{\omega}{\omega_n}\right)^2}{\sqrt{\left[1-\left(\frac{\omega}{\omega_n}\right)^2\right]^2 + \left(2\zeta\frac{\omega}{\omega_n}\right)^2}} \tag{8.46}$$

and where

$$\omega_n = \sqrt{\frac{K}{M}} \tag{8.47}$$

$$\phi = \tan^{-1}\frac{2\zeta\left(\frac{\omega}{\omega_n}\right)}{1-\left(\frac{\omega}{\omega_n}\right)^2}. \tag{8.48}$$

The force, which is transmitted from the casing into the housing (viewing the housing as grounded in this simple model), is then

$$F_T = K\left(\frac{m}{M}\right)e\left(\frac{\omega}{\omega_n}\right)^2 \frac{\sqrt{1+\left(2\zeta\frac{\omega}{\omega_n}\right)^2}}{\sqrt{\left[1-\left(\frac{\omega}{\omega_n}\right)^2\right]^2+\left(2\zeta\frac{\omega}{\omega_n}\right)^2}}. \qquad (8.49)$$

This equation indicates clearly that the transmitted force is proportional to the unbalance (me). Where the isolation design has zero damping ($\zeta = 0$), Equation 8.49 becomes, for $\omega/\omega_n \gg 1$,

$$F_T = K\left(\frac{m}{M}\right)e. \qquad (8.50)$$

This is the best one can do for normal, supercritical designs. This shows that it is imperative that the unbalance (me) be held as small as possible by balancing the rotor well. In summary, one must try to approach

$$\zeta = 0 \qquad (8.51)$$

and choose

$$\frac{\omega}{\omega_n} \gg \sqrt{2}, \qquad (8.52)$$

which means that the spring rate K should be

$$K \ll \frac{M\omega^2}{2}. \qquad (8.53)$$

Another theoretical possibility in a stationary compressor is to attach masses to the casing so that M becomes as large as possible. This, of course, is unacceptable in household refrigerating and air conditioning compressors where the total weight of a unit needs to be as small as possible.

Just to be complete, if we decide to operate subcritically, which means that

$$\frac{\omega}{\omega_n} \ll 1, \qquad (8.54)$$

we could approach, theoretically, for any amount of damping, zero force transmission. It would require that the spring rate would have to be very stiff:

$$K \gg M\omega^2. \tag{8.55}$$

This would practically require bolting the casing to the housing. This would not work for reciprocating piston compressors, where ω also represents the infinite number of higher harmonics (see Section 8.2.3).

8.2.3 Reciprocating Compressor Unbalance

A simple model that explains how reciprocating compressors produce forcing at multiples of the shaft speed is shown in Figure 8.7. The unbalanced effective mass of the crankshaft and an effective mass portion of the connecting rod are lumped together as mass m_1 as shown, and the other portion of the effective mass of the connecting rod is lumped with the piston mass as mass m_2, as shown.

The position of the casing mass M, which includes the unbalanced masses m_1 and m_2, is defined by x, the position of m_1 is defined by y_1, and the position

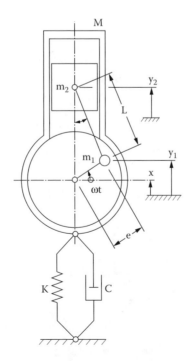

FIGURE 8.7 One-degree-of-freedom model of vibration isolation for a reciprocating compressor casing. The casing mass is M, the piston mass plus part of the connecting rod mass is m_1, and the unbalanced crank mass with part of the connecting rod mass is m_2.

of m_2 is defined by y_2, all measured from the position of the compressor when $t = 0$. We may write

$$y_1 = x + e \sin \omega t \qquad (8.56)$$

$$y_2 = x + e \sin \omega t + L(\cos \alpha - \cos \alpha_o) \qquad (8.57)$$

where α_o is the value of the angle α at $t = 0$.
From the geometry,

$$\cos \alpha = \sqrt{1 - \left(\frac{e}{L}\right)^2 \cos^2 \omega t} \qquad (8.58)$$

and when $t = 0$,

$$\cos \alpha_o = \sqrt{1 - \left(\frac{e}{L}\right)^2}. \qquad (8.59)$$

Therefore,

$$y_2 = x + e \sin \omega t + L \left[\sqrt{1 - \left(\frac{e}{L}\right)^2 \cos^2 \omega t} - \sqrt{1 - \left(\frac{e}{L}\right)^2} \right]. \qquad (8.60)$$

Because

$$\left(\frac{e}{L}\right)^2 \cos^2 \omega t < 1 \qquad (8.61)$$

we may use the series expansion

$$\sqrt{1+z} = 1 + \frac{z}{2} - \frac{z^2}{8} + \frac{3z^3}{48} - \frac{15z^4}{384} \ldots \text{etc.} \qquad (8.62)$$

Here, to keep the model simple, we will assume that only the first two terms of the series are important:

$$\sqrt{1+z} \cong 1 + \frac{z}{2}. \qquad (8.63)$$

Vibrations of Other Structural Components of a Compressor

This will allow us to investigate the importance of what are often called the primary and secondary unbalances of a reciprocating piston compressor (see Holowenko, 1955). We write, therefore,

$$y_2 = x + e \sin \omega t + \frac{L}{2}\left(\frac{e}{L}\right)^2 (1 - \cos^2 \omega t). \tag{8.64}$$

Because $\cos^2 \omega t = \frac{1}{2} - \frac{1}{2}\cos 2\omega t$, we get, after differentiating twice with respect to time

$$\ddot{y}_2 = \ddot{x} - e\omega^2 \left[\sin \omega t - \left(\frac{e}{L}\right) \cos 2\omega t \right]. \tag{8.65}$$

Next, drawing appropriate free body diagrams and utilizing Newton's second law of motion, we obtain the equation of motion of the system:

$$M\ddot{x} + C\dot{x} + Kx = (m_1 + m_2)e\omega^2 \sin \omega t - m_2 e\omega^2 \left(\frac{e}{L}\right) \cos 2\omega t. \tag{8.66}$$

This shows that the primary unbalance forces the system at the fundamental shaft frequency (or first harmonic) ω, and the secondary unbalance forces the system at the second harmonic frequency 2ω. If we had not neglected the higher-order terms in the series expansion Equation 8.62, we would also have derived forcing terms for the higher harmonics.

The steady-state response magnitude due to the first harmonic forcing term is

$$X_1 = \frac{\left(\frac{m_1 + m_2}{M}\right) e \left(\frac{\omega}{\omega_n}\right)^2}{\sqrt{\left[1 - \left(\frac{\omega}{\omega_n}\right)^2\right]^2 + \left(2\zeta \frac{\omega}{\omega_n}\right)^2}}. \tag{8.67}$$

The steady-state response magnitude due to the second harmonic forcing term is

$$X_2 = \frac{\left(\frac{m_2}{M}\right) e \left(\frac{2\omega}{\omega_n}\right)^2 \left(\frac{e}{L}\right)}{\sqrt{\left[1 - \left(\frac{2\omega}{\omega_n}\right)^2\right]^2 + \left(4\zeta \frac{\omega}{\omega_n}\right)^2}}. \tag{8.68}$$

If one wishes to obtain the total response, one must consider the phase angles:

$$\phi_1 = \tan^{-1} \frac{2\zeta\left(\dfrac{\omega}{\omega_n}\right)}{1-\left(\dfrac{\omega}{\omega_n}\right)^2} \tag{8.69}$$

$$\phi_2 = \tan^{-1} \frac{2\zeta\left(\dfrac{2\omega}{\omega_n}\right)^2}{1-\left(\dfrac{2\omega}{\omega_n}\right)^2}. \tag{8.70}$$

The total steady-state response is then

$$x = X_1 \sin(\omega t - \phi_1) + X_2 \cos(2\omega t - \phi_2), \tag{8.71}$$

and we can calculate the transmitted force F_T into the housing.

8.2.4 Isolating the Flexural Vibrations of the Casing or Housing

In the foregoing, the compressor casing, or the compressor housing as the case may be, was depicted as a rigid body with no flexure of its own. However, the conclusions also apply in an approximate sense to the isolation of any flexural motion of, say, the casing, if we consider each natural mode's modal mass and modal stiffness. The system model must be only slightly modified, as shown in Figure 8.8, where M is now the modal mass, and K is the modal stiffness of a natural mode. It is again our goal to minimize the force that is transmitted into the foundation, A.

The mathematical model parallels that of Section 8.2.1. The equation of motion becomes

$$M\ddot{x} + (C+c)\dot{x} + (K+k)x = F_o \sin \omega t. \tag{8.72}$$

The force transmitted into the foundation A will be

$$F_T = kx + c\dot{x}. \tag{8.73}$$

The steady-state solution of Equation 8.72 is of the form

$$x = X \sin(\omega t - \phi) \tag{8.74}$$

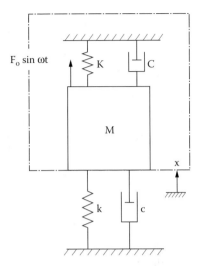

FIGURE 8.8 Very simplified one-degree-of-freedom model of the vibration isolation of a single, equivalent casing or housing natural mode. The natural mode is modeled in terms of its equivalent mass M, stiffness K, damping C, and forcing $F_o \sin \omega t$.

where

$$X = \frac{F}{\sqrt{(K+k-M\omega^2)^2 + (C+c)^2 \omega^2}} \quad (8.75)$$

$$\phi = \tan^{-1} \frac{(C+c)\omega}{(K+k) - M\omega^2}. \quad (8.76)$$

The force transmitted into the foundation A is, therefore,

$$F_T = X\sqrt{k^2 + (c\omega)^2} \sin(\omega t - \phi + \beta) \quad (8.77)$$

where

$$\beta = \tan^{-1}\left(\frac{c\omega}{k}\right). \quad (8.78)$$

The amplitude of the transmitted force is

$$\tilde{F}_T = X\sqrt{k^2 + (c\omega)^2} \quad (8.79)$$

or

$$\tilde{F}_T = \frac{F_o\sqrt{k^2+(c\omega)^2}}{\sqrt{\left[(K+k)-M\omega^2\right]^2+\left[(C+c)\omega\right]^2}}. \quad (8.80)$$

After some manipulation, the transmission ratio can be written as

$$\frac{\tilde{F}_T}{F_o} = \frac{\sqrt{\left(\frac{k}{K+k}\right)^2+\left(\frac{c}{C+c}\right)^2\left[2\zeta\left(\frac{\omega}{\omega_n}\right)\right]^2}}{\sqrt{\left[1-\left(\frac{\omega}{\omega_n}\right)^2\right]^2+\left[2\zeta\left(\frac{\omega}{\omega_n}\right)\right]^2}}. \quad (8.81)$$

Note that if we have a rigid body mode (M = m) where the modal stiffness K = 0, and the modal damping C = 0, this equation reduces to the formulation in Equation 8.39.

$$\omega_n = \sqrt{\frac{K+k}{M}} \quad (8.82)$$

$$\zeta = \frac{(C+c)}{2M\omega_n} \quad (8.83)$$

Thus, if a particular natural mode of the casing or the housing is excited by a harmonic equivalent modal force of amplitude F_o, the transmission ratio equation is similar to the single-degree-of-freedom case representing a rigid body motion. One comes to similar conclusions, namely that the isolation spring rate should be as small as feasible, and that the vibration isolation damping should approach zero (keep in mind that in this discussion it is assumed that the modal mass M and the modal stiffness cannot be changed and must be treated as a given).

8.3 SURGING IN COIL SPRINGS INTERFERING WITH VIBRATION ISOLATION

Surging of isolation springs may change the vibration isolation behavior as usually calculated using massless springs. The usual, but sometimes not justifiable, hypothesis is that springs do not contribute resonances of their own to a composite system. In the following, this restriction is lifted. The system model is considerably simplified, however, to capture the essence of the spring surge problem (see also Simmons and Soedel (1996), Hatch and Wollatt (2002), Kelly and Knight (1992b) and Yang et al., (1996). The results illustrate phenomena that are observed by engineers when they try to achieve noise control by changing spring rates.

For a real compressor, say, a refrigeration compressor supported inside a hermetic shell by three mounting springs, which may each deflect in three directions, the formulation is only a little more complicated, but the general conclusion that one reaches is the same.

8.3.1 A Simplified Spring Surge Model

In addition to the coil spring of mass M, stiffness K, and internal damping rate C, designated sub-system C, the system is divided into two other subsystems, B and D, as shown in Figure 8.9. B is an extremely simplified model of the compressor body and D is an extremely simplified model of one mode of a compressor shell represented by its modal mass and modal stiffness. Damping of subsystem C occurs in the spring itself; in other words, there is no *external* damper—the damping of the coil spring occurs via material damping or by means of a plastic sleeve stretched around the spring. This will allow for a continuous damping effect across the length L, rather than a net damping effect at the two end points of the spring.

In general, the receptance of a system is simply the ratio of harmonic displacement at one point to a harmonic force at another point. See Bishop and Johnson (1979) and Soedel (2004) for discussions of the receptance method. The receptances are, for systems B and D,

$$\beta_{11} = \beta_{22} = \beta_{12} = \frac{1}{(k_1 - m_1\omega^2) + j\omega C_1} \quad , \quad \delta_{33} = \delta_{44} = \delta_{43} = \frac{1}{(k_2 - m_2\omega^2) + j\omega C_2}$$

(8.84)

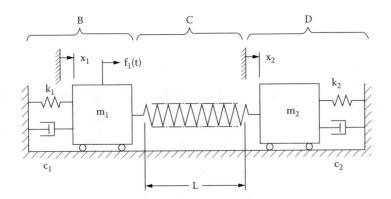

FIGURE 8.9 A simplified spring surge model. The spring (system C) connects an equivalent one-degree-of-freedom model of the compressor casing (system B) to an equivalent one-degree-of-freedom model of the housing (system D).

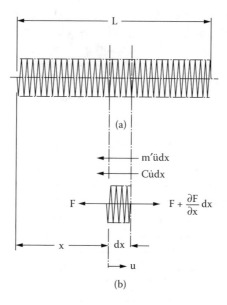

FIGURE 8.10 A coil spring. The number of windings is exaggerated to illustrate the continuum approach, but springs with only very few windings are also described by this model. (a) The spring is viewed as a continuum in the axial direction. (b) An element of length dx is cut from the coil spring.

where k_i, m_i, and c_i are the modal stiffnesses, the modal masses, and the modal damping constants, respectively.

The coil spring of Figure 8.10a has mass M, length L, stiffness K, and an internal damping coefficient C. The displacement in the x–direction is u (x, t). The force F, created in the spring, is related to deflection by

$$F = KL \frac{\partial u}{\partial x}. \tag{8.85}$$

Determining a constant mass per unit length $m' = dm/dx = M/L$, an element dx of the spring is acted on by forces shown in Figure 8.10b. A force balance and Newton's second law give

$$\frac{\partial^2 u}{\partial t^2} + \frac{C}{m'} \frac{\partial u}{\partial t} = \frac{KL}{m'} \frac{\partial^2 u}{\partial x^2}. \tag{8.86}$$

The displacement solution is of the general form

$$u(x,t) = (A_1 e^{-j\kappa_1 x} + B_1 e^{j\kappa_1 x}) e^{j\omega t}. \tag{8.87}$$

Vibrations of Other Structural Components of a Compressor

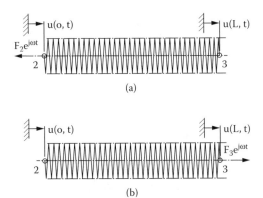

FIGURE 8.11 Boundary conditions for obtaining (a) receptances γ_{22} and γ_{32}, (b) receptances γ_{23} and γ_{33}.

Applying this to the equation of motion yields

$$\kappa_1 = \sqrt{\frac{m'\omega^2 - Cj\omega}{c^2 m'}}, \quad c^2 \equiv \frac{KL}{m'}, \tag{8.88}$$

Using this general expression for the equation of motion of the coil spring, the two boundary conditions of Figure 8.11a are introduced and we solve for the unknown constants A_1 and B_1. When A_1 and B_1 are known, the receptances at the left-hand endpoints γ_{22} and γ_{32} are evaluated:

$$\gamma_{22} = \frac{u(0,t)}{F_2 e^{j\omega t}} = \frac{e^{j\kappa_1 L} + e^{-j\kappa_1 L}}{j\kappa_1 KL(e^{j\kappa_1 L} - e^{-j\kappa_1 L})}, \quad \gamma_{32} = \frac{u(L,t)}{F_2 e^{j\omega t}} = \frac{2}{j\kappa_1 KL(e^{j\kappa_1 L} - e^{-j\kappa_1 L})}. \tag{8.89}$$

Similarly, receptances γ_{23} and γ_{33} are evaluated. The boundary conditions, however, are reversed and are shown in Figure 8.11b.

$$\gamma_{23} = \frac{u(0,t)}{F_3 e^{j\omega t}} = \frac{2}{j\kappa_1 KL(e^{j\kappa_1 L} - e^{-j\kappa_1 L})}, \quad \gamma_{33} = \frac{u(L,t)}{F_3 e^{j\omega t}} = \frac{e^{j\kappa_1 L} + e^{-k\kappa_1 L}}{j\kappa_1 KL(e^{j\kappa_1 L} - e^{-j\kappa_1 L})} \tag{8.90}$$

Note the symmetry of the sub-system C receptances: $\gamma_{22} = \gamma_{33}$, $\gamma_{32} = \gamma_{23}$.

The subsystems are generalized as block diagrams as shown in Figure 8.12a. To obtain the system A (the total system) receptances, it is necessary to break the system between the connections and generalize the force and displacements for each

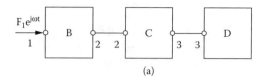

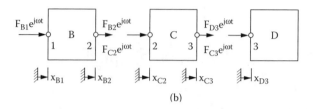

FIGURE 8.12 (a) Connection diagram for the total system, called A. (b) Subsystem definitions of displacements and forces.

subsystem. This is shown in Figure 8.12b. Setting up displacement expressions and following the procedure outlined in Soedel (2004) yields

$$\alpha_{31} = \frac{\beta_{22}\gamma_{23}\delta_{33}}{(\gamma_{22}+\beta_{22})(\delta_{33}+\gamma_{33})-\gamma_{23}^3}. \qquad (8.91)$$

Because of damping in some or all parts of the system, α_{31} is a complex number. To account for this, the magnitude and phase of the receptance will have to be considered when response behavior is analyzed. Complex receptances are discussed in Soedel (2004).

8.3.2 The Effect of Surging on the Response

The first observation to be made is the noticeable difference in a typical system response (receptance α_{31}) when spring surging is considered, as shown in Figure 8.13a. The natural frequencies of subsystems B and D were selected to be 6283 and 2000 rad/s, respectively. The spring dimensions and properties were selected to be typical for small refrigeration compressors.

Several observations can be made. First, notice that the external system spikes (due to the resonances of systems B and D) at $\omega = 6283$ rad/s and $\omega = 2000$ rad/s are magnified when surging occurs. Each of the dashed spikes that do not occur at $\omega = 2000$ or 6283 rad/s represents a surge frequency.

In addition to the overall increase at the system resonance, note the near coincidence effect of a spring surge resonance with the subsystem D natural frequency at 2000 rad/s. The spring surge peak is increased as a result of this near coincidence. Clearly, surge natural frequency intervals must be a consideration in the design of such a system. The separation between the coil spring resonance will clearly dictate

Vibrations of Other Structural Components of a Compressor

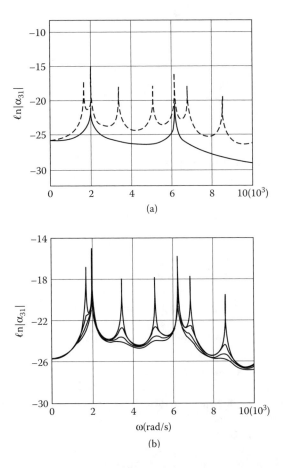

FIGURE 8.13 (a) System response in terms of the magnitude of the system receptance α_{31}, which is a measure of the vibration transmission due to the forcing on the compressor casing to the housing:—, when spring surging is neglected; ---, when spring surging is included in the model. (b) The effect of damping of the spring on the response curves; damping increases from top to bottom.

whether coincidence is likely to occur, and thus increase system response. Furthermore, it may be required to operate at a specific frequency or over a range of given frequencies, and in these cases, spring surging would reduce the operable ranges. If spring surging can be ignored, then a large range between the external subsystem natural frequencies exists in which to operate. However, if spring surging cannot be ignored, one must then consider how to appropriately space the intervals so as to avoid detrimental amplifications.

A final note about the system is that spring natural frequency harmonics will be present at the higher frequencies as well (i.e., those to the right of 6283 rad/s). On the other hand, when surging is ignored, the response behavior naturally decreases at theses higher frequencies (a false sense of security is created).

Of interest is the extent to which internal spring damping can reduce system response. A plastic sleeve around the coil spring might be one way to create internal damping. A typical response for various amounts of damping is shown in Figure 8.13b.

It is noted that damping of the coil spring appears to be a useful tool in reducing the contribution of the coil spring to the overall system receptance. However, coil spring damping does not seem to make a significant contribution to the damping of the external subsystem-controlled natural frequencies.

Also of interest is the influence of the spring rate value, K. First, consider a lower value of K. The classical hypothesis to test here is whether a lower spring rate will always reduce the response of system A. If one ignores spring surging, a case may be made that the most desirable K for response isolation is the lowest one possible. However, in the case of surging, it will be shown that this is not necessarily always true.

In Figure 8.14a, the solid line demonstrates that the intervals between surge natural frequencies are smaller for K reduced by 33% than for the original larger spring rate values (superimposed as the dashed curve). Thus, with a much smaller K, coincidence with other sub-system natural frequencies occurs more frequently and refutes the argument that a softer spring rate invariably reduces overall system response. (But there is a trend toward a lower average off-resonance response with decreasing spring rate.)

Next, consider a much higher subsystem C spring rate: K is increased by a factor of four. The comparison of this larger spring rate (solid line) with the reference value (dashed line) is given in Figure 8.14b. The result is characterized by a much higher mean receptance level, higher peaks (which are receptance features), but much wider surge natural frequency intervals. As a result of this, coincidence with system natural frequencies is less likely. The wider spring resonance intervals can be advantageous if the application calls for a system to operate at a specific frequency or in a specific frequency range. One would simply have to consider what design spring rate K would have the lowest mean receptance for the given operation range.

As one can observe, the receptance technique is an appropriate method for understanding how spring surging contributes to the overall response of a system. Also, computer simulations of actual spring isolation designs of actual compressors will lead to similar conclusions, which are that lower coil spring rates do create a somewhat lower mean system response, but coincidence of surging frequencies with system natural frequencies is more likely to happen due to the tighter intervals of the spring resonances. Selecting isolation designs and spring rates is ultimately an issue of the application and the range of operating frequencies of the compressor. Laboratory evaluation of the vibration isolation designs is highly desirable. The foregoing discussion does not suggest that one should not use soft springs, it simply attempts to explain why lowering isolation spring rates does not always produce beneficial results.

One of the major advantages of the receptance technique is that a wide variety of subsystems can be applied to the general expression for α_{31}. For example, one might wish to substitute a more complex subsystem into system A in place of subsystem D. The only requirement beyond what has previously been formulated

Vibrations of Other Structural Components of a Compressor

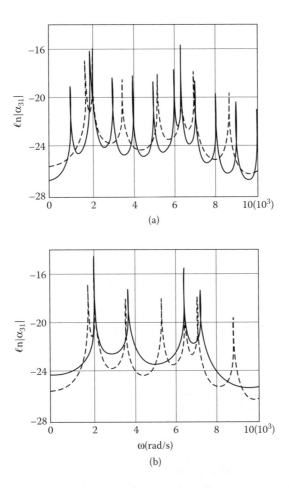

FIGURE 8.14 (a) Response curve when the spring rate K is reduced by 33% (solid curve), as compared to a reference K which produced the dashed curve. (b) Response when the spring rate K is increased by a factor of four (solid curve), as compared to the same reference K.

to make this substitution possible is that the receptance for subsystem D must be formulated or measured. The example that will be investigated here is a very simplified equivalent housing shell.

8.3.3 Surging and Housing Resonances

As mentioned, the system A receptance expression will remain unchanged. In order to substitute a real or even equivalent housing shell into the system, subsystem D must be replaced by the appropriate shell receptances. A typical housing shell receptance δ_{33} obtained analytically from a very simplified compressor housing model, is shown in Figure 8.15a. (Note that δ_{33} is very much a function of the location on the shell).

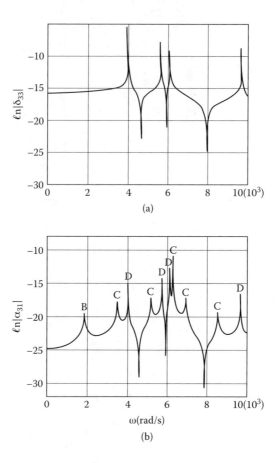

FIGURE 8.15 (a) Typical undamped receptance α_{33} of a housing shell. (b) Response α_{31} of the entire system (the compressor casing is still modeled as an equivalent one-degree-of-freedom system).

The lowest natural frequency occurs in this example for the $m = 1$, $n = 4$ mode and has a value of $\omega_{14} = 3939$ rad/s.

Next, this modified subsystem D receptance is substituted into the expression for the system A cross receptance α_{31}, with all other receptances remaining the same to yield the composite response plot of Figure 8.15b. The response resonances are labeled C, B, or D depending on whether they are primarily due to resonances of sub-systems C, B, or D, respectively. The difference here is that instead of having only one system D resonance as before, the shell introduces numerous resonances, each of which can be in coincidence or near coincidence with a surge frequency of the spring. Considered mode by mode, however, the system will behave in a manner similar to the simple cases discussed before.

The receptance method analysis has demonstrated the following:

1. Spring surging is often significant; it cannot categorically be neglected.
2. Internal or external damping of the surging coil spring is effective in damping surge resonance.
3. Surge frequencies should be detuned from other system natural frequencies if possible, or at least recognized as resonance contributions.
4. Lower coil spring rates reduce the average system response, but make coincidence more likely.
5. Higher coil spring rates may increase the spacing between surge resonances, and may result in certain situations in improved designs, even while increasing isolation spring rates is not recommended by the classical models of vibration isolation, and not necessarily advocated here.

9 Sound and Vibration of Compressor Valves

In the refrigeration industry, and also in the gas industry, where the compressors are usually hermetically sealed, valves are a noise source whose oscillatory energy is transmitted to the outside through pipes and the housing shell. Therefore, this interaction will usually have to be considered. In the air compressor industry, the discharge valves usually feed into a hermetic system and the problem is similar. Suction valves, on the other hand, are exposed to the atmosphere (except for a filter), and because gas pulsation waves or acoustic waves travel as easily against the flow as with the flow for typical, average flow velocities, they are important sources that transmit directly to the observer.

Valve noise is generated either by flow oscillations caused by the valves or by the metallic impact of the valve reed or plate on the seat or valve stop.

9.1 REMARKS ABOUT TYPES OF COMPRESSORS AND THEIR VALVING

Compressors without *automatic valves* still generate gas pulsations because their discharge flow, and to a somewhat lesser extent their suction flow, are intermittent. A screw compressor will discharge (back flow is defined here as part of the discharge) a given gas pocket in the interval during which this pocket is exposed to the port. The fundamental frequency of discharge f [Hz] is $f = z\, N/60$, where z = the number of male lobes and N = the rotational speed (RPM) of the male rotor. A scroll compressor will also discharge discrete gas pockets.

Actually, these *valveless* compressors indeed have valves; they simply belong kinematically to the category of *slider valve compressors*, except that the sliding is done by the scrolls or screws. (Therefore, no extra parts are needed.) These compressors operate perfectly only under particular design conditions, where the slider valve or port that is open at the moment the chamber pressure is equal to the discharge pressure, as shown, for example, by the 310 psia line in Figure 9.1a.

If the discharge pressure is less, say, 241 psia, the gas still compresses to 310 psia because the valve is not automatic, and the gas then discharges explosively. On the other hand, if the discharge pressure is 350 psia, the valve opens too soon at 310 psia, gas rushes back into the compression chamber, and is then pushed out. Obviously, any condition other than the design condition will produce larger harmonic components, as shown in Figure 9.1b.

Time-varying flow through ports (automatic valves or no automatic valves!) will produce gas pulsations (see Koai and Soedel, 1990a).

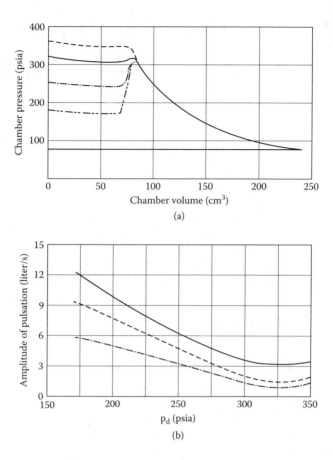

FIGURE 9.1 (a) Example of a pressure-volume diagram for a twin screw compressor with different discharge pressures:—, design pressure, $p_d = 310$ psia; ---, overpressure, $p_d = 350$ psi; $-\cdot-$, underpressure, $p_d = 240$ psi; $-\cdot\cdot-$, underpressure, $p_d = 170$ psi. (b) Amplitudes of flow pulsation harmonics for a screw compressor as a function of discharge pressure. The lowest pulsation levels are generated in the vicinity of the design pressure of $p_d = 310$ psia.

A second point is that given a certain volume of gas to be discharged at a given pressure condition and motor speed, it is in general better, from a noise as well as a valve (or port) design viewpoint, to do this in several discharges per rotation than in a single discharge. This is the basic advantage of the two- or multivane rotary compressor (and also of the screw and scroll compressor).

There is further similarity between the multivane rotary compressor and the screw compressor because the former can also operate without valves, except that it tends to operate in an underpressure condition. There will always be backflow into the trailing volume without loss of volumetric efficiency (but there will be a loss in thermal efficiency). Some air compressors work on this principle.

The purpose of automatic valves is basically to prevent back flow. They may introduce additional important harmonic components into the flow spectrum because they may modulate the flow by fluttering. Otherwise, the principle of gas pulsation generation is the same.

The apparent wide variety of automatic valves is misleading. Strictly speaking, we distinguish between spring-loaded valves and springless valves. The spring-loaded valves subdivide into two categories: (a) springs that are separate from the sealing element, for example, spring-loaded ring valves, and (b) springs that are built into the design, such as in flexible reeds. The purpose of the spring loading is mainly to assist proper valve closing in order to prevent backflow and its destructive effects, and to a lesser extent, help with sealing.

The multitude of reed shapes, and the fact that for larger compressors spring-loaded ring valves or strip valves dominate, is simply a fact of scaling. Cylinder volumes increase cubically while available surface area for valve porting increases only quadratically with size. For a small compressor, a single discharge hole with a beam-type reed suffices, while for a large compressor, the entire available surface area may have to be covered with ports, and accommodating flexible reeds would be a problem. For super-large compressors, special arrangements of valves need to be used to accommodate the limited space.

9.2 A SIMPLIFIED APPROACH TO UNDERSTANDING VALVE DESIGN

Because valves are one of the primary sound sources in compressors, and also one of the elements most likely to experience fatigue failure due to vibrations, the following outlines the steps of valve layout. Understanding the process of valve design allows us to understand why valves vibrate as they do and what influences this vibration. It also contributes to the understanding, possible reduction, and partial prevention of gas oscillations as a sound-generating mechanism, and also to an understanding of valve impact as a sound generating mechanism. This will clarify which design modifications resulting in sound and vibration reductions are permissible and which are not because they may lead to a deterioration of thermodynamic performance.

9.2.1 THERMODYNAMIC CONSIDERATIONS

Let us assume that we start from the beginning, with nothing to copy from and no idea about the required size of the valves. At a rotational speed Ω [rad/s], the compressor to be designed takes in gas at a pressure p_s [N/m^2] and at a temperature T_s [°K] and compresses it to a pressure p_d [N/m^2]. It is supposed to deliver a mass flow rate [kg/s].

Note that defining the specification this way bypasses the special conventions that have developed differently in the refrigeration industry than in the air compressor and gas compressor industry. The refrigeration industry prefers to specify conditions in terms of saturated gas temperatures and superheat (ASHRAE [American Society of Heating, Refrigerating and Air Conditioning Engineers], 1981). This makes some sense from the refrigeration system viewpoint but clouds the concept when we are concerned with compressor design. The conversion is made with the help of the pressure-enthalpy chart for the particular refrigerant. The suction density ρ_s is found by adding to the saturated suction gas temperature T_1, the superheat ΔT, to give the suction gas temperature T_s. Suction pressure and temperature determine ρ_s. To find the approximate discharge condition, an estimate of the compression process has to be made. A good first guess in refrigeration compressors and fast-running air and gas compressors (3600 rotations per minute and above) is to assume isentropic compression. The discharge density ρ_d is therefore found by moving along the constant entropy line to the discharge pressure P_d, which is determined by knowing the saturated discharge gas temperature T_2.

Slow speed, water-cooled air or gas compressors approach isothermal compression conditions. Using, in general, a polytropic coefficient n, the discharge temperature and density may be estimated from

$$T_d = T_s \left(\frac{P_d}{P_s} \right)^{\frac{n-1}{n}} \tag{9.1}$$

$$\rho_d = \rho_s \left(\frac{P_d}{P_s} \right)^{\frac{1}{n}} \tag{9.2}$$

The value of n should be bracketed by n = 1.0 for isothermal compression and n = k for isentropic compression. If it is found experimentally that n > k, it may be an indication that too much external heat is finding its way into the gas. This may be caused by heat transfer from an electromotor or diesel engine (in the case of an air compressor), or by high friction, or by leaking valves, and so on. It may also be caused by design reasons, for instance in high-side refrigeration compressors where the hot discharge gas surrounds the compressor cylinder. Even after having passed through an intercooler, the discharge gas can be expected to be of a higher temperature than the suction gas that surrounds a low-side compressor.

9.2.2 Indicator Diagrams, Valve Timing, and Flow Velocity Estimates

After these preliminaries, it is advisable to construct an idealized pressure-volume diagram to aid in the determination of valve timing. It will be assumed here that the required basic size of the swept compressor volume has been determined, and that the kinematic type of compressor has been selected, because this is not the subject of this discussion. However, it should be noted that when first sizing the compressor, a generous allowance for clearance volume should be made where its effect is of importance. Because the clearance volume will be a strong function of the valve design, a later revision in design dimensions may have to be made. Some kinematic types, like rotary vane compressors, do not have a clearance volume that re-expands and prevents fresh gas from filling the swept volume.

In order to lay out a valve, it is necessary to first determine the average flow velocity. This is determined by the suction and discharge conditions and by the valve timing.

Sound and Vibration of Compressor Valves

The latter is a strong function of the kinematic design and is achieved with the help of the idealized pressure-volume diagram. For a reciprocating piston compressor, a typical diagram is shown in Figure 9.2a. At position 1, the piston is at bottom dead center. Both valves are closed as the piston starts to compress the gas. Discharge pressure is reached at 2 and the discharge valve opens. Assuming that the valve is ideal (that it has no flow losses), the gas is pushed out under constant pressure until the top dead center position of the piston is reached at 3. Thus, the volume of discharge gas pushed out is $V_2 - V_3$.

To do the indicator plot, the well-known relationship

$$p = p_o \left(\frac{V_o}{V} \right)^n \tag{9.3}$$

can be used in this simplified model.

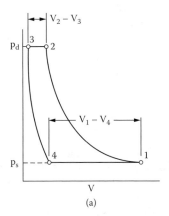

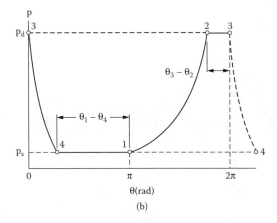

FIGURE 9.2 (a) Idealized pressure-volume diagram for a reciprocating compressor. (b) The same figure replotted as a pressure–crank angle diagram. For constant angular crank velocities, this diagram is proportional to the pressure-time diagram.

From the kinematics of the drive, we next establish a relationship between volume and time, or preferably shaft (crank) angle because the idealized indicator diagram is independent of shaft speed. This is shown in Figure 9.2b. The crank angle during which the suction valve is open is $\theta_1 - \theta_4$, and the discharge valve is open for $\theta_3 - \theta_2$.

These opening angles can then be converted to opening times, assuming a constant shaft speed Ω (rad/s):

$$t_1 - t_4 = \frac{1}{\Omega}(\theta_1 - \theta_4) \tag{9.4}$$

$$t_3 - t_2 = \frac{1}{\Omega}(\theta_3 - \theta_2) \tag{9.5}$$

While the idealized diagram of Figure 9.3 is always the same, for a given kinematic design and a given suction and discharge pressure, the duration of the valve opening is, obviously, inversely proportional to shaft speed. Thus, the average flow velocity of an ideal discharge valve of flow area A_d is

$$v_d = \frac{V_2 - V_3}{(t_3 - t_2)A_d} = \frac{Q_d}{A_d} \tag{9.6}$$

when Q stands for volume velocity.

It is important to realize that this is an average value. In reality, the volume pushed out per unit of time is not a constant, but is dependent on the kinematic design. Also, the inability of the valve to open suddenly is to be considered for more accurate calculations. This will be discussed later.

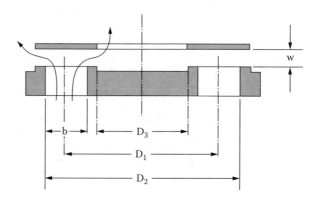

FIGURE 9.3 A typical plate valve (the valve springs are not shown). Parallel displacement w is assumed.

Volume V_3 is the clearance volume. This volume needs to be estimated first because it will be a function of valve design. Its presence affects the volumetric efficiency λ of the compressor. The volumetric efficiency is the ratio of the volume of gas entering through the suction valve (the mass at the suction condition is also the delivered mass at the discharge condition) to the swept volume of the piston. Because the clearance volume expands from 3 to 4, the gas entering the cylinder at suction condition is only $V_1 - V_4$. Thus

$$\lambda = \frac{V_1 - V_4}{V_1 - V_3}. \tag{9.7}$$

For a reciprocating piston compressor, as found in any text:

$$\lambda = 1 - \frac{V_3}{V_1 - V_3}\left[\left(\frac{P_d}{P_s}\right)^{\frac{1}{n}} - 1\right]. \tag{9.8}$$

There are other effects that influence volumetric efficiency (for instance, pressure drops in the suction valve, which delay closing, caused by pressure surges). Also, the amount of mass delivered will be reduced if the suction gas is heated when passing through the suction manifold. However at this point, it is best to ignore all effects except the clearance volume expansion, and obtain an average suction velocity of

$$v_s = \frac{V_1 - V_4}{(t_1 - t_4)A_s} = \frac{Q_s}{A_s} \tag{9.9}$$

where A_s is the effective flow area at maximum suction valve displacement. Again, all times and volumes are given by the proper kinematic relationships, which are obviously dependent on the type of compressor or crank device: reciprocating piston, rotary vane, stationary vane, and so on. For swing compressors, volumes and opening times are more difficult to estimate because the amplitude of oscillation varies with the load. Discharge valves are treated similarly.

9.2.3 Required Valve Port Areas

Based on experience, and the theoretical argument that pressure drops and flow losses in a valve are a function of the square of the flow velocity according to (ρ = mass density, Δp = pressure drop across valve):

$$\Delta p = \zeta \frac{\rho v^2}{2} = \zeta \frac{\rho M^2}{2} \quad c^2 = \zeta \frac{\rho}{2} kRTM^2. \tag{9.10}$$

The Mach number (the ratio of average speed to the average speed of sound) is important. Given a temperature and mass density, it is recommended that $M \leq 0.2$. Some authors distinguish between slow and fast, small and large compressors. The allowable flow velocity is, therefore,

$$v = Mc \qquad (9.11)$$

where the speed of sound is

$$c = \sqrt{kRT} \qquad (9.12)$$

and where k = the adiabatic coefficient, R = the gas constant, and T = the absolute temperature. These are average values for given discharge, or suction conditions. The required effective flow area is, therefore,

$$A = \frac{Q}{v} = \frac{Q}{Mc}. \qquad (9.13)$$

Note that v is the allowable flow velocity, averaged over the opening time of the suction or discharge valve. The first order of business is, therefore, to design a valve port arrangement that achieves this effective area. Introducing a contraction coefficient k_s, which may be assumed (because information will, in general, be lacking at this point) to be $k_s = 0.6$, the required geometric port area is

$$A_e = \frac{A}{k_s}. \qquad (9.14)$$

The same argument applies to both suction and discharge valve ports.

Because the volume of a compressor increases with the cube of a typical dimension, while the surface area available for valve ports increases only with the square, it becomes more and more difficult to find enough space for the valves as the size of a compressor increases, given that the compressor speed is held constant. Obviously, the flow area requirement also increases proportionally to the compressor speed. Thus, large and fast compressors are the most difficult for which to design valves. The smaller and slower the compressor, the easier valve design becomes as far as space constraints are concerned.

There are also certain designs for which it is innately easier to design valves. An example is the rotary vane compressor. As the number of vanes increases, the number of gas discharges per revolution increases, approaching a more and more continuous discharge. This has the effect of reducing the average flow velocity, given a certain port size and compressor speed. Actually, a discharge valve becomes unnecessary for compressor operation if one can tolerate inefficiency because already compressed gas will re-expand into each volume as it is opened to the discharge port.

Sound and Vibration of Compressor Valves

For a rotary, one-vane compressor, this would, of course, not work. For a two-vane compressor, pumping will continue even for a broken discharge valve, but the compressor will become very hot. For five vanes, it would probably be acceptable not to use a discharge valve, if one can tolerate other effects encountered in multivane compressors.

The purpose of this discussion is to point out that there is nothing magical about $M = 0.2$, and one may have to tolerate higher Mach numbers, but one should then not be surprised if the indicator diagram shows large valve losses. As always, design means compromise. The purpose of this discussion is to allow one to face difficulties with open eyes and a full understanding, not in ignorance, which usually manifests itself in a frantic "trial and error, see if you can fix it" procedure.

9.2.4 Allowable Valve Lift

Once the port area is established, it can be argued that the lift height is established by dividing the port area by the effectively available circumferences of the covering valve plate or reed. The term *effectively available* has been introduced because it is important at this point to sketch the reed design and to visualize a flow pattern around the reed. For example, for a ring valve (as shown in Figure 9.3, the available circumference is equal to

$$C = \pi(D_2 + D_3). \tag{9.15}$$

Note that D_3 is the smallest diameter for gas that passes through the inner annular gap, while D_2 is the smallest diameter for gas that passes through the outer gap. The smallest restriction usually controls the flow. A more refined treatment is possible taking into account the fact that the flow most likely will prefer the outer to the inner annular gap, but during the initial stages of valve layout, it is not necessary. Also, in some designs there is flow interference and the effectively available circumferential gap is less than the geometrically calculated one, resulting in an estimated lift height that is too small.

Thus, the average required lift height is

$$h = \frac{A_e}{C} = \frac{A_e}{\pi(D_2 + D_3)}. \tag{9.16}$$

For a flexible ring valve, the gap area is, of course, not simply the circumference times the lift height. Rather, h has to be interpreted as an average value.

9.2.5 Advantage of a Valve Stop

The best design probably results if one can introduce a stop for the valve to rest against during most of its opening time. If a stop cannot be introduced, the valve will overshoot its equilibrium position by almost 100% because the sudden opening and subsequent forcing behavior can be viewed as the response to a step function.

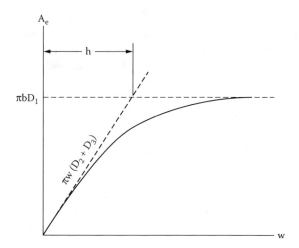

FIGURE 9.4 A typical effective flow area A_e plotted as a function of the parallel valve plate displacement w. The recommended stop height h is shown. For small displacements w, the ring gaps control the flow; for large w, the valve orifice is the controlling factor.

Based on this reason alone, flutter will be introduced because the valve, after having overshot its position, will not return slowly to equilibrium but will overshoot in the opposite direction at the same frequency. This causes a partial or full closing of the valve. Gas is partially prevented from escaping, the pressure builds up, and the valve will be fired again away from the seat, repeating the cycle. In this process, the valve may actually come in contact with the seat several times during each opening cycle.

When the thermodynamic process demands that the valve be closed, it may actually be caught at its maximum excursion from the seat and cannot close in time. Negative pressure differentials will build up very rapidly, slamming the valve into the seat with a high impact velocity and destructive effects.

In summary, a properly designed stop will not diminish the efficiency of the compressor because larger valve amplitudes than necessary will not increase the flow appreciably; in an approximate sense, the effective orifice area controls the flow for lifts equal to or greater than h, as illustrated in Figure 9.4. It will prevent flutter, and to a certain extent gas pulsations and noise. Because the valve will close at more or less the proper time, the thermodynamic efficiency is not degraded, and high-impact stresses are prevented.

9.2.6 Estimating the Flow Force on the Valve and Selection of the Effective Valve Spring Rate

The required effective stiffness or spring rate, provided by springs in the case of a spring-loaded plate valve, or by the flexural resistance in the case of a flexing reed valve, is at this stage determined by the maximum required lift height h. This height must be reachable by the action of the flow forces on the valve. While the nature of the flow forces are fairly complicated when viewed over the entire valve opening cycle, it can be argued that as a rough approximation we can estimate them from the momentum-impulse law, ignoring Bernoulli effects due to wide valve seats, streamline detachment and reattachment, and so on. Thus, the available average force to reach opening height h is

$$F = \rho A v^2 \qquad (9.17)$$

where A is the effective port area, and because the admissible velocity has been already given as

$$v = Mc \qquad (9.18)$$

we obtain (using $c^2 = kRT$ and $p/\rho = RT$)

$$F = kApM^2. \qquad (9.19)$$

In the case of a spring-loaded plate valve, the required effective spring rate K can now be determined from

$$K = \frac{F}{h}. \qquad (9.20)$$

In the case of flexible reeds, the designer needs to decide, at this point, the general design of his reed. The same is true for spring loaded ring valves. Because of the flexure, the force achieved by this simple approach represents an approximate value only, in terms of a resultant. If the reed simply covers a single circular hole, as is the case in valves for small compressors in the one horsepower or less category, the spring rate K represents the ratio of force over deflection of the reed at the center point of the hole. Thus, in the case of a cantilever-beam-type reed, we select the width b of the reed based on the seat area and obtain, from any handbook, that the valve thickness t should be

$$t = 2L\sqrt{\frac{K}{Eb}} \qquad (9.21)$$

where L = the length of the reed, b = the width of the reed, and E = the modulus of elasticity. Assuming that we have also selected the type of material, the two variables we can play with are the width b and length L. But because stresses may not exceed a certain level, we introduce the condition that the maximum stress may not exceed a certain value (or use any other failure theory). Assuming that the width b is determined by the porthole size, the stress condition will give us the length L.

9.2.7 Floating Valves or Spring Loaded Valves

Why do we need valve springs or flexible valves having a spring stiffness? Actually, we do not. There are floating valves, as for instance the multi-hat valve (Killman (1972a, b)) shown schematically in Figure 9.5, where many tiny hats float open with

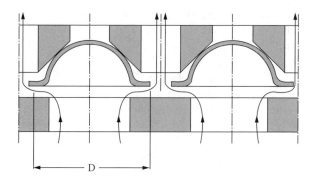

FIGURE 9.5 The multihat valve consists of a large number of small, hat-shaped valves that cover small circular ports. The diameters D are on the order of a few millimeters.

the gas stream, rest against a stop, and return as the flow reverses. To keep the flow reversal losses small because there is no spring that assists the return to the valve seat, each valve needs to be of the smallest mass possible, so there is negligible inertial resistance to opening or closing. If the mass is large, a time delay is caused by the inertial resistance, the reversed flow has time to develop, and the valve slams against its seat with high-impact velocity, but not before the reversing flow has caused a degradation in the volumetric and energy efficiencies. Again, the desire to assist a valve in closing at the proper time is the reason for spring loading or utilizing the stiffness of a flexing reed valve.

The requirements for quick opening are, again, a small effective mass; however, the spring effect is also undesirable to some extent. If the effective spring rate is too large, it may close the valve properly, but it will retard the opening, for which we pay with somewhat increased flow losses.

On the positive side, in a minor way, the spring effect will reduce the impact velocity against the stop, and in the case of preloading, perhaps increase the quality of seating. Reduced leakage means reduced losses.

9.2.8 Reed Valve Shapes

On balance, it is probably best to choose a valve design with a spring effect. If it is a flexible reed, we have to give some thought to what the covering reed should look like. Copying other designs is a time-honored occupation in engineering and should not be discouraged if one is honest about one's copying, copies from a successful rather than unsuccessful design (note it takes knowledge and judgment to identify what is successful), and does not infringe on someone's patent rights. Thus, one scenario is that we copied an existing design, but wonder about the scale if we had copied from a different size and type of compressor. The other scenario is that we are confined by space and cost and need to create an original reed shape. In both situations it pays to remember that failure can be interpreted as having pushed too much strain energy into a given mass of material. With other words, we should have as much material available as possible to store strain energy. The strain energy in the flexed reed should also be distributed as evenly as possible. This requires that we avoid areas of stress concentration.

Having decided on a shape, we can then determine the thickness. We do know K, the required effective spring stiffness. Assuming that the shape is too complicated to lend itself to calculation, we may determine the thickness t by a simple and perhaps even crude experiment. We cut our valve shape, scaled up if necessary, out of a sheet of plastic, support its boundaries as they would be supported in the compressor, and attach to it a weight at a location where the force resultant may be located, or several weights at locations of partial force resultants. The combined weight should not cause a deflection of more than 10% of the largest dimension of the valve, so that we are approximately in the linear range. In the case of a single force resultant, weight divided by deflection is the spring rate of the model. Next, we may scale the spring rate to the actual valve size and material by

$$\frac{K_1}{K_2} = \frac{E_1}{E_2}\left(\frac{t_1}{t_2}\right)^3 \frac{1-\mu_2^2}{1-\mu_1^2}\left(\frac{L_2}{L_1}\right)^2 \qquad (9.22)$$

where subscript 1 refers to the model, subscript 2 to the actual valve, and E = Young's modulus, μ = Poisson's ratio, t = the thickness, and L = the typical length dimension. Presumably, the simple experiment on the model gives us K_1 and we know E_1, t_1, μ_1 and L_1 of the model. We select the material for the actual valve reed, which gives us E_2 and μ_2 and we know L_2. Also, we know the spring rate K_2 we want. The thickness t_2 is then defined by

$$t_2^3 = t_1^3 \left(\frac{E_1}{E_2}\right)\left(\frac{K_2}{K_1}\right)\left(\frac{1-\mu_2^2}{1-\mu_1^2}\right)\left(\frac{L_2}{L_1}\right)^2. \qquad (9.23)$$

If a plate finite element program is easily available it may be used instead (see also Kristiansen, Soedel, and Hamilton, 1972; Soedel, 1971; and Soedel, 2004).

At this point we have selected a reed shape and know the thickness t that gives us the desired effective spring rate so that the reed can open under the action of the expected flow forces to a height (most likely defined by a stop) that limits the flow velocity. It is now time to estimate stress levels.

Because at this point of layout development we are most likely not in the position to do a dynamic analysis, it is recommended that a static stress analysis be done. Those who have a static plate finite element program (which they used to determine the thickness t) will also be able to obtain stresses with it. If the maximum static principal stress is in the neighborhood of the endurance strength of the material or larger, the reed needs to be changed. In the case of a stop, the static design stress should be, at this stage of initial design, about one-third of the endurance limit. If there is no stop, the nominal static stress should be approximately one-sixth of the endurance strength because the dynamic stress for the no-stop design can be expected to be twice the stress for the stop situation.

Obviously, these are all rules of thumb. It should be remembered that the actual dynamic situation in the compressor may be very different from the idealized static deflection situation, and thus a safety factor is needed. Also, fatigue limits are statistical averages, often based on samples of insufficient size.

9.2.9 SUCTION AND DISCHARGE VOLUME SELECTION

It is important that the designer of high-speed compressor valves (3600 rpm or more) keep in mind that any gas will resist velocity changes because of its mass.

Actually, at very high speeds the flow resistance will be dominated by the inertial effect; the classical valve flow resistance, which is proportional to the square of velocity and a steady flow concept, becomes less and less valid as the dominant model.

The flow in compressor manifolding is intermittent. When a discharge valve opens, the gas flowing from the cylinder has to push the gas already in the manifolding out of the way, so to speak. This is a problem whose importance increases with the compressor speed. At 3600 rpm, we have only 1/60 second per revolution, and only a small fraction of this is available for the gas mass in the cylinder to be emptied into the manifolding, accelerating in turn the gas already in the manifolding. The result is the development of a backpressure against which the compressor must work and the losses can be large. The reality is, of course, even more complicated because a pressure surge (oscillation) will be created, but this will be discussed later.

At this stage of the design process, one should make sure that the volume directly behind a discharge valve is as large as possible; as a minimum it should be equal to the cylinder volume for 3600 rpm compressors, but preferably it should be two or three times as large. The same is true for suction valves because the sudden filling of the cylinder depletes the supply of gas in the suction manifold and an underpressure is created against which the valve must work.

Never should the valve be asked to dump gas directly into a pipe or suck it directly from a pipe, unless the pipe cross-section is many times the effective valve flow area so that the pipe looks like an expansion volume to the valve.

The expansion volume acts like a collection tank or accumulator of gas, so that an over-or undersupply of gas can be stored temporarily, giving the gas molecules in the pipe sections of the manifold time to attain speed, thereby limiting the acceleration and thus the required over- or underpressure to overcome the dynamic resistance.

These measures will, in general, reduce the gas oscillations in the manifolding and piping and will therefore reduce the sound.

9.2.10 RELIABILITY CONSIDERATIONS

It is possible that the valve orifice area is so large that the valve experiences its largest stress when it is on its seat. The maximum pressure loading will, for this situation, always be the difference between discharge and suction pressure. Again, simplified models of this situation suffice at this stage of design.

Another consideration is impact velocity. Unfortunately, no good way has been developed so far to enable quick design estimates. The old definition of *setting velocity* seems not to be very useful. The setting velocity is the product of the compressor rotational speed ω (rads/s) and the design amplitude (stop) A (mm) of the valve reed or plate

$$v_s = \omega A, \tag{9.24}$$

and it is, for instance, recommended that it should not exceed 100 to 200 mm/s (see, for example, Chlumski, 1965; Frenkel, 1969; and Plastinin, 1984). It is a concept that manages to capture the idea that valve impact failure is related to impact speed, and thus somehow related to compressor speed and maximum valve amplitude, but ignores the fact that it is also related to the effective stiffness and mass of the valve, its opening history, and the gas forces. It essentially creates an amplitude criteria. For example, for a compressor turning with 3600 rotations per minute, $\omega = 377$ rad/s, and taking $v_s = 200$ mm/s as the allowable value, A is allowed to be less than one millimeter—clearly an absurd situation. Of course the quoted limits on the setting velocity have emerged from the experience of designers of large and rather slow compressors. Calculating v_s from actual high-speed refrigeration compressors, we find that it is in the neighborhood of 600 mm/s.

At this point of the design process, impact velocity is ignored as a design criterion, other than being aware of its potential importance. The introduction of a valve stop is useful because this measure will, in most cases, limit the seat impact velocity and is therefore a step in the right direction unless we are unfortunate enough to have created, in turn, a stop impact problem. We may wish to calculate an approximate minimum expected seat impact velocity if we know the natural frequency ω_n of the valve. The minimum expected impact velocity v_{min} on the seat at the time of valve closing is

$$v_{min} = h \omega_n \tag{9.25}$$

where h is the stop height. For example, a typical first natural frequency of a refrigeration reed valve might be $\omega = 2400$ rad/s and h = 3 mm. Thus, $v_{min} = 7200$ mm/s. Unfortunately, we have (even today) only a limited amount of data to tell us if this is a safe value. For more information, see Forbes and Mitchell (1974), Svenzon (1976), and Waltz and Soedel (1980).

9.2.11 Valve Material Selection

While not primarily important to sound and vibration, the material selection is important to the prevention of fatigue failure, and is thus related to valve vibrations.

The major criteria in the selection of valve steel are the endurance strength to bending fatigue, the impact resistance, and the corrosion resistance.

A high-carbon steel seems to offer good impact resistance, as far as that can be defined, as well as excellent endurance strength. If corrosion resistance is required, stainless steels of various types may have to be used. It is difficult to be more specific and still avoid discussion of various manufacturers. It is sufficient to say that all valve steel manufacturers will be very helpful in discussing the various options. For more information, see Cohen (1972), Dusil (1976), Dusil and Appell (1976), Dusil and Johansson (1978 a, b and 1980), Johansson and Persson (1976), Laub (1980), Smith (1978), Sprang et al. (1980), and Soedel (1984), for example.

There is a sentiment that at the beginning of the design process, not much is known about the required impact resistance, and it should, therefore, be ignored. The basic decision is whether corrosion resistance is required or not. If not, one should perhaps design the valve with a relatively low quality carbon steel on the argument that later unforeseen troubles could then be cured by a material switch. Of course sometimes the design constraints are so severe that a high-quality material needs to be considered immediately.

It should be remembered that the higher the endurance strength of a steel, the more random scatter in its properties can be expected. The statistical samples on which published S-N fatigue curves are based are probably too small to put much faith in a line below which supposedly only 2.3% of failures occur. In addition, a 2.3% failure rate is unacceptable for most compressor companies, but where is, for instance, the 0.2% line? It has become fashionable in practical engineering to extrapolate very small samples by fitting a distribution curve, Gaussian, Weibull, and so forth, to it. While this may be helpful in guiding the intuition of designers, it creates only an illusion of accuracy. Unfortunately, it would be unreasonable to expect the valve steel manufacturers to run samples large enough so that these extremes of the S-N distribution can be established with accuracy, and so we have to live with this approach. Thus, for lack of anything better, we extrapolate. For example, if the average fatigue limit is 125,000 psi for a high-grade steel, and the published 2.3% fatigue limit is 109,600 psi, then by extrapolation the 0.2% fatigue limit becomes, using Gaussian distribution tables, 95,600 psi. This means that at this stress level, probably 2 out of 1000 valves will have limited life and fail eventually.

The safe stress limit needs to be further reduced because it will decrease with increasing temperature. For example, unless published data is available, it would be advisable to reduce the allowable fatigue limit further by approximately 10% for discharge valves, to perhaps 86,000 psi.

Next, because in the initial design state where we are able to make only very approximate maximum stress estimates, one may want to choose a factor of safety, appropriately called a *factor of ignorance* at this state, of perhaps two, which in effect halves the fatigue limit (or halves the estimated allowable stress amplitude).

The philosophy of fatigue limit design is of course connected with the type of compressor application. In hermetic compressors, we have no choice but to design for unlimited endurance because a valve failure would be costly. For compressors where valves can be exchanged fairly easily, requirements can be somewhat relaxed.

9.2.12 MULTICYLINDER COMPRESSORS AND MULTISTAGE COMPRESSORS

The basic valve designs and most of the various advanced considerations are the same for a multicylinder compressor as they are for a single cylinder compressor. One of the exceptions is the gas pulsation in discharge and suction manifolds, which are more complicated. Because of this complexity, it is more difficult to detune manifolds so that resonances are avoided and a computer simulation is almost a must, even while simplifying approaches have been developed that do give some interesting insight. Again, the concern for valves is that these valve-generated pulsations may create large backpressures that interfere with flow and cause large energy or volumetric losses. The second important exception is noise, both the pulsation- and impact-generated varieties because the phasing between the various cylinders may make the harmonic content richer. Gas oscillation waves may cancel or reinforce each other.

It is beyond the scope of this text to discuss the thermodynamics of staging, except to state the obvious—that the interstage pressure is not determined by the valve design, but rather by the relative cylinder volume sizes of the different stages. Otherwise, valve design follows the same steps as for single-cylinder compressors. There is some simplification in the sense that the discharge valve of stage i can have more or less the same basic dimensions as the suction valve of state i + 1, except for features that are functions of the geometrical difference between discharge and suction valves, for example, the design of valve stops. There is also a difference due to the intercooler, which shrinks the hot discharge gas from stage i in volume before it enters state i + 1.

There is the possibility for pronounced gas oscillations in the piping between stages. This may be treated mathematically as one treats gas oscillations in a multicylinder compressor, as will be discussed later.

9.3 USEFUL VALVE CALCULATIONS

As the valve is being finished on the drawing board, based on the simple calculations of Section 9.2, we may notice that the flow passages, by geometric necessity, turn out to be more complicated than initially anticipated. It may therefore be of value to examine, on an approximate theoretical basis, the expected flow and force versus valve lift behavior. The experienced designer can draw preliminary conclusions as to the quality of the valve design from such curves. Again, properly designed valves may result in less vibration and therefore less sound, be it generated by gas pulsations or by valve impact.

9.3.1 Effective Valve Flow Areas

It is instructive to consider a simple model, courtesy of Schwerzler and Hamilton (1972), that allows calculation of the effective flow area as a function of displacement (see also Soedel, 1972). The following assumptions are made: (1) incompressible flow theory is assumed, (2) restrictions in the valving system can be considered as simple orifices between central volumes, (3) upstream and downstream conditions are stagnation conditions, and (4) each valve can be treated as a combination of orifices in series or parallel.

Because of condition (3), the velocity of approach factors are assumed to be unity and the mass rate of flow through each orifice is given by

$$m = KA\sqrt{2\rho_u(p_u - p_d)} \qquad (9.26)$$

where ρ_u is the upstream mass density, p_u is the upstream pressure, and p_d is the downstream pressure. The flow coefficients K have to be assumed, using as a guideline published literature concerning orifices that resemble the various valve

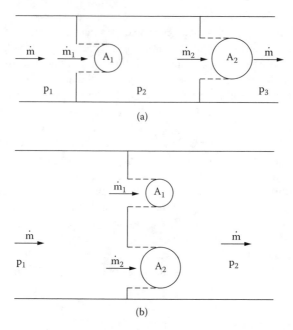

FIGURE 9.6 Valve orifices in (a) series and (b) parallel.

restrictions. Although this is the weakest link in this approach, it still provides insight. The effective orifice area is KA.

9.3.1.1 Orifices in Series

Let us now examine two orifices in series as shown in Figure 9.6a. For steady flow (and note that the implied assumption is again that we may use a quasi-steady approach to the nonsteady valve flow) we may write

$$\dot{m}_1 = \dot{m} \tag{9.27}$$

$$\dot{m}_2 = \dot{m} \tag{9.28}$$

$$(KA)_1 \sqrt{2\rho_1(p_1 - p_2)} = \dot{m} \tag{9.29}$$

$$(KA)_2 \sqrt{2\rho_2(p_2 - p_3)} = \dot{m} \tag{9.30}$$

$$(p_2 - p_3) = \frac{\dot{m}^2}{2(KA)_2^2 \rho_2}. \tag{9.31}$$

Also, because

$$p_1 - p_3 = \Delta p \tag{9.32}$$

where Δp is the pressure drop across the entire valve, we may write

$$(p_1 - p_2) - (p_3 - p_2) = \Delta p \tag{9.33}$$

or

$$(p_1 - p_2) = \Delta p - \frac{\dot{m}^2}{2(KA)_2^2 \rho_2}. \tag{9.34}$$

Substituting this into Equation 9.29 gives

$$\dot{m} = (KA)_1 \sqrt{2\rho_1 \left(\Delta p - \frac{\dot{m}^2}{2(K)_2^2 \rho_2} \right)}. \tag{9.35}$$

For approximately incompressible flow, $\rho_1 = \rho_2 = \rho$. Solving Equation 9.35 for $\dot{m}$ gives

$$\dot{m}^2 = \frac{2\rho\Delta p}{\frac{1}{(KA)_1^2} + \frac{1}{(KA)_2^2}}. \tag{9.36}$$

This result is not entirely unexpected intuitively. Designating $(KA)_e$ as the effective flow area, we may write

$$\frac{1}{(KA)_e^2} = \frac{1}{(KA)_1^2} + \frac{1}{(KA)_2^2} \tag{9.37}$$

and the mass flow rate is given by

$$\dot{m} = (KA)_e \sqrt{2\rho\Delta p}. \tag{9.38}$$

9.3.1.2 Parallel Orifices

If we apply a similar approach to two orifices in parallel, we obtain

$$\dot{m}_1 + \dot{m}_2 = \dot{m} \tag{9.39}$$

or

$$(KA)_1\sqrt{2\rho_1(p_1-p_2)} + (KA)_2\sqrt{2\rho_2(p_1-p_2)} = \dot{m}. \quad (9.40)$$

This gives (again assuming that $\rho_1 = \rho_2 = \rho$)

$$\dot{m} = (KA)_e\sqrt{2\rho\Delta p} \quad (9.41)$$

where $\Delta p = p_1 - p_2$ is the pressure drop across the valve and where

$$(KA)_3 = (KA)_1 + (KA)_2. \quad (9.42)$$

Note that any combination of more than two orifices can be handled in a similar manner.

9.3.1.3 Application Examples: Ring Valves

Let us, as a practical example, obtain the effective flow area for the ring valve, which was discussed earlier, and which is shown in Figure 9.3.

Note that all orifices are in the form of ring bands. Orifices A_2 and A_3 are in parallel, with

$$(KA)_2 = \pi D_2 w \quad (9.43)$$

and

$$(KA)_3 = \pi D_3 w. \quad (9.44)$$

The effective orifice area combining the two is

$$(KA)_{e23} = (KA)_2 + (KA)_3 = K\pi w(D_2 + D_3). \quad (9.45)$$

K is really a function of the deflection w, but an average approximation is used; for preliminary work, perhaps even $K = 1$.

The flow area $(KA)_1$ is

$$(KA)_1 \cong K\pi D_1 b \quad (9.46)$$

Thus, the effective flow area for the total valve is given by the following (because $(KA)_1$ and $(KA)_{e23}$ are in series)

$$\frac{1}{(KA)_e^2} = \frac{1}{(KA)_{e23}^2} + \frac{1}{(KA)_1^2} \quad (9.47)$$

or

$$(KA)_e = \frac{(KA)_{e23}}{\sqrt{1 + \frac{(KA)_{e23}^2}{(KA)_1^2}}}. \tag{9.48}$$

The effective flow area can now be plotted as a function of w as shown in Figure 9.4. We see from Equation 9.48 that for small values of w, the equation is essentially

$$(KA)_e = \cong (KA)_{e23} = K\pi(D_2 + D_3)w \tag{9.49}$$

and defines the slope of the tangent line to the effective flow area curve at the $w = 0$ position. As w increases, there comes a point where the effective flow area has converged to the port area $(KA)_1$. Equation 9.48 gives us

$$(KA)_e \cong (KA)_1 = K\pi D_1 b. \tag{9.50}$$

This is the value of the horizontal line, which is the tangent line to the effective flow area curve as the valve deflection w becomes large.

Figures 9.7a and 9.7b show an example of a more complicated ring valve design, together with the equivalent orifice interpretation (see also Hamilton, 1974). As the valve plate lifts off the seat, the ringlike orifices, $(KA)_2$ and $(KA)_5$, increase proportionally to the valve lift. On the other hand, the ring-like orifice $(KA)_4$, between the valve stop and the rising valve plate decreases proportionately to the valve lift. When the valve plate comes to rest against the stop, the orifice area $(KA)_4$ approaches zero and the entire flow pass $(KA)_2 \rightarrow (KA)_3 \rightarrow (KA)_4$ is shut off.

From a thermodynamic efficiency viewpoint, this is an undesirable design, and we would be inclined to fix it by drilling a series of portholes into the valve stop. However, considering a possible cushioning effect that reduces the impact velocity of the valve plate on the valve stop, this design may possibly make some sense. In any case, this example illustrates that viewing valve passages as equivalent orifices in series or in parallel, however approximate, is a valuable tool.

9.3.2 Effective Valve Force Areas

At the present state of the art, a simple, uniformly good theoretical approach for flow forces on the valves has not been developed yet, even while some attempts have been fairly successful in special cases. Considerable engineering judgment is required to create a reasonably simple model of the forces acting on the valves. The following concept of a valve force area based on experiments was first introduced by Wambsganss and Cohen (1967), and the following calculation approach follows Schwerzler and Hamilton (1972); see also Wambsganss (1964), and Wambsganss, Coates, and Cohen (1967). Using the valve of Figure 9.3 as the example, the approach

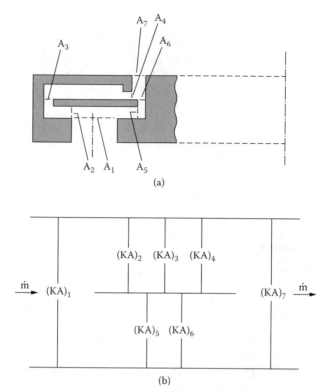

FIGURE 9.7 (a) Ring valve with complicated stop design: the flow has to pass through six critical restrictions, labeled A_1 to A_6. (b) Equivalent orifice arrangement.

is to visualize pressures on it as shown in Figure 9.8. The force on the valve is given approximately by

$$F = p_2 A_2' + p_1 A_1' + p_3 A_3' - p_4 A_4' \qquad (9.51)$$

where

$$A_1' = \frac{\pi}{4}\left(D_2^2 - D_4^2\right) \qquad (9.52)$$

$$A_2' = \frac{\pi}{4}\left(D_5^2 - D_2^2\right) \qquad (9.53)$$

$$A_3' = \frac{\pi}{4}\left(D_4^2 - D_3^2\right) \qquad (9.54)$$

$$A_4' = \frac{\pi}{4}\left(D_5^2 - D_3^2\right) \qquad (9.55)$$

Sound and Vibration of Compressor Valves

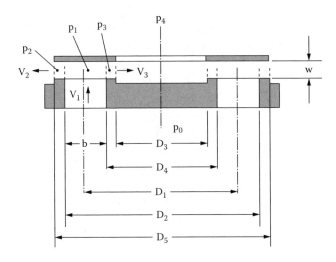

FIGURE 9.8 Pressures acting on a typical ring valve subjected to parallel displacement w.

The static pressure p at each point in the ring gap approximately equals the total pressure p_o, less the local dynamic head

$$p_1 = p_o - \rho \frac{v_1^2}{2} \tag{9.56}$$

$$p_2 = p_o - \rho \frac{v_2^2}{2} \tag{9.57}$$

$$p_3 = p_o - \rho \frac{v_3^2}{2}. \tag{9.58}$$

In the port region, we have an additional pressure term due to the change in flow direction. Thus, approximately,

$$p_1 = p_o + \rho v_1^2 - \frac{\rho v_1^2}{2} = p_o + \frac{\rho v_1^2}{2}. \tag{9.59}$$

The velocities are given by

$$v_1 = \frac{\dot{m}}{(KA)_1 \rho} \tag{9.60}$$

$$v_2 \cong \frac{\dot{m}}{(KA)_2 \rho} \tag{9.61}$$

$$v_3 \cong \frac{\dot{m}}{(KA)_3 \rho}. \tag{9.62}$$

Note that the approximations for v_2 and v_3 are only permissible for a ring valve with fairly narrow seating. This formulation neglects flow separation.

Neglecting compressibility, the mass flow rate is given approximately by

$$\dot{m} = (KA)_e \sqrt{2\rho \Delta p} \qquad (9.63)$$

where $\Delta p = p_0 - p_4$, the pressure differential across the entire valve, and where $(KA)_e$ is the effective flow area of the entire valve.

Thus, the force on the valve is approximately

$$F = p_0(A_2' + A_1' + A_3') - p_4 A_4' - A_2' \frac{\rho}{2} \frac{\dot{m}^2}{(KA)_2^2 \rho^2} - A_3' \frac{\rho}{2} \frac{\dot{m}^2}{(KA)_3^2 \rho^2} + A_1' \frac{\rho}{2} \frac{\dot{m}^2}{(KA)_1^2 \rho^2}. \qquad (9.64)$$

Because

$$A_4' = A_2' + A_1' + A_3' \qquad (9.65)$$

and

$$(KA)_1 \cong A_1', \qquad (9.66)$$

we get

$$F = \Delta p A_4' + \frac{\dot{m}^2}{2\rho}\left[\frac{1}{(KA)_1} - \frac{A_2'}{(KA)_2^2} - \frac{A_3'}{(KA)_3^2}\right] \qquad (9.67)$$

or substituting Equation 9.63 we obtain

$$F = \Delta p \left[A_4' + (KA)_e^2 \left(\frac{1}{(KA)_1} - \frac{A_2'}{(KA)_2^2} - \frac{A_3'}{(KA)_3^2} \right) \right]. \qquad (9.68)$$

The bracketed quantity is the effective force area $B(w)$. This force area can now be plotted as a function of w. First of all, when $w = 0$ and there is no flow, $F = \Delta p A_1'$ or $F = \Delta p A_4'$, depending on the seating. The actual value is probably between. When flow develops, we obtain, for small values of w, setting all K equal to unity as a first approximation and utilizing Equation 9.49,

$$F = \Delta p \left[A_4' + \pi^2 (D_2 + D_3)^2 w^2 \left(\frac{1}{\pi D_1 b} - \frac{A_2'}{\pi^2 D_2^2 w^2} - \frac{A_3'}{\pi^2 D_3^2 w^2} \right) \right]. \qquad (9.69)$$

Sound and Vibration of Compressor Valves

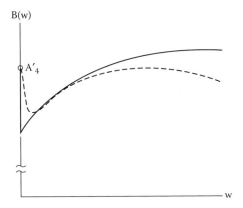

FIGURE 9.9 Typical effective force area as a function of displacement: —, approximate theory; ---, actual behavior.

This gives us, at $w = 0^+$,

$$F = \Delta p \left[A_4' - (D_2 + D_3)^2 \left(\frac{A_2'}{D_2^2} + \frac{A_3'}{D_3^2} \right) \right]. \tag{9.70}$$

This drop in effective force area is due to the Bernoulli effect. It is overpredicted here because the viscosity of the gas is not considered (Figure 9.9).

For large values of w, $(KA)_e = \pi D_1 b$ and we obtain approximately

$$F = \Delta p (A_4' + \pi D_1 b). \tag{9.71}$$

In reality, this is an upper bound that is never reached; the force rapidly drops with deflection sooner or later because the assumed ninety-degree deflection of the velocity jet by the valve becomes less and less true as w increases. Also, it is less and less true that the stagnation pressure in the gaps is equal to the cylinder pressure. But as a rough approximation, the model helps our understanding of how important certain design parameters are.

On a practical level, the analysis shows that the Bernoulli effect can be reduced by making the areas A_2' and A_3' as small as possible; in other words, the seat areas should be small relative to the port areas. For ring valves, this is usually the case.

If the valve seats are relatively wide, the sharp edge orifice model employed here cannot any longer be used.

The significance of the Bernoulli effect to valve vibrations is that the valve will be sluggish in opening. This delay causes a pressure buildup, which causes the valve, when it is finally forced off the seat, to attain larger vibration amplitudes or higher impact velocities against the valve stop. This delay is therefore undesirable.

Another model of valve opening delay was introduced by Wambsganss and Cohen (1967). They postulated a *sticktion force*, which must be overcome before the valve can open. Sticktion is thought to be due to an oil film that is present on the valve seat in refrigerating, air conditioning, and heat pump compressors (see also Giacomelli and Giorgetti, 1974).

The Bernoulli effect was studied by Killmann (1972a, b), who was motivated by a valve design that used a multitude of tiny, floating hats, with no springs. For different approaches to valve flow and valve forces, see also Fleming et al. (1982), and Boswirth (1980a, b and 1982).

9.3.3 Measuring Effective Valve Flow and Force Areas

Because flow and force areas are relatively important from a valve vibration, sound generation, and flow loss viewpoint, and because the simple theoretical models are very approximate, there is an advantage to measuring these properties directly. The experimental approaches discussed in the following were introduced by Wambsganss and Cohen (1967).

9.3.3.1 Effective Valve Flow Areas

The purpose of this measurement is to establish an equivalent square-edged orifice area for the valve assembly as a whole. For spring-loaded plate valves, the valve is displaced parallel to the seat plate by a lift height w. The effective (equivalent) flow area is established by placing the valve assembly into a steady air stream whose velocity is known. Wambsganss and Cohen (1967) mounted the valve assembly on a standard orifice flow meter (ASME [American Society of Mechanical Engineers], 1959) as shown in Figure 9.10. The effective flow area is, for practical purposes, independent of the type of gas used in the experiment. The type of gas enters through the mass density and flow velocity achieved in a given

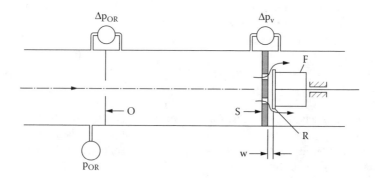

FIGURE 9.10 Effective flow area measurement: S = valve seat, F = adjustable fixture, R = valve plate or reed, O = orifice. To determine the flow rate, p_{OR} and Δp_{OR} are measured; the pressure differential across the valve, Δp_v, is also measured, as w is adjusted to various settings.

compressor. It should be noted that the valves may be scaled, and may also be mock-ups, made of plastic.

For poppet valves and spring-loaded ring valves, the valve plate is displaced parallel to the seat by a distance w, and the effective flow area is determined from

$$A_v = \frac{\dot{m}}{p_u \sqrt{\frac{2k}{(k-1)RT_u}} \sqrt{\left(\frac{p_d}{p_u}\right)^{\frac{2}{k}} - \left(\frac{p_d}{p_u}\right)^{\frac{k+1}{k}}}} \quad (9.72)$$

where, in metric units, m is the measured mass flow rate [kg/s], p_u is the upstream pressure, p_d is the downstream pressure, both in [N/m²], T_u is the upstream temperature [°K], R is the gas constant [Nm/°Kkg], and k is the ratio of the specific heats. The effective flow area is usually plotted as a function of the displacement, w. For a well-designed valve, there will be a region where the effective flow area is proportional to the valve lift, and the slope will be proportional to the effective circumference of the ports. After a transition region, the curve asymptomatically approaches a horizontal line, which will be equal to the effective flow area of the ports alone.

For a well-designed valve, the stop height should neither be smaller nor larger than the transition valve of w. If it is less than the transition value, then the port size is either unnecessarily large or the flow is artificially restricted by the stop. In other words, it does not make sense for a valve plate not to lift close to the value of w, at which the maximal flow area can be achieved. Neither does it make sense to have it lift beyond this point because little additional flow area is gained. The negative tradeoffs, such as large strains, high impact velocities, are not justified.

The experiment becomes less clear cut if effective flow areas for flexible reed valves are to be measured. There are two ways they may be approached. For situations where the deflection shape of the reed in the compressor under operating conditions is not known, it was advocated by Wambsganss and Cohen (1967) to still use the parallel displacement test and calculate what the effective flow area would be for a deflected shape. This approach is useful for computer simulation, but not always for a practical test where the overall valve design is to be evaluated. In such a case, it is perhaps best to estimate the deflection shape in the compressor and hold the valve reed in positions of increasing lift by set screws, as done by Elson, Soedel, and Cohen (1976). Many times, the expected deflection is close to a static deflection, the forces being supplied by adjustment screws over the port locations. In this way, a curve can be generated that has a more practical meaning. It represents the actual effective flow area as a function of the amount of overall reed deflection.

From a diagnostics viewpoint, there are various insights that can be gained from these flow experiments. For instance, if the slope in the proportionality region is much less than theoretically expected, it may mean that ports interfere with each other. If the curve is reluctant to reach the horizontal limiting line, it may mean that there is another flow restriction in the valve assembly that has nothing to do with the flexing valve plate. Also, the question of whether it is of value to streamline ports can be investigated in this way.

A drawback of the experiment is that it is difficult to do for large valves because a large compressed gas supply may be needed to maintain steady flow conditions during the time that is necessary to do each measurement. Because in compressible flow it is the pressure ratio across the valve that determines velocity, it may be advisable to use a vacuum tank and vacuum pump with the supply pressure being at ambient conditions. This way, even supersonic velocities may be achieved. The reason it is possible to use steady-state flow measurements in the first place is, of course, the fact discussed in Wambsganss and Cohen (1967) that as a reasonable approximation, the unsteady flow process of a compressor valve during compressor operation can be treated as quasi-steady. Limitations to this were discussed by Trella and Soedel (1974), for example. As compressors become faster in the future, this simplification may become less permissible. The reader can also consult Fleming and Brown (1982), Killman (1972a), Richardson, Gatecliff, and Griner (1980), and Weiss and Boswirth (1982).

9.3.3.2 Effective Valve Force Areas

While one would expect that the integrated force acting on a valve is a function of lift height and pressure differential across the valve, it is interesting that the data in most cases indeed collapses into the concept of an effective area, which is a function of valve lift, and which gives the force if multiplied by the pressure differential:

$$F = B(w)\Delta p \qquad (9.73)$$

This type of behavior was predicted by the previously discussed simplified theoretical approach.

The implied assumption is that the force area is not a function of Δp. In cases where this simplification does not hold, the relationships among force, pressure, and lift height will have to be mapped out. Note that some investigators (for example, Boswirth, 1982) use the concept of drag coefficient.

The measurement is usually done in a parallel displacement mode, as shown in Figure 9.11. When the data is collapsed, a typical curve as shown in Figure 9.9 results. This data can be applied directly to the understanding of spring-loaded ring valves, but has to be adapted to flexing reed valves by viewing the flexed reed as a summation of small parallel plate elements (Wambsganss and Cohen, 1967). A direct measurement of the force on a flexing reed is, of course, very difficult because a reaction force is provided by the boundary, which has to be taken into account.

No matter what method is used to obtain the effective force area, the expected result is that for zero deflection, the effective force area is a function of the port area covered by the valve. This area will then, in general, decrease with increasing lift, unless a strong Bernoulli effect is present for small displacement, so that the negative pressure in the gap between the valve and valve seat overrides the positive pressure over the port, creating a dip in the curve. In Figure 9.9, the dip in the example curve is due to this effect. For large seat areas, this effect is more pronounced than for small seat areas. A narrow seat design is, in general, preferable. In certain

Sound and Vibration of Compressor Valves

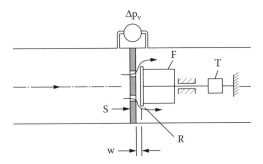

FIGURE 9.11 Effective force area measurement. The same flow measurement arrangement as in Figure 9.12 is used. The pressure differential across the valve, Δp_v, and the total reaction force on the valve plate or reed using transducer T is measured, as w is adjusted.

designs, as, for instance, in free-floating valves, where many individual small discs or hats cover many small circular ports, it is difficult to keep the ratio of seat to port area small, and the effect needs to be monitored (Killman, 1972a,b).

9.4 VALVE DYNAMICS

In the following, typical models of the vibration of discharge and suction valves are discussed. These models are useful for computer simulation, but also allow us to glean important information about the vibration behavior of valves.

Compressor valves in small refrigeration, air, or gas compressors are typically flexible reed valves. In larger compressors, spring-loaded nonflexing ring valves are typically used. In very small compressors, poppet valves are sometimes found.

9.4.1 Poppet Valves

For the purpose of introducing this subject, let us first consider a one-degree-of-freedom poppet valve where the valve plate is rigid, of mass M, and supported by springs, whose effective spring constant is K. We will also assume that there is zero preset in the spring, which is true for practically all automatic valves because preloading would unnecessarily delay the valve opening, which is undesirable from both thermodynamic efficiency and valve vibration and impact viewpoints. Subscripts d and s in Figure 9.12 refer to discharge and suction valves, respectively. However, the following derivation will be done without these subscripts because the underlying principles are common to both valves. Damping in the real valves, caused mainly by a combination of flow resistance and material damping, is assumed to be expressible by equivalent viscous damping factors C, which are estimated and adjusted during the actual simulation. The poppet valve model also applies to spring-loaded, nonflexing ring valves. The earliest work on poppet valve dynamics can be found in Hort (1922). See also Tsui, Oliver, and Cohen (1972), MacLaren (1972), Trella and Soedel (1974), Wollatt (1972), and Plastinin (1984), for example.

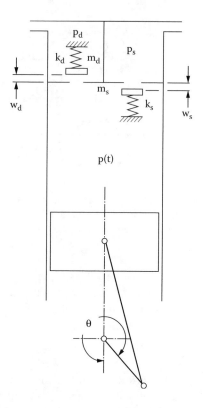

FIGURE 9.12 One-degree-of-freedom suction and discharge poppet valve model.

The equation of motion becomes, utilizing the free body diagram of Figure 9.13,

$$M\ddot{w}(t) + C\dot{w}(t) + Kw(t) = F(t) \qquad (9.74)$$

where $w(t)$ = the valve displacement [mm], $\dot{w}(t)$ = the valve velocity [mm/s^2], $\ddot{w}(t)$ = the valve acceleration [mm/s^2], M = the valve mass [Ns^2/mm^4], K = the effective spring stiffness [N/mm], C = the effective damping [Ns/mm], and $F(t)$ = the valve lift force [N].

Equation 9.74 may also be written as

$$\ddot{w}(t) + 2\zeta\omega_n\dot{w}(t) + \omega_n^2 w(t) = \frac{F(t)}{M} \qquad (9.75)$$

where $\omega_n = \sqrt{\frac{K}{M}}$, the natural frequency of the valve in rad/s, and $\zeta = \frac{C}{2M\omega_n}$, the damping coefficient, which is also the ratio of the actual damping constant C to the critical damping constant $C_c = 2M\omega_n$. Typically, valves operate with $\zeta = 0.1$ to 0.2.

Sound and Vibration of Compressor Valves

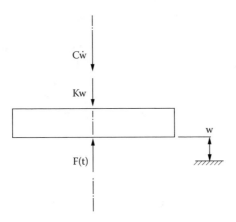

FIGURE 9.13 Free body diagram of a spring-loaded poppet valve plate with damping.

The way the valve lift force is obtained was discussed earlier. A condition that has to be introduced is that

$$w(t) \geq 0, \tag{9.76}$$

which assures that the valve is not driven into its seat. There is another condition on $w(t)$ if a valve stop is to be considered. In this case

$$w(t) \leq w_s \tag{9.77}$$

where w_s is the stop distance of valve travel. Also, the introduction of a coefficient of restitution, ϵ, has been advocated by some:

$$\dot{w}(t)_{\text{(after impact)}} = -\epsilon \, \dot{w}(t)_{\text{(before impact)}}. \tag{9.78}$$

However, in most applications it turns out experimentally that the coefficient of restitution ϵ can be assumed to be zero, which means that all kinetic energy is changed by way of oil film displacement (refrigerating compressors) into heat energy during impact, and there is no rebound. The bouncing of valves one experimentally sees is typically due to gas forces.

While the actual flow forces on, say, a discharge valve are usually a function of time and have to be determined by interactive computer simulations, one can investigate certain idealized situations. Referring to Figure 9.2, the idealized pressure-volume diagram of a reciprocating compressor, let us assume that there is no valve stop and the force on the valve is

$$F(t) = B(w)\Delta p \, U(t - t_2) \tag{9.79}$$

where B(w) is the effective force area (Section 9.3.2), assumed here to be constant, B(w) = B; during an idealized cylinder discharge process, Δp, which is the pressure differential across the valve, and also constant. Furthermore, the step function signifies that at top dead center, we assume that the cylinder pressure immediately reaches, at point 2 (time t_2), $p_d + \Delta p$, where p_d is the discharge pressure, and p_d and Δp stay constant until point 3 (time t_3) when the force on the valve abruptly ceases, and ideally the valve should return to its seat. Therefore, Equation 9.75 becomes

$$\ddot{w}(t) + 2\zeta\omega_n \dot{w}(t) + \omega_n^2 w(t) = \frac{B\Delta p U(t-t_2)}{M}. \tag{9.80}$$

The initial conditions, at $t = t_2$, are that the valve mass is at the seat and at rest:

$$w(t = t_2) = 0 \tag{9.81}$$

$$\dot{w}(t = t_2) = 0. \tag{9.82}$$

Taking the Laplace transformation of Equation 9.80 gives

$$w(s) = \left[\frac{F(s)}{(s+\zeta\omega_n)^2 + \omega_n^2(1-\zeta^2)} \right] \frac{B\Delta p}{M} \tag{9.83}$$

where F(s) is the Laplace transformation of F(t):

$$F(s) = \mathcal{L}(F(t)). \tag{9.84}$$

Because it is very difficult to highly damp valves, typical valves of the damping ratio ζ are from practically zero to 0.2. This means that valves are subcritically damped (critical damping would translate into $\zeta = 1$). Therefore, using convolution when taking the inverse Laplace transformation gives

$$w(t) = \frac{B\Delta p}{\gamma_n M} \int_0^t F(\tau) e^{-\zeta\omega_n(t-\tau)} \sin\gamma_n(t-\tau) d\tau \tag{9.85}$$

where

$$\gamma_n = \omega_n\sqrt{1-\zeta^2}. \tag{9.86}$$

Therefore, in this particular case,

$$F(\tau) = U(\tau - t_2)$$

Sound and Vibration of Compressor Valves

and we obtain

$$w(t) = \frac{B\Delta p}{\gamma_n^2 M}\left\{1-\zeta^2 - \sqrt{1-\zeta^2}\ e^{-\zeta\omega_n(t-t_2)}\cos[\gamma_n(t-t_2)-\phi_n]\right\} \quad (9.87)$$

where

$$\phi_n = \tan^{-1}\frac{\zeta}{\sqrt{1-\zeta^2}}. \quad (9.88)$$

Let us now assume that the valve has negligible damping, so that we may set $\zeta = 0$. Equation 9.87 becomes

$$w(t) = \frac{B\Delta p}{\omega_n^2 M}\{1-\cos[\omega_n(t-t_2)-\phi_n]\} \quad (9.89)$$

or, since $\phi_n = 0$:

$$w(t) = \frac{B\Delta p}{\omega_n^2 M}[1-\cos\omega_n(t-t_2)] \quad (9.90)$$

While these solutions (Equations 9.87 and 9.90) describe the behavior of the valve only in a very approximate sense, they illustrate several important points.

First, Equation 9.87 shows that if there were enough damping to make the vibrations die out, the valve displacement would reach a steady state-value of approximately

$$w_{s.s.} = \frac{B\Delta p}{\omega_n^2 M}, \quad (9.91)$$

which is equal to $(\omega_n^2 M = K)$:

$$w_{s.s.} = \frac{B\Delta p}{K}. \quad (9.92)$$

So ideally we should design the valve stop w_s to be equal to this steady-state value:

$$w_s = w_{s.s.} = \frac{B\Delta p}{K}. \quad (9.93)$$

Earlier we established another condition for w_s based on the effective flow area curve. This means that either B or K should be adjusted to satisfy both conditions.

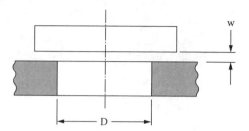

FIGURE 9.14 Flow area geometry of a poppet valve.

For a poppet valve with a circular porthole as shown in Figure 9.14, we obtain as an effective flow area (two orifices in series as in Equation 9.48)

$$(KA)_e = \frac{(KA)_2}{\sqrt{1 + \frac{(KA)_2}{(KA)_1}}} \qquad (9.94)$$

where $(KA)_1$ is the adjusted cross-sectional area of the circular port. Here, for simplicity of argument, we ignore the contraction coefficient, setting $K = 1$, and get

$$(KA)_1 = \frac{D^2 \pi}{4}. \qquad (9.95)$$

Next, $(KA)_2$ is the adjusted cross-sectional area of the gap formed by the edge of the hole and the displacement as:

$$(KA)_2 = \pi D w. \qquad (9.96)$$

For small valve displacement w,

$$(KA)_e \cong \pi D w. \qquad (9.97)$$

For large valve displacement w,

$$(KA)_e \cong \frac{\pi D^2}{4} = \text{constant}. \qquad (9.98)$$

Therefore, there is no point in selecting a stop height that is larger, in an approximate sense, than the value for which

$$\pi D w = \frac{\pi D^2}{4} \qquad (9.99)$$

Sound and Vibration of Compressor Valves

or

$$w = w_s = \frac{D}{4}. \tag{9.100}$$

Equating, for this poppet valve example, Equations 9.93 and 9.100 gives

$$\frac{B\Delta p}{K} = \frac{D}{4}. \tag{9.101}$$

Therefore, if we could estimate the product $B\Delta p$, which is the average force acting on the valve, we should design the spring rate to be

$$K = \frac{4B\Delta p}{D} \tag{9.102}$$

Because of the difficulty of estimating $B\Delta p$, Equation 9.102 is of limited usefulness, but the thought process that led to this equation should be of value.

Next, examining Equation 9.90, which describes the solution of the zero damping case, we see that if we do not introduce a valve stop, the maximum oscillating valve lift would be twice the steady-state valve deflection:

$$w_{max}(t) = \frac{2B\Delta p}{\omega_n^2 M} = \frac{2B\Delta p}{K}. \tag{9.103}$$

This illustrates why it is important to have a valve stop that limits this large excursion.

Another issue is that the valve should be closed as soon as the cylinder pressure starts its rapid decrease at $t = t_3$ (see Figure 9.2). If there is no valve stop, this will only be possible in an approximate sense if the natural frequency ω_n of the valve is such that (see Equation 9.90)

$$\cos \omega_n (t_3 - t_2) = 1 \tag{9.104}$$

or

$$\omega_n (t_3 - t_2) = 0, \pi, 2\pi, \ldots, m\pi \tag{9.105}$$

where $m = 0, 1, 2, \ldots$ or

$$\omega_n = \frac{m\pi}{(t_3 - t_2)} \tag{9.106}$$

and because, in the case of our poppet valve, $\omega_n = \sqrt{K/M}$,

$$\sqrt{K/M} = \frac{m\pi}{(t_3 - t_2)}. \tag{9.107}$$

Assuming that the mass M is a given quantity, and $(t_3 - t_2)$ is given for a particular operating condition, K should be selected such that it satisfies

$$K = \frac{m^2\pi^2}{(t_3 - t_2)^2} M \tag{9.108}$$

for the nearest choice of m = 1, 2, 3, ..., which also satisfies the other selective criteria. Note that m stands for the number of oscillations (or bounces). The problem with this spring rate selection criterion is that it will work only for the operating condition where $(t_3 - t_2)$ is approximately fixed. For air conditioning compressors, where the discharge pressure p_d varies according to the outdoor temperature demand, $(t_3 - t_2)$ is not constant.

This is another reason that a valve stop should be used, because we will then not have to worry about the timing of $(t_3 - t_2)$.

Finally, let us examine the situation where the valve, during the discharge process, lies against the valve stop, having a constant displacement w_s. If properly designed, it is held barely against the stop by an assumed flow force $B(w_s)(\Delta p)_c$, where $(\Delta p)_c$ is assumed to be a constant value. At t_3, the cylinder pressure drops rapidly (see the pressure-time diagram of Figure 9.3). Let us assume that the pressure differential across the discharge valve can be approximately described as

$$\Delta p = (\Delta p)_c \left[1 - \frac{a}{(\Delta p)_c}(t - t_3) \right]. \tag{9.109}$$

The value of a is the slope at time t_3 of the pressure-time diagram of Figure 9.3, with units of [N/m²s]. In this simple model,

$$t \geq t_3 \tag{9.110}$$

when $t = t_3$, $\Delta p = (\Delta p)_c$ and the valve lies against the stop. The pressure $\Delta p = p - p_d$, where p is the cylinder pressure and p_d is the discharge pressure, rapidly becomes negative. This occurs when

$$\frac{a}{(\Delta p)_c}(t - t_3) = 1 \tag{9.111}$$

or

$$t = \frac{(\Delta p)_c}{a} + t_3. \tag{9.112}$$

This means that the valve, which is fully open at $t = t_3$, should be closed at $t = \frac{(\Delta p)_c}{a} + t_3$ to avoid a backpressure buildup, which would slam the valve into the seat with a higher velocity than ideally expected. Therefore, the question is: how long does it take the valve to return to the seat and how does this compare to the closing criterion?

For simplicity sake, let us set damping of the valve to zero. We must solve Equation 9.75 in the form

$$\ddot{w}(t) + \omega_n^2 w(t) = \frac{F(t)}{M} \tag{9.113}$$

where

$$F(t) = B(w)\Delta p = B(w)(\Delta p)_c \left[1 - \frac{a}{(\Delta p)_c}(t - t_3)\right]. \tag{9.114}$$

This means that when $t - t_3 = 0$, the force will be $B(w)(\Delta p)_c$, which will be just enough to hold the valve against the stop of height $w_s = \frac{B(w)(\Delta p)_c}{K} = \frac{B(w)(\Delta p)_c}{M\omega_n^2}$. During the time the valve is returning to its seat, $B(w)$ will not be constant, but for the sake of illustration, let us assume that it is approximately constant, $B(w) = B$. Thus, we must solve

$$\ddot{w}(t) + \omega_n^2 w(t) = \frac{B(\Delta p)_c}{M}\left[1 - \frac{a}{(\Delta p)_c}(t - t_3)\right] \tag{9.115}$$

with the initial conditions that at $t = t_3$,

$$w(t = t_3) = w_s \tag{9.116}$$

$$\dot{w}(t = t_3) = 0. \tag{9.117}$$

The solution of Equation 9.115 can be shown to be

$$w(t) = \frac{B(\Delta p)_c}{K}\left[1 - \frac{a}{(\Delta p)_c}(t - t_3) + \frac{a}{\omega_n(\Delta p)_c}\sin\omega_n(t - t_3)\right]. \tag{9.118}$$

The initial condition that when $t - t_3 = 0$,

$$w(t_3) = w_s = \frac{B(\Delta p)_c}{K} \qquad (9.119)$$

is satisfied. The valve displacement $w(t) = 0$, which means the valve has returned to its seat, occurs when

$$1 - \frac{a}{(\Delta p)_c}(t - t_3) + \frac{a}{\omega_n(\Delta p)_c}\sin\omega_n(t - t_3) = 0. \qquad (9.120)$$

We may solve for the time $(t - t_3)$ of valve closing by expanding the sine function in a series, keeping only the first two terms:

$$\sin\omega_n(t - t_3) = \omega_n(t - t_3) - \frac{\omega_n^3(t - t_3)^3}{3!} + \cdots \qquad (9.121)$$

Equation 9.120 becomes, therefore,

$$1 - \frac{a\omega_n^2(t - t_3)^3}{6(\Delta p)_c} = 0 \qquad (9.122)$$

or

$$t - t_3 = \sqrt[3]{\frac{6(\Delta p)_c}{a\omega_n^2}}. \qquad (9.123)$$

To keep the closing time $(t - t_3)$ as short as possible, we must select $\omega_n^2 = K/M$ as large as possible (we have no influence on the slope a) by making the mass M of the valve as small as possible or by selecting a reasonably large spring rate K (also keeping in mind, of course, the other design criteria demands on K).

This simple model is based on rather severe assumptions, but it manages to verify and illustrate the conventional wisdom of the design community, which has always known that a spring will aid the proper return of a valve to its seat, unless the mass M can be made very small (springless floater valves typically have a very small mass M (see, for example, Killman, 1972a, b).

The simple model also allows us to estimate impact velocities, which excite the structural elements of the valve assembly and transmit vibrations into the compressor casing and through the isolation springs and suction and discharge pipes into the compressor housing (shell).

When the valve closes, at seat impact, we obtain, by differentiation of Equation 9.118,

$$v_c = -\frac{Ba}{K}[1 - \cos\omega_n(t - t_3)] \qquad (9.124)$$

where $(t - t_3)$ is given approximately by Equation 9.123. We see that the impact velocity decreases with an increasing spring rate K, but not linearly because the natural frequency ω_n is also a function of K. Also, as the slope amplitude "a" of the cylinder pressure-time diagram becomes larger, so does the impact velocity. This is one of the explanations as to why some members of the refrigeration and air conditioning industry ran into valve failure problems when they changed from four-pole induction motors for small compressors to two-pole induction motors, doubling compressor shaft speeds. Assuming that the idealized pressure-volume diagram remains more or less unchanged (there will be some differences in heat transfer, and also in the pressure differential across the valves if the effective flow areas are not changed), this means that the slope magnitude "a" of the pressure-time diagram approximately doubled, and therefore the impact velocities doubled because the valve designs were left unchanged at the time of the changeover.

The valve will also impact the valve stop shortly after the valve opens. The stop is impacted when w(t) of Equation 9.90 is equal to the stop height w_s. Thus

$$w_s = \frac{B\Delta p}{\omega_n^2 M}[1 - \cos\omega_n(t - t_2)] \tag{9.125}$$

or

$$(t - t_2) = \frac{1}{\omega_n}\cos^{-1}\left(1 - \frac{w_s K}{B\Delta p}\right). \tag{9.126}$$

The velocity is obtained by differentiating Equation 9.90:

$$\dot{w}(t) = \frac{B\Delta p}{M\omega_n}\sin\omega_n(t - t_2). \tag{9.127}$$

If the stop height w_s is designed according to Equation 9.93, Equation 9.124 becomes

$$(t - t_2) = \frac{1}{\omega_n}\frac{\pi}{2} \tag{9.128}$$

and the velocity of the stop impact is

$$v_o = \frac{B\Delta p}{M\omega_n} = \frac{B\Delta p}{\sqrt{KM}} \tag{9.129}$$

or, in terms of the stop height:

$$v_o = w_s \omega_n = w_s \sqrt{\frac{K}{M}}. \tag{9.130}$$

For example, if we know what an allowable impact velocity is, say, $v_o = 6$ m/s, and the natural frequency of the valve is, say, $\omega_n = 1200$ rad/s (small refrigeration compressor), then the stop height should not be larger than $w_s = v_o/\omega_n = 0.005$ m.

9.4.2 Flexible Reed or Plate Valves

As in the previously discussed one-degree-of-freedom poppet valve, a valve reed (beam) or plate without a stop or backup plate will oscillate, no matter what, for at least the beginning of the opening phase because the rapidly changing pressure will act like a step or steep ramp function, causing a transient oscillation that may or may not decay quickly, depending on damping and possible instabilities. These instabilities, or tendencies to flutter, perpetuate the initially introduced oscillation, causing the valve to perhaps slam against its seat or at least causing it to continue to oscillate at an amplitude of the same order of magnitude as the initially introduced one.

From a very simplified viewpoint, a valve tends to exhibit this type of continuing oscillation to a greater degree if the effective flow force versus displacement curve has a positive slope somewhere in the lift domain. For well-behaved valves, the flow force decreases continually with lift (negative slope), which means that as the valve reed or plate wishes to return from its overshot position right after opening, it will be opposed by an increasing force as displacement decreases. For a badly behaved valve, one with a positive slope, the returning valve meets with less force as it returns to the seat. A positive slope occurs typically for valves that have a fairly wide seat because the Bernoulli effect, causing a negative pressure over the seat, may in this case start to dominate the positive total pressure (Killman, 1972a, b). This simplified reasoning does not always apply because the nonlinear interactions among valve flow, cylinder thermodynamics, suction or discharge gas oscillations, and valve motion make the motion in certain situations very complex and requires a simulation.

The frequency of valve flutter is typically close to the fundamental natural frequency of the valve, modified somewhat by the presence of flow. For spring-loaded poppet or ring valves, the natural frequency in [Hz] is

$$f = \frac{1}{2\pi} \sqrt{\frac{K}{M}} \qquad (9.131)$$

where K is the effective spring stiffness and M is the mass. Flexible reed valves are of the plate or beam type. The following discussion also applies to plates that are structural elements of the compressor and its housing (end caps). We start with the shell equations of motion presented previously (see also Soedel, 2004).

9.4.2.1 Equations of Motion

A plate is a shell of zero curvature:

$$\frac{1}{R_1} = \frac{1}{R_2} = 0. \qquad (9.132)$$

The third equation of motion, relating transverse forces, uncouples and reduces to, after substituting the strain-displacement relationships, and so forth,

$$D\nabla^4 u_3 + \rho h \ddot{u}_3 = q_3 \tag{9.133}$$

where

$$\nabla^2(\cdot) = \frac{1}{A_1 A_2}\left[\frac{\partial}{\partial \alpha_1}\left(\frac{A_2}{A_1}\frac{\partial(\cdot)}{\partial \alpha_1}\right) + \frac{\partial}{\partial \alpha_2}\left(\frac{A_1}{A_2}\frac{\partial(\cdot)}{\partial \alpha_2}\right)\right]. \tag{9.134}$$

In Cartesian coordinates, $A_1 = A_2 = 1$, $\alpha_1 = x$, $\alpha_2 = y$, and

$$\nabla^2(\cdot) = \frac{\partial^4(\cdot)}{\partial x^4} + 2\frac{\partial^4(\cdot)}{\partial x^2 \partial y^2} + \frac{\partial^4(\cdot)}{\partial y^4}. \tag{9.135}$$

In polar coordinates, $A_1 = 1$, $A_2 = r$, $\alpha_1 = r$, $\alpha_2 = \theta$, and

$$\nabla^2(\cdot) = \frac{\partial^2(\cdot)}{\partial r^2} + \frac{1}{r}\frac{\partial(\cdot)}{\partial r} + \frac{1}{r^2}\frac{\partial^2(\cdot)}{\partial \theta^2}. \tag{9.136}$$

Cutting a narrow strip of constant width b from a rectangular plate gives the equation of motion for a transversely vibrating beam: setting $\frac{\partial}{\partial y} = 0$ (no twisting) and multiplying the resulting equation by the width b of the strip gives

$$EI\frac{\partial^4 u_3}{\partial x^4} + \rho A \ddot{u}_3 = q_3' \tag{9.137}$$

where $D \to EI = \frac{Eh^3 b}{12}$, $\rho h \to \rho A$, $q_3 \to q_3'$ (load per unit length). The boundary conditions reduce also. For the platelike reed valve, we must specify bending moments and Kirchhoff shear at the edges, or slope and displacement. The same is true for the ends of a beam-like valve reed.

9.4.2.2 Natural Frequencies and Modes

Natural frequencies of plates are obtained by setting forcing and damping to zero and obtaining the free vibration equation

$$D\nabla^4 u_3 + \rho h \ddot{u}_3 = 0. \tag{9.138}$$

At a natural frequency,

$$u_3(\alpha_1, \alpha_2, t) = U_3(\alpha_1, \alpha_2)e^{j\omega t}. \tag{9.139}$$

This gives

$$DV^4 U_3 - \rho h \omega^2 U_3 = 0. \tag{9.140}$$

From this equation, and appropriate boundary conditions, we obtain natural frequencies and modes in the form of analytical functions. They could, of course, also be obtained by other means such as the finite element method or by experiments.

For valve reeds that resemble clamped cantilever beams of uniform width, the boundary conditions are, for example, at the clamped end (x = o):

$$u_3(0,t) = 0 \quad \text{or} \quad U_3(0) = 0 \tag{9.141}$$

$$\frac{\partial u_3}{\partial x}(0,t) = 0 \quad \text{or} \quad U_3'(0) = 0 \tag{9.142}$$

and at the free end (x = L)

$$\frac{\partial^2 u_3}{\partial x^2}(L,t) = 0 \quad \text{or} \quad U_3''(L) = 0 \tag{9.143}$$

$$\frac{\partial^3 u_3}{\partial x^3}(L,t) = 0 \quad \text{or} \quad U_3'''(L) = 0. \tag{9.144}$$

The free equation of motion, after eliminating time, becomes

$$EI\, U_3''''(x) - \rho A \omega^2 U_3(x) = 0 \tag{9.145}$$

and has the general solution

$$U_3(x) = A \sin\left(\frac{\lambda x}{L}\right) + B \cos\left(\frac{\lambda x}{L}\right) + C \sinh\left(\frac{\lambda x}{L}\right) + D \cosh\left(\frac{\lambda x}{L}\right) \tag{9.146}$$

where

$$\omega = \frac{\lambda^2}{L^2} \sqrt{\frac{EI}{\rho A}}. \tag{9.147}$$

Substituting this into the four boundary conditions results in the frequency equation

$$\cos \lambda \cosh \lambda + 1 = 0 \tag{9.148}$$

which has the roots $\lambda_1 = 1.875$, $\lambda_2 = 4.694$, $\lambda_3 = 7.855$, ... , etc.

Thus, the lowest (fundamental) natural frequency of a cantilevered reed valve in Hz is given by

$$f_1 = \frac{(1.875)^2}{2\pi L^2} \sqrt{\frac{EI}{\rho A}} \tag{9.149}$$

Sound and Vibration of Compressor Valves

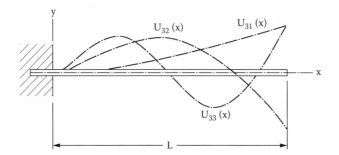

FIGURE 9.15 The first three natural modes of a cantilever beamlike valve reed.

where L = the length, E = Young's modulus, I = the area moment, ρ = mass density, and A = the cross-sectional area.

The natural modes are (λ = 1, 2, 3, ...)

$$U_{3k}(x) = \cosh\left(\frac{\lambda_k x}{L}\right) - \cos\left(\frac{\lambda_k x}{L}\right) - \alpha_k\left[\sinh\left(\frac{\lambda_k x}{L}\right) - \sin\left(\frac{\lambda_k x}{L}\right)\right] \quad (9.150)$$

where

$$\alpha_k = \frac{\cosh\lambda_k + \cos\lambda_k}{\sinh\lambda_k + \sin\lambda_k}. \quad (9.151)$$

The first three natural modes are shown in Figure 9.15.

9.4.2.3 Response of Beam or Platelike Valve Reeds to Forcing

To investigate the forced response of a platelike reed, we solve

$$D\nabla^4 u_3 + \lambda \dot{u}_3 + \rho h \ddot{u}_3 = q_3. \quad (9.152)$$

The solution is expected to be in the form of a modal series (see also the chapters on housing and casing vibrations):

$$u_3(\alpha_1, \alpha_2, t) = \sum_{k=1}^{\infty} \eta_k(t) U_{3k}(\alpha_1, \alpha_2) \quad (9.153)$$

where the natural modes $U_{3k}(\alpha_1, \alpha_2)$ satisfy all boundary conditions, therefore the forced response $u_3(\alpha_1, \alpha_2, t)$ satisfies all boundary conditions.

Substituting this into the equation of motion, we utilize the orthogonality condition, which for a platelike reed is (see Soedel, 2004)

$$\iint_{\alpha_2\,\alpha_1} U_{3k}U_{3p}A_1A_2 d\alpha_1 d\alpha_2 = \begin{cases} 0, & p=k \\ N_k, & p \neq k \end{cases} \tag{9.154}$$

where

$$N_k = \iint_{\alpha_2\,\alpha_1} U_{3k}^2 A_1 A_2 d\alpha_1 d\alpha_2. \tag{9.155}$$

We obtain

$$\ddot{\eta}_k + 2\zeta_k \omega_k \dot{\eta}_k + \omega_k^2 \eta_k = F_k \tag{9.156}$$

where

$$F_k = \frac{1}{\rho h N_k} \iint_{\alpha_2\,\alpha_1} q_3 U_{3k} A_1 A_2 d\alpha_1 d\alpha_2 \tag{9.157}$$

$$\zeta_k = \frac{\lambda}{2\rho h \omega_k}. \tag{9.158}$$

The general solution expression for η_k is identical to the one given for the forced response of a shell. It is

$$\eta_k(t) = e^{-\zeta_k \omega_k t}\left\{\eta_k(0)\cos\gamma_k t + [\eta_k(0)\zeta_k\omega_k + \dot{\eta}_k(0)]\frac{\sin\gamma_k t}{\gamma_k}\right\}$$

$$+ \frac{1}{\gamma_k}\int_0^t F_k(\tau)e^{-\zeta_k\omega_k(t-\tau)}\sin\gamma_k(t-\tau)\,d\tau \tag{9.159}$$

where $\gamma_k = \omega_k\sqrt{1-\zeta_k^2}$. The initial conditions are given by

$$\eta_k(0) = \frac{1}{N_k}\iint_{\alpha_2\,\alpha_1} u_3(\alpha_1,\alpha_2,0)U_{3k}(\alpha_1,\alpha_2)A_1A_2 d\alpha_1 d\alpha_2 \tag{9.160}$$

$$\dot{\eta}_k(0) = \frac{1}{N_k}\iint_{\alpha_2\,\alpha_1} \dot{u}_3(\alpha_1,\alpha_2,0)U_{3k}(\alpha_1,\alpha_2)A_1A_2 d\alpha_1 d\alpha_2. \tag{9.161}$$

When the valve first opens (designating time to be zero here, or adapting it, for, say, a discharge valve, to $t - t_2 = 0$), it will be at zero displacement and zero velocity.

Thus, $u_3(\alpha_1, \alpha_2, 0) = 0$ and $\dot{u}_3(\alpha_1, \alpha_2, 0) = 0$; therefore, $\eta_k(0) = 0$ and $\dot{\eta}_k(0) = 0$. For this situation, Equation 9.159 consists of the convolution integral only.

For example, for a step change in pressure differential across the valve, assuming that the pressure distribution $\Delta p^*(\alpha_1, \alpha_2)$ does not change:

$$q_3(\alpha_1, \alpha_2, t) = \Delta p^*(\alpha_1, \alpha_2) U(t - t_2). \tag{9.162}$$

Equation 9.159 becomes

$$\eta_k(t) = \frac{F_k^*}{\gamma_k^2}\left\{1 - \zeta_k^2 - \sqrt{1-\zeta_k^2}\, e^{-\zeta_k \omega_k(t-t_2)} \cos[\gamma_k(t-t_2) - \phi_k]\right\} \tag{9.163}$$

where

$$F_k^* = \frac{1}{\rho h N_k} \int_{\alpha_2}\int_{\alpha_1} \Delta p^*(\alpha_1, \alpha_2) U_{3k}(\alpha_1, \alpha_2) A_1 A_2 \, d\alpha_1 d\alpha_2 \tag{9.164}$$

$$\gamma_k = \omega_k \sqrt{1 - \zeta_k^2} \tag{9.165}$$

$$\phi_k = \tan^{-1}\frac{\zeta_k}{\sqrt{1-\zeta_k}}. \tag{9.166}$$

This means that each natural mode of the platelike reed responds in a manner similar to that of the one-degree-of-freedom poppet valve responding to a step change in differential pressure, except that the total response $u_3(\alpha_1, \alpha_2, t)$ is the summation of all these one-degree-of-freedom responses, one for each natural mode.

For practical purposes, the infinite summation of Equation 9.153 is limited to only a few modes. If valve flutter or impact is of concern, as few as one or two modes have been used. On the other hand, if valve stresses are to be estimated, ten modes or more may be needed. The reason is that stresses are proportional to second derivatives of $U_{3k}(\alpha_1, \alpha_2)$, which means that the model series will converge slower. Small deflection errors in the $u_3(\alpha_1, \alpha_2, t)$ distribution, which do not appreciably influence the prediction of the deflection response behavior of the valve reed, will cause large errors in the stress prediction.

The results for a transversely vibrating beam are the same as for the plate, except that $A_1 = 1$, $A_2 = 1$, $\alpha_1 = x$, $\alpha_2 = y$. The integration over $\alpha_2 = y$ gives the width of the reed if it is uniform:

$$\int_{\alpha_2} A_2 \, d\alpha_2 = b. \tag{9.167}$$

This is so because for a beamlike reed, the natural modes will not be a function of y. (If a beamlike valve reed experiences torsional components of vibration, it cannot be treated any longer as a beam, but may have to be viewed as a plate.)

We solve Equation 9.137. Equations 9.153 to 9.155 become

$$u_3(x,t) = \sum_{k=1}^{\infty} \eta_k(t) U_{3k}(x) \qquad (9.168)$$

$$\int_x U_{3k} U_{3p} dx = \begin{cases} 0, & p = k \\ N_k, & p \neq k \end{cases} \qquad (9.169)$$

and

$$N_k = \int_x U_{3k}^2 dx. \qquad (9.170)$$

Equation 9.157 becomes

$$F_k = \frac{1}{\rho A N_k} \int_x q_3' U_{3k} dx \qquad (9.171)$$

and Equations 9.160 and 9.161 become

$$\eta_k(o) = \frac{1}{N_k} \int_x u_3(x,o) U_{3k}(x) \, dx \qquad (9.172)$$

$$\dot{\eta}_k(o) = \frac{1}{N_k} \int_x \dot{u}_3(x,o) U_{3k}(x) \, dx. \qquad (9.173)$$

All other equations stay the same.

For example, if the loading on the beamlike reed is

$$q_3'(x,t) = \Delta p^*(x) U(t - t_2) \qquad (9.174)$$

where $\Delta p^*(x) = b(\Delta p)$ is a force per unit beam length, Equation 9.163 stays the same, but Equation 9.164 becomes

$$F_k^* = \frac{1}{\rho A N_k} \int_x \Delta p^*(x) U_{3k}(x) \, dx. \qquad (9.175)$$

9.4.2.4 Approximate Flow Forces on Reed Valves

For simplicity, we will use Cartesian coordinates ($A_1 = 1$, $A_2 = 1$, $d\alpha_1 = dx$, $d\alpha_2 = dy$), and assuming that we have a platelike reed, the forcing function $q_3(x,y,t)$ of Equation 9.152 can be approximately calculated from the flow through the valve ports as first established by Wambsgauss and Cohen (1967), and extended by Schwerzler and Hamilton (1972).

Sound and Vibration of Compressor Valves

Flow forces result from the pressure difference across the valve and the flow through the valve. An approximate theoretical approach to the prediction of these forces for simple valve ports, and an experimental approach, were discussed in Section 9.3.

For a poppet valve, the total force $F(t)$ acting on the valve can be expressed adequately by

$$F(t) = B(w)\Delta p(t) \tag{9.176}$$

where $B(w)$ = the effective force area, and $\Delta p(t)$ = the pressure differential across the valve.

Note that the assumption has been made that steady-state experiments or theoretical models are indicative of what happens during non-steady conditions. Numerous simulations have more or less substantiated this assumption.

Equation 9.176 is sufficient for the one-degree-of-freedom valve of Equation 9.74. However, for cases where the valve reed is not displaced parallel to the valve seat as the valve opens, we have to convert the basic information of $B(w)$ obtained for valve displacement parallel to the valve seat, into a useable form. Let us illustrate this with several examples.

First, consider a valve reed that covers a single, relatively small porthole. In this case we may assume that the force acting on the valve reed in its flexed position is equal to the valve reed in its parallel position (Figure 9.16). Thus, we may assume, for one hole at location (x_1, y_1) and of the effective cross-sectional area ΔA_1, that

$$q_3(x, y, t)\Delta A_1 = B(w(x_1, y_1))\Delta p(t) \tag{9.177}$$

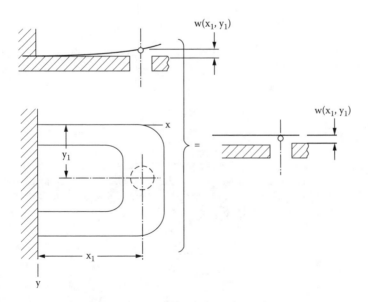

FIGURE 9.16 Single-port valve.

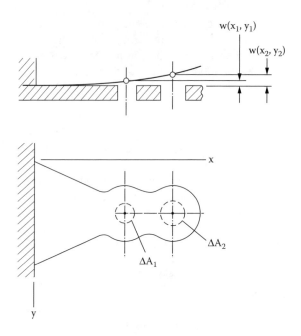

FIGURE 9.17 Two-port valve.

where $B(w(x_1,y_1))$ is the effective force area of the reed when the valve is displaced $w(x_1,y_1)$ parallel to the seat, and $\Delta p(t)$ is the pressure differential across the valve. Equation 9.157 becomes

$$F_k = \frac{1}{\rho h N_k} U_{3k}(x_1,y_1) B(w(x_1,y_1)) \Delta p(t) . \tag{9.178}$$

Next, let us consider the case where we have two individual portholes of cross-sectional areas ΔA_1 and ΔA_2 at locations (x_1, y_1) and (x_2, y_2) as shown in Figure 9.17. In this case, Equation 9.75 becomes

$$F_k = \frac{1}{\rho h N_k}[U_{3k}(x_1,y_1)q_3(x_1,y_1,t)\Delta A_1 + U_{3k}(x_2,y_2)q_3(x_2,y_2,t)\Delta A_2] \tag{9.179}$$

where $q_3(x_1,y_1,t)$ and $q_3(x_2,y_2,t)$ are equivalent pressures. Assuming that for the valve reed displaced parallel to its seat at height $w(x_1,y_1)$, the ratio of the forces over each porthole is equal to the ratio of each port area to the total port area $A = \Delta A + \Delta A_2$, we may write

$$q_3(x_1,y_1,t)\Delta A_1 = B(w(x_1,y_1))\Delta p(t)\frac{\Delta A_1}{A} \tag{9.180}$$

$$q_3(x_2,y_2,t)\Delta A_2 = B(w(x_2,y_2))\Delta p(t)\frac{\Delta A_2}{A} , \tag{9.181}$$

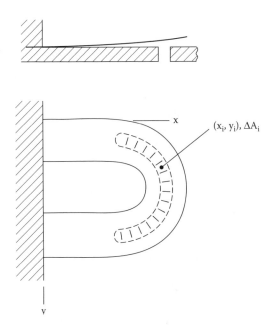

FIGURE 9.18 Multiport or continuous-port valve. Port segment locations are identified by (x_i, y_i).

and Equation 9.179 becomes

$$F_k = \frac{\Delta p(t)}{A\rho h N_k}[U_{3k}(x_1,y_1)B(w(x_1,y_1))\Delta A_1 + U_{3k}(x_2,y_2)B(w(x_2,y_2))\Delta A_2]. \tag{9.182}$$

Finally, let us consider k portholes, or for that matter a continuous port as, for example, shown in Figure 9.18. We divide the port into N small segments of areas ΔA_i at locations (x_i, y_i). This is admittedly a severe approximation, but seems to give results that are not too unreasonable. We are still in the realm of simple models, which is desirable because we wish to understand valve behavior, however approximately. For a later full simulation, the use of finite element-based computational fluid dynamics methods can always be attempted, but we are able to obtain a generalized approximation for F_k:

$$F_k = \frac{\Delta p(t)}{A\rho h N_k} \sum_{i=1}^{N} U_{3k}(x_i, y_i)B(w(x_i, y_i))\Delta A_i \tag{9.183}$$

where $A = \sum_{i=1}^{n} \Delta A_i$.

In summary, we obtain $U_{3k}(x,y)$ and $B(w(x,y))$ from experiments or theory. The ΔA_i are design choices and $\Delta p(t)$ is a simulation input or an assumed value for the sake of design exploration.

9.4.2.5 Natural Frequencies and Modes by Experiment

Forced response predictions require knowledge of the natural frequencies ω_k and the mode shapes $U_{3k}(x,y)$, using Cartesian coordinates in this discussion. Theoretically, k ranges from 1 to infinity, but for practical purposes only the first few modes, perhaps up to k = 3 (in certain cases even only k = 1), are necessary.

In the following we will briefly discuss a procedure that is commonly used to obtain natural frequency and mode information for valve reeds.

If a valve reed is excited with a force that varies sinusoidally in time at a frequency that coincides with the reeds of kth natural frequency, then the reed will vibrate in a characteristic shape that is called the *kth natural mode*. The spatial distribution of the force is of no consequence. A typical mode of a reed valve is shown in Figure 9.19. The shape shown in this figure is frozen in time. While the characteristic shape remains the same, the amplitude of motion changes from the positive maximum to the negative maximum to positive maximum every $2\pi/\omega_k$ seconds. Certain points or lines on the reed will exhibit zero motion and are called nodes. If sand is applied to the reed vibrating in one of its natural modes, it will collect at these nodes and will be thrown off elsewhere. The resulting lines are called *Chladni figures*. This technique is useful because it enables one to look at the node pattern and estimate the natural modes without any measurements. Laser time-averaging techniques can also be used to make node lines visible.

For more precise work, however, the valve reed has to be vibrated at each of its natural frequencies of interest and the characteristic shape has to be measured. For this

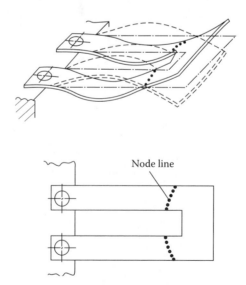

FIGURE 9.19 Node line of the second natural mode of a valve reed.

Sound and Vibration of Compressor Valves

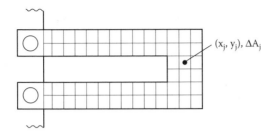

FIGURE 9.20 Locations of mode measurements are identified by (x_j, y_j).

purpose the valve reed is attached to an experimental fixture in the same way it is attached inside the compressor. It is important that the boundary conditions the reed experiences in the compressor are duplicated in the fixture, as much as possible. The excitation device can be an electromagnet.

The electromagnet driven by the signal generator is used to excite the valve while a proximity transducer is used to measure the valve vibration amplitude and phase.

The simplest way a natural frequency can be detected is by observing the displacement amplitude. When this amplitude reaches a local maximum and if sand applied to the reed produces a clearly defined Chladni figure, one has surely found one of the natural frequencies of the valve. Most of the time a natural frequency will also announce itself acoustically. While the reed is vibrating at one of its natural frequencies, the associated natural mode is measured by scanning the entire reed surface with a proximity transducer. The difference between minimum and maximum amplitude is measured for each grid point of the reed (Figure 9.20). Care has to be taken to record the result with the correct plus (+) or minus (−) signs because the measured values are twice the amplitude and are magnitudes only. The + or − signs are, however, easily taken care of if one starts at a particular point with (+) and changes the sign every time a node line is crossed, or one can use the observed 180° phase shift of the transducer output that occurs every time a node line is crossed as a (+) or (−) indicator. Care has to be taken to keep the forcing constant during this procedure, which includes keeping the electromagnet at the same location and distance with respect to the valve reed.

The result will be a natural frequency value ω_k and an array of amplitude data (M points) describing the natural modes $U_{3k}(x, y)$ for each k of interest. Note that the absolute amplitudes of each natural mode are unimportant; only the relative amplitudes along the reed surface constitute the mode. The natural mode is a mode shape.

As can be seen from Equations 9.153, 9.157, and 9.155, constant multiplication factors of natural modes will cancel out. This is the reason why one does not have to worry about the magnitude of the electromagnetic driving force, as long as it remains unchanged during the measurements and does not drive the valve reed up into amplitudes where nonlinear effects in the reed or the measuring system affect the results.

To utilize the data in the computer simulation, either mathematical expressions are fitted to the measurement arrays, which allow us to evaluate, for instance, $\int_x \int_y U_{3k}^2 dx\, dy$ of Equation 9.154 directly. We write

$$\int_x \int_y U_{3k}^2(x,y)dx\, dy = \sum_{j=1}^M U_{3k}^2(x_j, y_j)\Delta A_j. \tag{9.184}$$

Equation 9.183 becomes in this case

$$\ddot{\eta}_k + 2\zeta_k \omega_k \dot{\eta}_k + \omega_k^2 \eta_k = \frac{\Delta p(t) \sum_{i=1}^N U_{3k}(x_i, y_i) B(w(x_i, y_i))\Delta A_i}{A\rho h \sum_{j=1}^M U_{3k}^2(x_j, y_j)\Delta A_j}. \tag{9.185}$$

Note that the (x_i, y_i) locations and elements ΔA_i (N of them) were picked according to porting requirements, while the (x_j, y_j) locations M of them are picked according to reed geometry requirements. Thus, if the (x_i, y_i) and (x_j, y_j) locations do not coincide, one must also measure the natural mode shape at the (x_i, y_i) locations. Many investigators have used the approach to reed valve dynamics developed by Wambsganss and Cohen (1967), or have refined parts of it. See, for example, Wambsganss, Coates and Cohen (1967), Adams, Hamilton, and Soedel (1974), Boyle et al. (1982), Bredesen (1974), Brown et al. (1982), Bukac (2002), Collings and Weadock (2004), Doige and Cohen (1972), Elson, Soedel, and Cohen (1976), Friley and Hamilton (1976), Futakawa et al. (1978), Gatecliff et al. (1980), Gatecliff and Lady (1972), Hamilton (1974), Joergensen (1980), Joo et al. (2000), Khalifa and Liu (1998), Lenz (2000), Libralato and Contarini (2004), Luszczycki (1978 a, b), Ma and Bae (1996), Machu et al. (2004), MacLaren (1972), MacLaren and Tramschek (1972), MacLaren et al. (1978), Madsen (1976) who investigated plastic deformations; and Matos et al. (2002), McLaren et al. (1982), Moaveni et al. (1972), Papastergiou et al. (1980), Papastergiou et al. (1982a, b), Payne and Cohen (1971), Reddy and Hamilton (1976), Richardson et al. (1980), Schwerzler and Hamilton (1972), Simonich (1978), Soedel (1982), Tramschek and MacLaren (1980), Upfold (1972), Wollatt (1972, 1980, 1982).

9.4.3 Pumping Oscillation

Every compressor valve produces flow oscillations, even if it does not flutter. This is due to the fact that in a positive displacement compressor, flow cannot be steady. The valves open, dump mass into the discharge system or take it from the suction system, close, and after one cycle, open again. The precise nature of how the valve opens, when it closes, and so on, is only a perturbation (except when strong valve flutter is present). The fundamental frequency component will be equivalent to the shaft speed of the compressor, assuming that there are not several compressors connected to a common system, causing beat phenomena. In multicylinder compressors,

Sound and Vibration of Compressor Valves

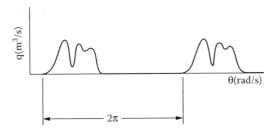

FIGURE 9.21 Periodic volume velocity of period 2π.

if the opening times of, let us say the discharge valve, are distributed such that they are equally phased with regard to each other, the apparent fundamental frequency may be a multiple of the shaft speed, but if one looks closely enough at an experimental spectrum, one will discover that it is actually at the shaft speed frequency because tolerances will make the valves slightly different.

The periodic discharge (or intake) can be viewed as being composed of an infinite sum of harmonic waves. Mathematically we can decompose the volume flow rate curve q(t) through a given valve as a function of time, into an infinite number of Fourier components, all multiples of the fundamental frequency.

Dividing the mass flow rate $\dot{m}$ by the average suction or discharge density ρ_o gives us the volume velocity q, which passes through the suction or discharge valve

$$q = \frac{\dot{m}}{\rho_o} \tag{9.186}$$

where $\dot{m}$ = the suction or discharge mass flow rate (N sec/m), and ρ_o = the average suction or discharge mass density (N sec^2/m^4). Results look typically as shown in Figure 9.21. The Fourier series description of it is given by

$$q = a_o + \sum_{n=1}^{\infty} a_n \cos n\theta + b_n \sin n\theta \tag{9.187}$$

where

$$a_o = \frac{1}{2\pi} \int_0^{2\pi} q \, d\theta \tag{9.188}$$

$$a_n = \frac{1}{\pi} \int_0^{2\pi} q \sin n\theta \, d\theta \tag{9.189}$$

$$b_n = \frac{1}{\pi} \int_0^{2\pi} q \sin n\theta \, d\theta. \tag{9.190}$$

q has been expressed in terms of crank angle θ because the mass flow rate of the computer simulations is usually in terms of crank angle. Realizing that for a constant compressor speed Ω in rad/sec,

$$\theta = \Omega t \qquad (9.191)$$

we may write

$$q = a_o + \sum_{n=1}^{\infty} a_n \cos n\Omega t + b_n \sin n\Omega t \qquad (9.192)$$

where

$$a_o = \frac{1}{T} \int_0^T q \, dt \qquad (9.193)$$

$$a_n = \frac{2}{T} \int_0^T q \cos n\Omega t \, dt \qquad (9.194)$$

$$b_n = \frac{2}{T} \int_0^T q \sin n\Omega t \, dt \qquad (9.195)$$

and where the period T is

$$T = \frac{2\pi}{\Omega}. \qquad (9.196)$$

Let us now introduce the so-called impedances. It is of convenience to use complex notation. We see from Equation 9.187 that all volume velocity input components are of the form

$$q = Q\,e^{j\omega t} \qquad (9.197)$$

where

$$j = \sqrt{-1} \qquad (9.198)$$

$$\omega = n\Omega \qquad (9.199)$$

and where, for the cosine terms

$$Q = a_n \qquad (9.200)$$

$$e^{j\omega t} = R(e^{j\omega t}) = \cos n\omega t \qquad (9.201)$$

and for the sine terms

$$Q = b_n \qquad (9.202)$$

$$e^{j\omega t} = I(e^{j\omega t}) = \sin n\omega t . \qquad (9.203)$$

The input impedance is the ratio of pressure to volume velocity at the input location, let us say, at $x = 0$:

$$Z_o = \frac{p(o,t)}{Q(o,t)} . \qquad (9.204)$$

If Q is harmonic in time, p will be harmonic in time and we can write

$$p(o,t) = P(o)e^{j\omega t} . \qquad (9.205)$$

Thus,

$$Z_o = \frac{P(o)}{Q(o)} \qquad (9.206)$$

or

$$P(o) = Z_o Q(o) . \qquad (9.207)$$

This means that we can write the pressure p at the input location as

$$p(o,t) = Z_o(o)a_o + \sum_{n=1}^{\infty} [a_n Z_o(n\Omega)\cos n\Omega t + b_n Z_o(n\Omega)\sin n\Omega t] . \qquad (9.208)$$

This formula describes the back pressure experienced by the valves because of the presence of gas pulsations. This backpressure must be used in compressor simulation programs to calculate an improved flow distribution q. New Fourier coefficients are calculated using Equations 9.193 to 9.195. Substituting these into Equation 9.208 gives a further improved backpressure p. This iteration is repeated several times until p has converged.

If the objective is only to model the gas pulsation amplitudes and not the pressure feedback, the procedure described can stop at the first go around (no iteration) and will give reasonable results for noise calculation purposes. However, if the energy loss due to gas pulsations is to be determined accurately, the iteration process becomes important. From a simplified viewpoint, the backpressure should be as small as possible over the opening time of a valve because the energy needed to overcome it represents an efficiency loss.

The iteration model was first developed by Elson and Soedel (1974) and extended by Kim and Soedel (1990 a). It was found to converge rapidly in most cases. In the few cases where reasonable convergence was not achieved, a slowing down of the iteration process, by using only a certain average of the new and old predictions, always eventually resulted in a satisfactory convergence. Sometimes it is possible that the solution will not converge because the system is not modeled realistically. For instance, if damping in the discharge system is neglected and a resonance occurs, the impedance approach will give infinite pressure amplitudes, which are clearly not realistic and will obviously not lead to a convergence of the iteration. An unrealistic valve model, for instance one where the valve displacement is allowed to increase without bound, may lead to similar difficulties.

Physically, each of the harmonic components acts like a volume source pulsing at its particular harmonic of the fundamental frequency. These volume source pulsations will excite the various resonances of the discharge or suction system. These resonances occur at the natural frequencies of the discharge gas system or intake gas system, slightly modified by damping or by the fact that there is a mean velocity (which may produce a usually negligible Doppler effect). It is important to stress that the gas oscillations will always occur at multiples of the compressor running speed. It is only the pulsation amplitude that is determined by how close any one of the excitation harmonics is to a resonance frequency.

Often the gas pulsation amplitudes in the manifolding of a hermetic system are quite large, perhaps on the order of 100 to 180 dB at certain frequencies. These pulsations may cause the pipes and the hermetic shell to vibrate, and again there is the opportunity for a resonance, except that it is now structural. The surface vibrations of the pipes or shell radiate the noise, which we hear.

In an air compressor with open-air intake, the harmonic volume sources act directly on the atmosphere, with no manifolding to intervene. Usually the manifolding structure will cause a transmission loss unless the metal of the structure is excited. This may occur at a structural resonance condition where excitation is quite easy. Thus, in general, open intake valves create noise, which is many times louder than when it has to go through piping. It may be possible to use the intake filter as a muffler.

Because our options are limited at the source of this type of noise, except for the elimination of possible valve flutter, efforts will have to be directed toward either preventing structural resonances, damping them out, or preventing the harmonic volume sources from exciting the gas resonances in the manifolding system and hermetic housing interior by using gas pulsation mufflers.

It was found, for instance, that the first volume the valve experiences when it dumps gas into the discharge system or takes it from the suction system is very important. This volume acts, primarily speaking, as a temporary storage tank (an accumulator), tending to equalize the periodic pressure that the periodic volume flow wants to create.

9.4.4 Valve Stops and Damping

Because flutter modulates the flow, we expect a strong sound component at the flutter frequency to appear in the frequency spectrum, for example, as shown in Figure 9.22a for an air compressor intake system exposed to the surroundings (see also Trella and Soedel, 1972). In the case of hermetic systems, Tojo et al. (1980) showed that a strong sound component at the flutter frequency will also be present if it interacts

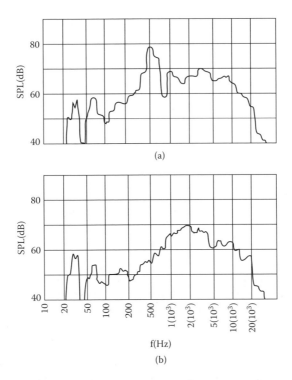

FIGURE 9.22 Typical sound pressure level spectra of an air compressor where the suction valve is exposed to the atmosphere: (a) the valve does not have a valve stop and flutters, (b) a valve stop that suppresses flutter is introduced.

with an efficient sound radiator like a shell. Because the flutter frequency is only a weak function of the compressor speed, we can often identify it by running the compressor at different speeds, as long as we do not mistake it for one of the gas resonances, which of course will not change significantly with compressor speed.

As already mentioned, the easiest way to eliminate flutter and its noise is to introduce a stop, as illustrated in Figure 9.22b (see also Dhar and Soedel, 1978a, b). Experimental evidence of how a motion limiter can suppress flutter was, for example, given by Luszczycki (1978a, b). Figure 9.23 shows a typical suction valve lift curve where the stop height is held constant, but the operating conditions vary. It also illustrates the important point that valve stops are often only optimal for a certain operating condition.

Valve stops should be designed as discussed earlier—they should be at a height at which further opening would not contribute to a significant increase in effective flow area. This is the energy-efficient way of designing a stop. From a flutter viewpoint, the stops should be at the lift height the valves would want to occupy if flutter is completely damped out. In a good design, these requirements do not conflict at the design stage, but we have to remember that there is a multitude of other operating conditions, especially in refrigeration compressors, where the valve does not even reach the stop. Thus, we may have flutter in certain operating conditions, even while

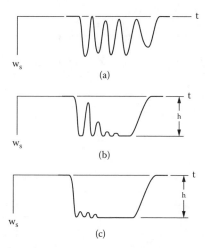

FIGURE 9.23 Typical suction valve displacement for different operating conditions. In the top figure, the valve plate does not contact the valve stop because of insufficient flow force (h = stop height) at that operating condition.

we have suppressed it effectively at others. A second way of suppressing flutter is to dampen it out. A penalty may be somewhat slower opening or closing. Unfortunately, it is often difficult to introduce the required amount of damping. Typical material damping of metals is much too low. Sandwich-constructed valve reeds are suggested, where a viscoelastic layer is sandwiched between two metal reeds. The viscoelastic material must be able to withstand the hostile environment of the compressor. It was suggested that perhaps a sandwich of two steel layers with a copper layer in between could be used for flexing reeds. The hysteresis loss of copper during deformation provides the increased damping. Another possibility is to introduce mechanical friction as the valve reed flexes. This is sometimes provided by backup reeds.

A purely fluid mechanical approach to flutter, centered probably around a redesign of the seat (especially reducing the seat width if it is too large), may also help in some cases.

For suction valves, it is often very difficult to introduce stops, and the reed plate is allowed to flutter. The introduction of an effective muffler is then a must, and it will be discussed in the following sections.

9.4.5 Simulation of Valve Motions

The equations presented so far are valuable in their own right for understanding the parameters that govern valve vibrations. But there are interactive processes going on in a compressor that may make it desirable, or even necessary, for the sake of accuracy of prediction, to do a compressor valve motion simulation on the computer. Here, we will discuss an approach pioneered by Wambsganss and Cohen (1967), which is still the basis for more modern simulations. There were other valuable simulation attempts before Wambsganss and Cohen, but none were as successful in predicting valve motion.

Sound and Vibration of Compressor Valves

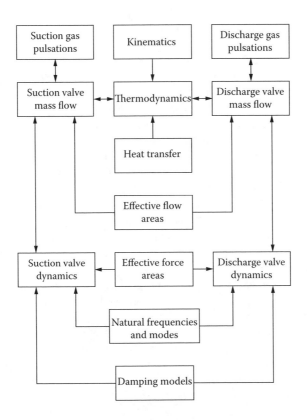

FIGURE 9.24 Typical structure of a compressor simulation.

Also contributing to this approach in the early years were Wambsganss, Coates, and Cohen (1967), Payne and Cohen (1971), MacLaren, Tramschek, and Kerr (1974), Benson and Ucer (1972), Brablik (1972), MacLaren, and Tramschek (1972), Schwerzler and Hamilton (1972), Soedel, Padilla-Navas, and Kotalik (1973), Elson and Soedel (1974), and others. Many of the references cited at the end of Section 9.4.2 contain contributions to the development of valve motion computer simulations.

While an accurate prediction of valve motion is the real goal, the simulation is constructed philosophically around the prediction of the cylinder pressure as a function of cylinder volume. This is depicted in Figure 9.24 as the box labeled *thermodynamics*, with the box labeled *kinematics* supplying the cylinder volume as a function of time. The pressure in the cylinder is a function of the mass that is already in the cylinder, plus any gas mass taken in during the suction process, and minus any gas mass that flows out through the suction, respectively, discharge valves. In Figure 9.24 this is depicted by the boxes labeled *suction valve mass flow* and *discharge mass flow*. The mass flow in or out of the cylinder is a function of the instantaneous suction or discharge valve lift, which is depicted by the boxes labeled *suction valve dynamics* and *discharge valve dynamics*. The mass flows are not only a function of the valve lifts and thus the instantaneous flow area, but also of the pressure differentials across the valves. Thus, the instantaneous cylinder pressure must be known, but the instantaneous suction and discharge pressure must be simulated. This is depicted by the boxes labeled *suction gas pulsations* and *discharge gas pulsations*. (Gas pulsations will be discussed later in chapter 10.) Thus, these five boxes are the heart of the simulation.

The other line boxes depict auxiliary calculations or experiments to establish effective flow and force areas, natural frequencies and modes, and inputs such as damping and polytrophic indices.

Experiments that measure cylinder pressure to establish polytrophic indices, suction and discharge manifold pressures to establish gas damping coefficients, and valve motion to establish valve damping in a prototype compressor are, in the absence of fundamental modeling, often necessary to match the simulation to the experimental evidence. This is typically done at one operating condition. It is then argued that these coefficients do not change appreciably under different operating conditions, and this is typically verified for the compressor that is being studied. It is also argued that these coefficients do not appreciably change for perturbations of the compressor design under investigation, and therefore the effect of changing certain geometric design parameters (valve shape, port design, muffler configurations, and so on) can be predicted.

As simulation models evolve, fewer auxiliary experiments are necessary. The same with prototype measurements. If a realistic heat transfer model is attached to the simulation, the polytrophic process model can be replaced by a first law of thermodynamics model, and an experimental determination of a polytrophic index becomes unnecessary. Of course, auxiliary experiments may have to be added for the determination of heat transfer coefficients.

9.5 STRESSES IN VALVES

While a discussion of valve stresses may go beyond what might be considered under the heading of sound and vibrations of compressors, it is included here (1) because of the importance of valve reliability and (2) because valve motion is a vibration and sound source and the study of stress waves helps us understand it.

In valve stresses, one distinguishes between various types of stresses, such as bending or flexural stresses, impact stresses, stress waves, plate stresses, hoop stresses, impact stresses of the bending kind, static seating stresses, and so on. One may ask why is this so—is stress not simply stress? Indeed, this is true; any point in a valve reed or plate is, in general, subjected to a three-dimensional stress state. The various terms mainly reflect the approximations that are involved in theoretically obtaining these stresses, using perhaps wording that reflects how these stresses are produced or the way they are interpreted. This is illustrated in Figure 9.25. These distinctions are by no means trivial, especially because approximations, if they are good approximations, provide physical insight.

9.5.1 BENDING STRESSES

The approximation involved in the bending stress model of a valve reed or plate is typically that transverse stresses are zero. Strictly speaking, this is only true at a free surface point. It is not true at an impact contact point. However, in a time frame

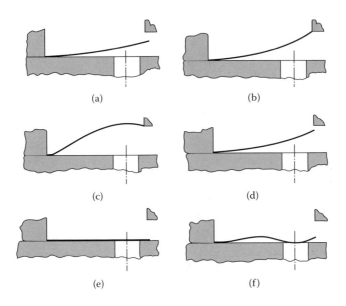

FIGURE 9.25 Examples of stress and deformation states in a vibrating reed valve: (a) bending stresses during opening motion, (b) impact stress when the reed contacts the motion limiter, (c) reversed bending as the reed bends beyond the motion limiter, (d) bending stresses during downward motion, (e) impact stresses as the reed contacts the valve seat, (f) static bending stresses; the reed is partially deflected into the valve port.

that is long enough that three-dimensional effects due to impact, where the surface load is high, have disappeared, we may assume transverse stresses everywhere to be zero. This simplifies any mathematical model greatly because the stress–strain relationships, which in three dimensions are

$$\varepsilon_x = \frac{1}{E}[\sigma_x - \mu(\sigma_y + \sigma_z)] \qquad (9.209)$$

$$\varepsilon_y = \frac{1}{E}[\sigma_y - \mu(\sigma_x + \sigma_z)] \qquad (9.210)$$

$$\varepsilon_z = \frac{1}{E}[\sigma_z - \mu(\sigma_x + \sigma_y)] \qquad (9.211)$$

$$\varepsilon_{xy} = \frac{\tau_{xy}}{G} \qquad (9.212)$$

$$\varepsilon_{xz} = \frac{\tau_{xz}}{G} \qquad (9.213)$$

$$\varepsilon_{yz} = \frac{\tau_{yz}}{G} \qquad (9.214)$$

become with this simplification, and also setting $\varepsilon_{xz} = \varepsilon_{yz} = 0$,

$$\varepsilon_x = \frac{1}{E}(\sigma_x - \mu\sigma_y) = -\alpha_3 \frac{\partial^2 u_3}{\partial x^2} \tag{9.215}$$

$$\varepsilon_y = \frac{1}{E}(\sigma_y - \mu\sigma_x) = -\alpha_3 \frac{\partial^2 u_3}{\partial y^2} \tag{9.216}$$

$$\varepsilon_{xy} = \frac{\tau_{xy}}{G} = -\alpha_3 \frac{\partial^2 u_3}{\partial x \partial y}. \tag{9.217}$$

See Soedel (2004) for details. The stress approximation is illustrated in Figure 9.26. Equations 9.215 to 9.217 also show how the strains are related to plate valve deflections u_3, measured or calculated; α_3 is the distance normal to the neutral surface.

In the case of flexible reeds, which resemble beams, one simplifies further by assuming that σ_y and τ_{xy} are negligible, x being defined as the beam axis. In this case, we simply have

$$\varepsilon_x = \frac{\sigma_x}{E} = -\alpha_3 \frac{\partial^2 u_3}{\partial x^2}. \tag{9.218}$$

In spring-loaded ring valves, the major flexing elements are the springs. They may be classical coil springs or they may be of a complicated shape. Impact stresses are not of concern when sizing coil springs; rather, the springs are usually designed

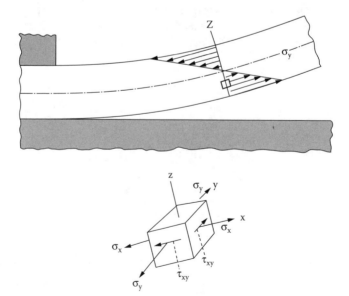

FIGURE 9.26 Typical assumptions when analyzing or measuring bending stresses of vibrating reeds: $\sigma_z = 0$, $\tau_{xz} = 0$, $\tau_{yz} = 0$; only σ_x, σ_y and τ_{xy} are considered.

for a maximum static or dynamic deflection. However, spring-loaded valve plates do break, even while bending-type flexures seem relatively small at first glance. This is due to the fact that these plates do bend because the spring supports are spaced around the ring and are not continuous.

Please note that simulations that are sufficient to predict valve deflections with acceptable accuracy are often insufficient to predict stresses mainly, again, because of the fact that stresses are proportional to the curvature of flexure, which amounts to saying that they are proportional to the second derivative of displacement. Thus, if modal expansion is used in stress analysis, many more natural modes are required than is usual for displacement predictions. Simple forcing models may have to be expanded. If finite elements are being used, the mesh may have to be relatively small.

When reeds deviate from the cantilever beam and become cantilever plates, which means that bending normal to the cantilever axis cannot be ignored, a finite element analysis is usually used to obtain natural modes and frequencies, which can then be used in the modal expansion technique to obtain bending deflections and stresses for the forced situation.

Because of the accuracy problems mentioned above, it is recommended that bending stress predictions be verified by direct strain measurements in an operating compressor using strain gauges. From a historical viewpoint, Lowery and Cohen (1962, 1963, 1971) and Gluck, Ukrainetz, and Cohen (1964) were the first to successfully measure reed valve strain, and thus stress, in an operating, fractional-horsepower refrigeration compressor. Previously, strain gauges were able to survive the hostile environment on, for example, a compressor discharge valve for only a few seconds—not enough time to take useful measurements.

9.5.2 Impact Stresses

Impact stresses occur in reed- or plate-type valves locally at the point where the reed tip contacts the stop, and in the seat zone when the reed impacts the valve seat at closure. The reed tip impact stress during stop contact is of importance, because a well-designed stop is at half of the displacement magnitude that the reed would reach without the stop, and thus the impact velocity may be fairly high.

Seat impact is even more important. This has long been recognized, and an empirically established, so-called *setting velocity* has been formulated; see, for example, Chlumski (1965), for impacting rigid ring valves. This setting velocity, v_s, has been defined as the amplitude of valve displacement as determined by the stop multiplied by the compressor rotational frequency in [rad/s]:

$$v_s = \frac{\pi}{30} NH \tag{9.219}$$

where N = the crank speed of the compressor [RPM] and H = the maximum valve displacement in inches. While it has been recommended that it should not exceed 4 to 9 in/sec, one should note that the setting velocity is not even close to the true impact velocity of a valve reed or plate. It is a correlation parameter based on the fact that some time ago (before World War II) investigators correlated valve failure with amplitude of motion and compressor speed. Because for typical modern high-speed refrigeration compressors, $v_s =$ 15 to 24 in/sec, exceeding the earlier allowable limit considerably, a more detailed examination of the true allowable impact velocities is desirable.

Impact fractures show cracks in or close to an impact zone. Classical fatigue failure analysis of the bending type would expect a cantilever reed to break close to its clamped boundary or at some other area of bending stress concentration, but not at the impact point.

Impact stresses in valves are broadly defined as stresses caused by valve reeds or rings impacting a stop or seat. This definition is often insufficient because it does not distinguish between impact stresses that occur in the valve, particularly in the vicinity of the impact region, during and shortly after impact (the first phase), and impact stresses that are plate- or beam-type bending stresses, which are caused by impact, but emerge in their developed form only some time after the impact process (the second phase).

The essential difference is one of time scale, which in turn requires different philosophies of interpretation and mathematical description. The impact phenomenon during the first phase takes place during a time span that is typically measured in microseconds. Simplifications that are used in the bending theory do not apply, and wave-type theories have to be used. While a valve strikes its seat, highly localized normal forces arise, which serve to prevent interpenetration of the valve and seat. Relative tangential motion of the valve and seat is resisted by frictional forces, which are in proportion to the normal forces. These surface impact forces then propagate away from the impact point at high speed, traveling through both the valve and seat as combinations of compression or tension (P) waves and shear (S) waves. An example of wave progression is given in Figure 9.27 (see Kim and Soedel, 1986, 1988).

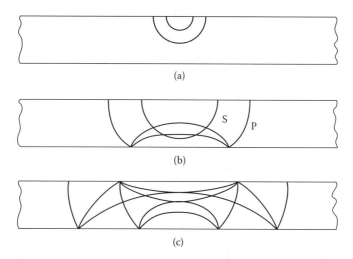

FIGURE 9.27 Typical pressure (P) and shear (S) wave fronts shortly after a point impact: (a) $t = 2\Delta t$, (b) $t = 5\Delta t$, (c) $t = 8\Delta t$, where Δt is on the order of one microsecond or less. The P-wave generates after reflection from the bottom free surface both a P- and an S-wave, as does the S-wave.

Sound and Vibration of Compressor Valves

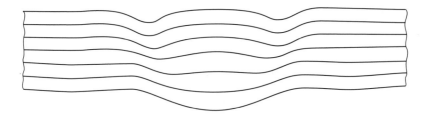

FIGURE 9.28 Typical transverse plate deflection shortly after impact.

See also the work of Boswirth (1980b) for a shear wave interpretation. On the surface, these wave forms combine to give rise to a surface wave called the *Rayleigh wave*. There is also a fourth wave, commonly called a *P-generated S-wave*, whose front is a line that is located at the intersection of the P-wave front with the reed surface and is tangential to the S-wave front. A typical transverse deflection shortly after impact is shown in Figure 9.28.

Eventually, the waves combine to form the bending deflection of the plate after impact. The impact stress analysis during the second phase covers the time after impact contact is completed and ignores local effects in the impact zone. In this case, the common bending simplifications of thin beam or plate theory can be made. Typically, the bending stresses away from the tip, but caused by the tip of a valve reed hitting the seat or a stop, are of this type. It is common that one finds these referred to as impact stresses, but it would be better to call them bending stresses caused by impact, or impact stresses of the second phase.

There is, of course, a smooth transition from the impact stresses of the first phase to those of the second phase. At first, all stress activity is confined to the local zone in the vicinity of the impact, with stress waves spreading out, reflecting from free surfaces, and so on. Some of these waves are compressive or tensile, and others are of a shear type. The rest of the valve reed is still stress free. As the stress waves spread out farther into the rest of the reed at their respective wave speeds, wave components that reflect between the two free surfaces decay rapidly, and only stress waves parallel to the reed surface remain. These eventually create the dynamic bending stress distribution.

Impact stresses of the first phase, as they occur in valves, have been analyzed under the assumption that in certain situations shear and Rayleigh waves may be ignored and that compression-tension type waves move as plane waves (colinear impact) through the valve plate and seat (for example, see Soedel, 1974; Pandeya and Soedel, 1978; Kim and Soedel, 1986; and Paczuski, 2004).

In this model, there is no appreciable velocity gradient transverse to the impact direction and the valve hits the seat squarely, all points contacting at the same time. This is a good assumption in the case of rigid ring plate valves where impact velocities are more or less uniform, except possibly at the edges of ports or when the plate tumbles. In the case of a flexing reed, impact velocities are approximately equal to the product of the time derivative of the first mode participation factor and the first mode, and are therefore not uniform. However, as a first approximation, colinear impact can be assumed.

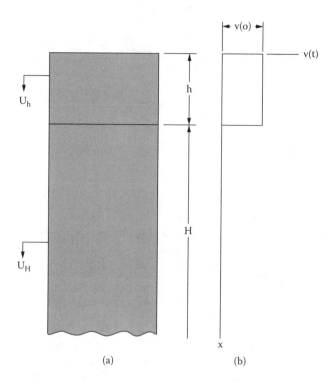

FIGURE 9.29 (a) Valve reed or plate in contact with the valve seat during impact. (b) Initial velocity distribution.

While the treatment includes both the flexible reed and the spring-loaded rigid valve plate, both impacting the seat, it is awkward to carry both terms throughout the paper. The term *valve plate* will therefore (in the following) also mean *valve reed*, and *impact velocity* will mean, in the case of the reed, *tip velocity* or *maximum velocity*.

Let the thickness of the valve plate be h and the thickness of the seat be H. The plate impacts the seat with velocity $v_h(o)$. The seat is at rest with $v_H(o) = 0$. The coordinate origin is the upper edge of the plate, away from the impact point (Figure 9.29). The governing equation for the plate is, after impact contact has been made, according to section 7.1.7,

$$\frac{\partial^2 u_h}{\partial t^2} = c_h^2 \frac{\partial^2 u_h}{\partial x^2}. \tag{9.220}$$

The governing equation of the seat is, after impact contact has been made,

$$\frac{\partial^2 u_H}{\partial t^2} = c_H^2 \frac{\partial^2 u_H}{\partial x^2}, \tag{9.221}$$

Sound and Vibration of Compressor Valves

where u_h and u_H are the elastic displacement of the plate and seat during impact. In the following, the meaning of the various symbols will be: h = thickness of valve plate or reed [in], H = thickness of valve seat [in], v(o) = velocity of valve plate at start of impact [in/sec], u = elastic displacement [in], t = time [sec], x = coordinate [in], c = wave speed [in/sec], f, f′ = waves traveling in a positive direction where the prime means *derivative*, g, g′ = waves traveling in a negative direction, E = Young's modulus [lb/in^2], σ = stress [lb/in^2], ρ = mass density [lb sec^2/in^4], and the subscript h means valve plate and the subscript H means valve seat.

The solutions to these two-wave equations are

$$u_h(x,t) = f_h(x - c_h t) + g_h(x + c_h t) \tag{9.222}$$

$$u_H(x,t) = f_H(x - c_H t) + g_H(x + c_H t) . \tag{9.223}$$

Stresses are given by

$$\sigma_h(x,t) = E_h \frac{\partial u_h}{\partial x} = \rho_h c_h^2 [f_h'(x - c_h t) + g_h'(x + c_h t)] \tag{9.224}$$

$$\sigma_H(x,t) = E_H \frac{\partial u_H}{\partial x} = \rho_H c_H^2 [f_H'(x - c_H t) + g_H'(x + c_H t)] . \tag{9.225}$$

Velocities are given by

$$v_h(x,t) = \frac{\partial u_h}{\partial t} = c_h [-f_h'(x - c_h t) + g_h'(x + c_h t)] \tag{9.226}$$

$$v_H(x,t) = \frac{\partial u_H}{\partial t} = c_H [-f_H'(x - c_H t) + g_H'(x + c_H t)] \tag{9.227}$$

where

$$c_h = \sqrt{\frac{E_h}{\rho_h}} \tag{9.228}$$

$$c_H = \sqrt{\frac{E_H}{\rho_H}} . \tag{9.229}$$

The functions f_h or f_H represent waves traveling in the positive direction with velocity c_h or c_H, while the functions g_h or g_H represent waves traveling in the opposite directions. Primes indicate derivatives with respect to the arguments of these functions. For convenience, these arguments are dropped in the following. Initial velocity conditions are

$$c_h(-f_h' + g_h') = v_h(o) \tag{9.230}$$

for $t = 0$ and $0 \leq x \leq h$,

$$-f'_H + g'_H = 0 \tag{9.231}$$

for $t = 0$ and $h \leq x \leq H$, and reflecting the fact of a zero initial stress state before impact,

$$f'_H + g'_H = 0 \tag{9.232}$$

$$f'_h + g'_h = 0. \tag{9.233}$$

From Equation 9.230 and 9.231, we obtain the initial magnitudes of the traveling waves:

$$f'_h = -\frac{v_h(o)}{2c_h} \tag{9.234}$$

$$g'_h = \frac{v_h(o)}{2c_h} \tag{9.235}$$

$$f'_H = 0 \tag{9.236}$$

$$g'_H = 0. \tag{9.237}$$

Because a free boundary cannot support a normal stress,

$$f'_h + g'_h = 0 \tag{9.238}$$

at $x = 0$, but at any time t. This gives

$$f'_h = -g'_h \tag{9.239}$$

and establishes the fact that at the free end, wave components reflect with equal amplitude, but opposite in phase.

At the impact location during impact contact, stresses on the valve plate boundary have to equal stresses at the valve seat boundary. Also, particle velocities have to be identical. Thus, at $x = h$,

$$\rho_h c_h^2 (f'_h + g'_h) = \rho_H c_H^2 (f'_H + g'_H) \tag{9.240}$$

$$c_h(-f'_h + g'_h) = c_H(-f'_H + g'_H). \tag{9.241}$$

These equations give

$$f'_H = \frac{f'_h}{2}\left[\frac{\rho_h}{\rho_H}\left(\frac{c_h}{c_H}\right)^2 + \frac{c_h}{c_H}\right] + \frac{g'_h}{2}\left[\frac{\rho_h}{\rho_H}\left(\frac{c_h}{c_H}\right)^2 - \frac{c_h}{c_H}\right] \qquad (9.242)$$

$$g'_H = \frac{f'_h}{2}\left[\frac{\rho_h}{\rho_H}\left(\frac{c_h}{c_H}\right)^2 - \frac{c_h}{c_H}\right] + \frac{g'_h}{2}\left[\frac{\rho_h}{\rho_H}\left(\frac{c_h}{c_H}\right)^2 + \frac{c_h}{c_H}\right]. \qquad (9.243)$$

For identical materials ($\rho_h = \rho_H$, $c_h = c_H$)

$$f'_H = f'_h \qquad (9.244)$$

$$g'_H = g'_h. \qquad (9.245)$$

The physical interpretation of the general case is that, for instance, a component f'_h approaching the impact boundary will split into a transmitted wave component f'_H and a reflected component g'_h due to the change in material properties. In the case of identical materials, no reflection takes place.

The boundary condition at the other side of the valve seat cannot be clearly defined because the valve seat is part of the total cylinder head. It is, however, reasonable to assume that

$$H \gg h \qquad (9.246)$$

and that therefore all waves traveling in the positive x–direction are dispersed and do not return.

Let us investigate the case when f'_h approaches the impact boundary. It will split into a reflected component g'_h and a transmitted component f'_H. The value of g'_H is zero. Thus, the reflection will be

$$g'_h = f'_h \frac{\rho_H c_H - \rho_h c_h}{\rho_H c_H + \rho_h c_h} \qquad (9.247)$$

and the transmitted component will be

$$f'_H = f'_h \frac{2\rho_h c_h}{\rho_H c_H + \rho_h c_h}\left(\frac{c_h}{c_H}\right). \qquad (9.248)$$

The peak impact stress amplitudes in the valve reed and the valve seat are therefore, by properly adding forward moving and reflecting wave components (Soedel, 1974).

$$\sigma_{h\,max} = \sigma_{H\,max} = -\frac{\rho_h c_h}{1 + \frac{\rho_h c_h}{\rho_H c_H}} v(o) \qquad (9.249)$$

The developing impact stress wave is shown in Figure 9.30 for the case of identical materials.

Equation 9.249 is plotted in Figure 9.31. We recognize, taking the case of equal material for valve plate and seat as the point of departure, that in order to reduce

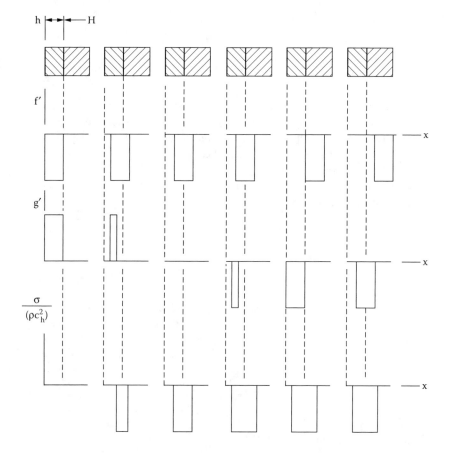

FIGURE 9.30 Pressure wave propagation in the x-direction: f' propagates in a positive direction; g' propagates first in a negative direction, reflects from the free boundary of the valve reed or plate, and then travels, with the opposite sign, also in the positive direction. The two waves add up to $f' + g' = \sigma/(\rho c_h^2)$.

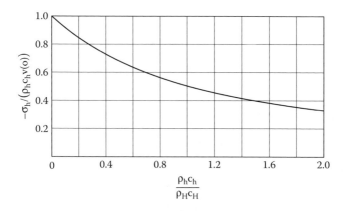

FIGURE 9.31 Compressive stress during impact. The region to the left of $\rho_h c_h/(\rho_H c_H) = 1$ is a *hard* valve seat, to the right it is a *soft* valve seat.

impact stresses in a given valve plate, we should make the seat more elastic and reduce the mass density of the seat material such that

$$\rho_H c_H < \rho_h c_h, \qquad (9.250)$$

and of course make the impact velocity v(o) as small as possible.

It does not matter, as far as impact stress amplitudes are concerned, whether the valve plate is of steel impacting a brass seat or a brass valve plate is impacting a steel seat, if the impact velocity is the same. Note further that the thickness of the valve plate does not enter into the stress relationship. Merely reducing the thickness of the valve plate will therefore not change the impact stresses, again if the impact velocity is the same, which it may not.

It is also possible to plate the seat with a softer material. This was investigated, for example, by Pandeya and Soedel (1978). Such a soft seat cushion reduces the impact stresses both in the valve plate and the valve seat. Unfortunately the problem now is that failure in the seat cushion has to be prevented. In refrigeration compressors, oil that is transported to the valve seat may act as a cushion.

It must be noted that the colinear impact model, even while very helpful to the understanding of how stresses develop during impact, describes the most benign impact condition. It is much more damaging if the valve reed impacts the valve seat obliquely (see, for example, Svenzon (1976) and Dusil and Johansson (1980)).

For additional information, see also Futakawa and Namura (1980), Graff (1975), and Zukas (1982).

10 Suction or Discharge System Gas Pulsations and Mufflers

Due to the intermittent nature of valve flow, gas pulsations are generated in the discharge and suction systems. This chapter discusses mathematical models of these pulsations and the analysis and design of gas pulsation mufflers. A considerable amount of work has been done on this topic. For additional information, see, for example, Akashi et al. (2000), Akella et al. (1998), Brablik (1972, 1974), Chen and Huang (2004), Cossalter et al. (1994), Dahr and Soedel (1978a–b), Elson and Soedel (1972a, 1974), Gatley and Cohen (1970), Groth (1953), Johnson et al. (1990), Kim and Soedel (1992b), Kim (1992), Kim and Soedel (1987, 1988a–b, 1989a–c, 1990 a–c, 1996), Koai and Soedel (1990a–b), Lai and Soedel (1996a–e, 1997, 1998), Lee et al. (1985), Lee and Soedel (1985), Lee et al. (2000), Liu and Soedel (1992, 1994), Ma and Min (2000), MacLaren et al. (1974), Mutyala and Soedel (1976), Nieter and Kim (1998), Park and Adams (2004), Park et al. (1994), Qvale et al. (1972), Roy and Hix (1998), Rauen and Soedel (1980a–b), Singh and Soedel (1974, 1975, 1976, 1978a–b, 1979), Soedel and Soedel (1992), Soedel (1976a–b, 1978, 1992), Soedel et al. (1973), Soedel et al. (1976), Song and Soedel (1998), Suh et al. (1998), Suh et al. (2000), Svendsen (2004), Yoshimura et al. (2002), and Zhou and Kim (1998, 2002).

10.1 REDUCING VALVE NOISE

10.1.1 Dissipative Mufflers

On the simplest conceptual level we find the dissipative muffler. The flow is forced through a number of small holes or some porous material, and oscillations caused by the valve are damped out. An air filter on an air compressor acts as a sort of free dissipative muffler for the intake noise. We get something for nothing—a situation which unfortunately does not happen very often in engineering.

However, dissipative mufflers are not common because deposits may eventually clog them, and even in their virgin state they are an impediment to flow and cause at least some energy loss. Reactive mufflers are more popular. In general, one can divide them into tuned side-branch resonators and low-pass filter mufflers, or combinations of the two.

10.1.2 Side-Branch Resonators

The side-branch resonator works on the principle of the dynamic absorber. It is tuned to an objectionable frequency, or in some of its forms to an objectionable frequency band. It works such that its sympathetic resonance will take oscillatory energy out of the main branch system and transfer it to the side-branch resonator, where it will eventually dissipate into heat. It is not a concept that always works well for refrigeration compressors because the operating conditions change so much, and with them gas temperatures and resonances. Neither does it always work well for automotive compressors, be they air conditioning compressors or positive displacement turbochargers, because the operating speed changes continuously. Of course, one could in principle use a microprocessor to change the tuning frequency of a side-branch resonator as the shaft speed changes or as the operating temperatures change.

10.1.3 Low Pass Filter Mufflers

For compressors of variable speed or variable operating conditions, low pass filter reaction mufflers are recommended. The word *filter* comes from electrical engineering where in circuits we have low and high pass filters. The analogy of vibration isolation applies. The suspension system of an automobile is a low pass filter. It passes very low road frequencies, but attenuates higher road input frequencies up to a certain limit. The same is true for a low pass filter muffler. Above its design frequency, it isolates the periodic noise source, which in our case is the valve, from the rest of the system. There is, of course, a limit to what such a muffler can do, depending on its size and also on the fact that the muffler geometry introduces resonance frequencies of its own at which the muffler obviously will not attenuate. Again, the analogy to a car suspension holds because at higher road input frequencies, resonances of the tire and the suspension structure will make attenuation difficult. A low pass filter muffler does, however, provide a broad frequency range over which it will be effective, and is therefore not very sensitive to speed and temperature changes.

In its simplest form, a low-pass filter is a flow-through Helmholtz resonator as shown in Figure 10.1. Its natural frequency f_n should be designed according to

$$f_n \sqrt{2} < f \qquad (10.1)$$

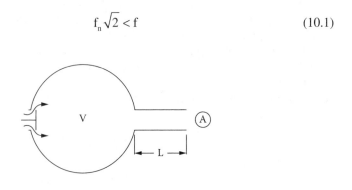

FIGURE 10.1 Flow-through Helmholtz resonator.

where f is the lowest frequency at which we wish to have the attenuation start. The natural frequency itself is given by

$$f_n = \frac{c}{2\pi}\sqrt{\frac{A}{LV}} \quad (10.2)$$

where c is the speed of sound, A is the cross-section of the neck, L is the length of the neck, and V is the volume. At the frequency $\sqrt{2}\,f_n$, there is no attenuation or amplification. As we move with increasing frequency through the excitation spectrum, we see that attenuation becomes larger and larger until we reach a limit where the wavelength becomes so short that the Helmholtz resonator does not act as a simple resonator, but has resonances inside its neck or volume. At this point and beyond, attenuation will exist in bands only, and we should really use the acoustic wave equation to obtain insight into what is going on.

Please note that there is no conflict between the Helmholtz resonator theory and the wave equation theory; it can be shown that the former is a limiting case of the latter.

While one can, as a first-order engineering approximation, look at the valve and the muffler separately, they do of course interact in the sense that the muffler will produce a backpressure on the valve, which may interfere with flow, and many computer simulations treat this interaction. From a practical viewpoint, the backpressure is negligible if we succeed in making the volumes immediately behind the discharge valve and immediately in front of the suction valve large. As a rule of thumb, they should be at least two times larger than the swept cylinder volume. Ideally, these volumes are also the volumes of flow-through Helmholtz resonator reaction mufflers and have to be large simply because we need to make f_n as low as possible for noise attenuation. There is, therefore, no conflict, except that often there is no space available for do it. It is important that enough space be provided for these volumes in the initial design stage.

Note also that this type of muffler requires a second so-called *decoupling volume* behind the neck, if it is connected to a pipe.

Often, several of these mufflers are arranged in series. If the spacing between any two of these mufflers is a pipe section longer than a quarter of a wavelength at the highest frequency of interest, wave theory must be used for calculations, but the Helmholtz concept is still useful for the initial design process.

10.1.4 Impact Noise

Each valve impacts at least once per cycle—namely, when it impacts the valve seat during closing. If there is a stop or backup plate, at least another impact occurs, and if the valve flutters, it may impact several times during its opening cycle.

Mufflers are not effective in dealing with valve impact. Each impact sets up stress waves that travel through the compressor structure. Eventually, as these waves are reflected back and forth through the compressor structure, they will interact with the various boundaries in such a way that they excite the various natural frequencies and mode of the structure.

If impact noise radiation in the vicinity of impact is of interest, a wave travel model will most likely have to be analyzed. But if the total noise radiation of the compressor structure due to impact is of concern, a modal expansion approach may be more informative. In the chapter on the vibration of the compressor casing, such a model was discussed and it was pointed out that vibration response amplitudes, and thus sound pressure levels radiated from the casing or from the housing, are proportional to the momentum transferred from the impacting valve to the casing. This momentum, while distributed, is roughly proportional to the average impact velocity of the valve and the mass of the valve.

What is probably most annoying about impact noise is that it has the potential for a very wide frequency content, starting with noise components at the fundamental compressor speed, and having, most likely, significant components over the entire range of hearing (up to 6000 Hz). Because it is a structure-borne sound, its higher frequency components will travel relatively easily through metallic support springs into shells and foundations.

One measure against the transmission of waves due to impact is to introduce an impedance mismatch at the spring supports or valve seat plate in the form of a rubber or plastic pad, which interrupts metallic contact.

At the source itself, what we can perhaps do is to try to avoid repetitive impacts during the valve opening cycle, try to keep impact velocities and valve masses small, use mismatched valve plate and seat materials (see Soedel, 1974), plastic valves where feasible, or isolate the valve seat area from the compressor head in some way.

10.1.5 Noise Due to Turbulence

Noise due to turbulence is of course always present, but difficult to predict and quantify analytically. To suppress it, valve passages should be as streamlined as possible and perhaps polished. This includes removing all sharp edges, misaligned flow cross-sections, protruding gaskets, and so on, and trying to prevent flow separation. It is important to keep flow velocities low. Noise due to turbulence contributes to the overall sound mainly in the higher frequency ranges, perhaps 3000 to 6000 Hz. It may become amplified by resonances of the gas in the valve passages, yet reaction mufflers do not seem able to deal with it very effectively. Typically, dissipative muffler elements seem to be needed.

10.2 REACTIVE MUFFLERS USING THE HELMHOLTZ SIMPLIFICATION

10.2.1 The Helmholtz Resonator

Any geometric shape that consists of a volume and a short neck, and is filled with a compressible gas or fluid can be viewed as a Helmholtz resonator. If a volume-neck combination is made to vibrate, it behaves essentially as if the gas in the neck is an incompressible plug that vibrates as a whole on the spring provided by the compressible gas in the volume. We can analyze the resulting oscillation if we assume that the compression process in the volume is approximately linear.

Suction or Discharge System Gas Pulsations and Mufflers

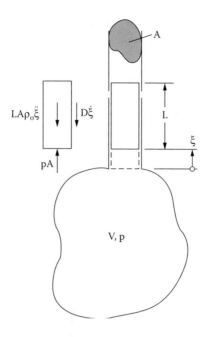

FIGURE 10.2 Helmholtz resonator model.

Let us derive the equation of motion of a Helmholtz resonator. The forces that act on the plug of gas in the neck are shown in Figure 10.2. A positive displacement ξ will cause a volume change,

$$dV = A\xi \tag{10.3}$$

where A is the cross-sectional area of the neck.

Thus the change in pressure is

$$\Delta p = -K_o \frac{A\xi}{V_o} \tag{10.4}$$

where K_o is the bulk modulus of the gas in N/m².

Let us in the following, switch notation, replacing Δp by p, where p is understood to be the pressure change with respect to a mean pressure. This is purely for convenience of notation. Multiplying both sides by the area A gives the dynamic pressure force,

$$pA = -\frac{K_o A^2}{V_o}\xi. \tag{10.5}$$

Note that the total pressure p_T is the average pressure p_o plus the oscillating pressure p

$$p_T = p_o + p. \tag{10.6}$$

Because of energy dissipation due to laminar or turbulent friction, we postulate an equivalent viscous damping coefficient D. The inertia of the plug of gas causes it to resist motion with a force $LA\rho_o\ddot{\xi}$. Thus we get

$$LA\rho_o\ddot{\xi} + D\dot{\xi} + \frac{K_o A^2}{V_o}\xi = 0. \tag{10.7}$$

This is the equation of a one-degree-of-freedom oscillator and the Helmholtz resonator behaves as such. The equation can also be written as

$$\ddot{\xi} + 2\zeta\omega_n\dot{\xi} + \omega_n^2\xi = 0 \tag{10.8}$$

where

$$\omega_n = c_o\sqrt{\frac{A}{LV_o}} \tag{10.9}$$

$$\zeta = \frac{D}{2LA\rho_o\omega_n} \tag{10.10}$$

and where D = the damping coefficient [N sec/m], ζ = the damping factor, ω_n = the natural frequency in rad/sec, and A = the cross-sectional area of the neck [m²]. The length L is an effective length and is given by

$$L = L_G + \sqrt{\frac{\pi A}{2}} \tag{10.11}$$

where L_G = the geometric length of the neck [m]. The approximate end correction takes into account that the gas acceleration in the vicinity of the neck ends are not negligible.

A limitation on the use of Helmholtz resonator theory is the wavelength λ of the highest frequency f_{max} of oscillation, which has to be predicted. It is suggested that for accurate results, the largest element dimension, a, of the resonator should be less than 1/4 of this wavelength λ. Because

$$\lambda = \frac{c_o}{f_{max}} \tag{10.12}$$

we get

$$a < \frac{c_o}{4f_{max}}. \tag{10.13}$$

For less accurate results, this requirement can be relaxed somewhat.

Necks that are not cylindrical are treated by formulating an equivalent cylinder of cross-sectional area A

$$A = \frac{L_G}{\int_0^{L_G} \frac{1}{A(x)} dx} \tag{10.14}$$

where $A(x)$ = the actual geometric area, and x = the coordinate used to define $A(x)$ between $x = 0$ and $x = L_G$.

10.2.2 THE HELMHOLTZ RESONATOR APPROACH APPLIED TO COMPRESSORS

Let us start with the simplest case imaginable. For instance, an air compressor discharge system consists of a cavity into which the discharge valve dumps a certain mass of gas once every cycle. This cavity is connected by a relatively short passage to a large collecting tank that is so large that the pressure in it can be assumed to be constant. This is sketched in Figure 10.3. (The suction side is similar.) This case is analogous to what is called, in vibrations, a *base excited oscillator.*

Let us introduce the subscript d to identify that the development described next is for the discharge side. The free body diagram of the discharge neck mass gives

$$p_d A_d - L_d A_d \rho_{od} \ddot{\xi}_d - D_d \dot{\xi}_d = 0. \tag{10.15}$$

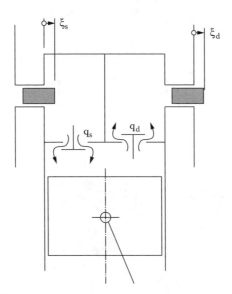

FIGURE 10.3 Air compressor that takes in air through a Helmholtz suction muffler and discharges the compressed air into a large tank through a Helmholtz discharge muffler.

However, the volume change dV_d is now

$$dV_d = \xi_d A_d - \int_0^t q_d dt \tag{10.16}$$

where q_d = the volume velocity through the discharge valve [m³/sec].
Thus

$$p_d = -K_{od}\frac{\xi_d A_d}{V_{od}} + \frac{K_{od}}{V_{od}}\int_0^t q_d dt \tag{10.17}$$

and the equation of motion becomes

$$L_d A_d \rho_{od} \ddot{\xi}_d + D_d \dot{\xi}_d + \frac{K_{od} A_d^2}{V_{od}}\xi_d = \frac{K_{od} A_d}{V_{od}}\int_0^t q_d dt. \tag{10.18}$$

Note that

$$q_d = \frac{\dot{m}_d}{\rho_{od}} \tag{10.19}$$

is the volume velocity through the discharge valve, where $\dot{m}_d$ = the mass flow rate through the valve [N sec/m = kg/sec] and ρ_{od} = the mean mass density in volume V_{od} [N sec²/m⁴ = kg/m³]. The equation of motion can be brought into a more convenient form, for computer simulation:

$$\ddot{\xi}_d + 2\zeta_d \omega_{nd} \dot{\xi}_d + \omega_{nd}^2 \xi_d = G_d \int_0^t \dot{m}_d dt \tag{10.20}$$

where

$$G_d = \frac{c_{od}^2}{L_d \rho_{od} V_{od}} \tag{10.21}$$

$$\omega_{nd} = c_{od}\sqrt{\frac{A_d}{L_d V_{od}}} \tag{10.22}$$

$$\zeta_d = \frac{D_d}{2L_d A_d \rho_{od} \omega_{nd}}. \tag{10.23}$$

Suction or Discharge System Gas Pulsations and Mufflers

This equation can now be solved if the mass flow rate through the valve is known as a function of time. The oscillation pressure is obtained once ξ_d is known.

$$P_d = \frac{c_{od}^2}{V_{od}} \int_0^t \dot{m}_d dt - \frac{\rho_{od} c_{od}^2 A_d}{V_{od}} \xi_d. \tag{10.24}$$

The total discharge pressure experienced by the valve is therefore the cylinder pressure on one side and

$$P_{Td} = P_{od} + P_d \tag{10.25}$$

on the other side.

Very similar development leads to equations for the suction system, if the suction system takes in air from a large tank of mean pressure p_{os}, or from the ambient:

$$\ddot{\xi}_s + 2\zeta_s \omega_{ns} \dot{\xi}_s + \omega_{ns}^2 \xi_s = G_s \int_0^t \dot{m}_s dt \tag{10.26}$$

$$G_s = \frac{c_{os}^2}{L_s \rho_{os} V_{os}} \tag{10.27}$$

$$\omega_{ns} = c_{os} \sqrt{\frac{A_s}{L_s V_{os}}} \tag{10.28}$$

$$\zeta_s = \frac{D_s}{2 L_s A_s \rho_{os} \omega_{ns}} \tag{10.29}$$

$$P_s = \frac{\rho_{os} c_{os}^2 A_s}{V_{os}} \xi_s - \frac{c_{os}^2}{V_{os}} \int_0^t \dot{m}_s dt \tag{10.30}$$

$$P_{Ts} = P_{os} + P_s. \tag{10.31}$$

10.2.3 Steady-State Harmonic Response to a Harmonic Volume Velocity Input and Design Criteria

Here, we are interested in a simplified model. Most likely, the volume velocity q will be periodic. According to Equation 9.192, it can be written

$$q = u_o + \sum_{n=1}^{\infty} (a_n \cos n\Omega t + b_n \sin n\Omega t). \tag{10.32}$$

It is of general interest, therefore, to evaluate the steady-state harmonic response of the Helmholtz resonator gas at its exit to a harmonic input volume velocity $Q_n e^{jn\Omega t}$, which stands for both the $a_n \cos n\Omega t$ (the real part) and $b_n \sin n\Omega t$ (the imaginary part).

Differentiating Equation 10.20 with respect to time and defining the volume velocity of gas plug displacement as

$$q_d = A_d \dot{\xi}_d \tag{10.33}$$

gives

$$\ddot{q}_d + 2\zeta_d \omega_{nd} \dot{q}_d + \omega_{nd}^2 q_d = \omega_{nd}^2 Q_n e^{jn\Omega t} \tag{10.34}$$

where ω_{nd} and ζ_d are given by Equations 10.22 and 10.23. The steady-state solution is expected to be

$$q_d = \tilde{Q}_d e^{jn\Omega t} \tag{10.35}$$

where

$$\tilde{Q}_d = Q_d e^{-j\phi_d}. \tag{10.36}$$

Substituting Equation 10.35 in Equation 10.34 gives

$$\tilde{Q}_d = \frac{\omega_{nd}^2 Q_n}{\left[\omega_{nd}^2 - (n\Omega)^2\right] + 2\zeta_d \omega_{nd}(n\Omega)j} \tag{10.37}$$

or

$$\frac{Q_d}{Q_n} = \frac{1}{\sqrt{\left[1-\left(\dfrac{n\Omega}{\omega_{nd}}\right)^2\right]^2 + 4\zeta_d^2\left(\dfrac{n\Omega}{\omega_{nd}}\right)^2}} \tag{10.38}$$

and

$$\phi_d = \tan^{-1} \frac{2\zeta_d\left(\dfrac{n\Omega}{\omega_{nd}}\right)}{1-\left(\dfrac{n\Omega}{\omega_{nd}}\right)^2}. \tag{10.39}$$

Suction or Discharge System Gas Pulsations and Mufflers

Note that the letter n in ω_{nd}, the natural frequency of the Helmholtz resonator, has nothing to do with the multiplier n of the Ω term, which refers to the harmonic number in the Fourier series (Equation 10.32).

For the flow through Helmholtz resonator to attenuate a particular gas oscillation component Q_n coming from the valve, we require that

$$\frac{Q_d}{Q_n} < 1. \qquad (10.40)$$

According to Equation 10.38, this will be the case if

$$\frac{n\Omega}{\omega_{nd}} < \sqrt{2} \qquad (10.41)$$

or if

$$\omega_{nd} < \frac{n\Omega}{\sqrt{2}}. \qquad (10.42)$$

For example, if we are limited to a certain muffler volume V_{od}, and the cross-sectional area cannot be smaller than a certain A_d (from a flow loss viewpoint, A_d should certainly not be less than the fully opened port area of the valve), what is the necessary corrected length L_d? Utilizing Equation 10.22 in Equation 10.42 gives

$$c_{od}\sqrt{\frac{A_d}{L_d V_{od}}} < \frac{n\Omega}{\sqrt{2}} \qquad (10.43)$$

or

$$L_d > \frac{2A_d}{V_{od}} \frac{c_{od}^2}{(n\Omega)^2}. \qquad (10.44)$$

This means that it is easier to attenuate higher harmonics n of the fundamental frequency Ω because the L_d do not have to be as long. The longest L_d results if we wish to attenuate all harmonics, even $n = 1$. It must be remembered that the Helmholtz simplification only holds if L_d is appreciably less than the shortest wavelength of interest; see Equation 10.13, where

$$f_{max} = \frac{n\Omega}{2\pi}. \qquad (10.45)$$

Otherwise, the Helmholtz simplification is invalid. Thus, not only does L_d have to satisfy Equation 10.44, but also

$$L_d < \frac{\pi c_{od}}{2n\Omega} \qquad (10.46)$$

in order for Equation 10.44 to be valid in the first place.

If the restriction of Equation 10.46 is more and more violated by selecting longer and longer necks, resonances (standing waves) of the gas in the necks come into play, which cause the overall attenuation predicted by Equation 10.38 to have certain frequency bands where standing waves of the gas in the neck may actually cause amplifications. A more complicated theory is then needed for prediction, which will be discussed in some of the later sections. But this does not invalidate the Helmholtz theory predictions entirely. They can still be used for rough estimates.

Note the similarity of this discussion to the discussion in the vibration isolation chapters.

10.2.4 Discharge System with Two Resonators in Series

A second resonator is often added in series to act as a low pass filter muffler (note that the first resonator can also be designed as a low pass filter resonator muffler). Let us investigate this by obtaining the governing equations of motion.

From the free body diagrams of Figure 10.4, we get

$$A_{1d}(p_{1d} - p_{2d}) - L_{1d}A_{1d}\rho_{od}\ddot{\xi}_{1d} - D_{1d}\dot{\xi}_{1d} = 0 \qquad (10.47)$$

$$A_{2d}p_{2d} - L_{2d}A_{2d}\rho_{od}\ddot{\xi}_{2d} - D_{2d}\dot{\xi}_{2d} = 0. \qquad (10.48)$$

The oscillatory pressure p_{1d} is the pressure increase in volume V_{old} above the mean pressure p_{od}, and is, as before

$$p_{1d} = \frac{c_{od}^2}{V_{old}} \int_0^t \dot{m}_d dt - \frac{\rho_{od} c_{od}^2 A_{1d}}{V_{old}} \xi_{1d}. \qquad (10.49)$$

The pressure differential p_{2d} is the pressure increase in volume V_{o2d} above the mean pressure and is

$$p_{2d} = -K_{od}\frac{dV_{o2d}}{V_{o2d}}. \qquad (10.50)$$

Because

$$dV_{o2d} = A_{2d}\xi_{2d} - A_{1d}\xi_{1d} \qquad (10.51)$$

$$K_{od} = \rho_{od}c_{od}^2 \qquad (10.52)$$

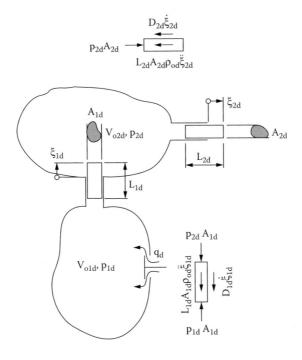

FIGURE 10.4 Discharge system with two Helmholtz resonators in series.

we get

$$p_{2d} = \frac{\rho_{od} c_{od}^2}{V_{02d}} (A_{1d}\xi_{1d} - A_{2d}\xi_{2d}). \tag{10.53}$$

The equations of motion become, therefore,

$$L_{1d}A_{1d}\rho_{od}\ddot{\xi}_{1d} + D_{1d}\dot{\xi}_{1d} + \rho_{od}c_{od}^2 A_{1d}^2 \left(\frac{1}{V_{o1d}} + \frac{1}{V_{o2d}}\right)\xi_{1d}$$

$$- \frac{\rho_{od}c_{od}^2 A_{2d}A_{1d}}{V_{o2d}}\xi_{2d} = \frac{c_{od}^2 A_{1d}}{V_{01d}} \int_0^t \dot{m}_d dt \tag{10.54}$$

and

$$L_{2d}A_{2d}\rho_{od}\ddot{\xi}_{2d} + D_{2d}\dot{\xi}_{2d} + \frac{\rho_{od}c_{od}^2 A_{2d}^2}{V_{02d}}\xi_{2d}$$

$$- \frac{\rho_{od}c_{od}^2 A_{1d}A_{2d}}{V_{o2d}}\xi_{1d} = 0. \tag{10.55}$$

In matrix form, this can be written as

$$\rho_{od}\begin{bmatrix} L_{1d}A_{1d} & 0 \\ 0 & L_{2d}A_{2d} \end{bmatrix}\begin{Bmatrix} \ddot{\xi}_{1d} \\ \ddot{\xi}_{2d} \end{Bmatrix} + \begin{bmatrix} D_{1d} & 0 \\ 0 & D_{2d} \end{bmatrix}\begin{Bmatrix} \dot{\xi}_{1d} \\ \dot{\xi}_{2d} \end{Bmatrix}$$

$$+ \rho_{od}c_{od}^2 \begin{bmatrix} A_{1d}^2\left(\dfrac{1}{V_{old}} + \dfrac{1}{V_{o2d}}\right) & -\dfrac{A_{1d}A_{2d}}{V_{o2d}} \\ -\dfrac{A_{1d}A_{2d}}{V_{o2d}} & \dfrac{A_{2d}^2}{V_{o2d}} \end{bmatrix}\begin{Bmatrix} \xi_{1d} \\ \xi_{2d} \end{Bmatrix}$$

$$= \begin{Bmatrix} \dfrac{c_{od}^2 A_{1d}}{V_{old}}\int_0^t \dot{m}_d dt \\ 0 \end{Bmatrix}.$$

(10.56)

It can be shown that all matrices must be symmetrical or must be capable of being brought into a symmetric form.

10.2.5 DISCHARGE SYSTEM WITH ANECHOIC PIPE

In high-side refrigeration compressors, the compressor discharges into a pipe that is connected to the condenser. Thus, the condenser is a boundary condition. It was found that in this case it is not entirely unreasonable to model the pipe as an infinite pipe, assuming that pressure waves will not be reflected from the condenser, but will be dissipated. The suction side is effectively decoupled from the evaporator piping by the relatively large volume of the shell. In low-side compressors, this is reversed. The suction pipe is modeled as anechoic and the discharge is uncoupled from the condenser by the shell.

How to model a pipe as infinite (anechoic in acoustic terminology) will, at this point, not be derived because discussion of the wave equation follows in a later section. It will be shown that the oscillation pressure at the entrance of an infinite pipe is related to the oscillating velocity component of gas molecules by (see, for example, Soedel, 1974a and Lai and Soedel, 1996b)

$$p = c_o \rho_o \dot{\xi} \qquad (10.57)$$

where p = the oscillating pressure at the pipe entrance, and $\dot{\xi}$ = the velocity at the pipe entrance.

Let us work out the example shown in Figure 10.5. In this case the pressure in volume V_o is

$$p_d = \dfrac{C_{od}^2}{V_{od}}\int_0^t \dot{m}_d dt - \dfrac{\rho_{od}c_{od}^2 A_d}{V_{od}}\xi_d. \qquad (10.58)$$

Suction or Discharge System Gas Pulsations and Mufflers

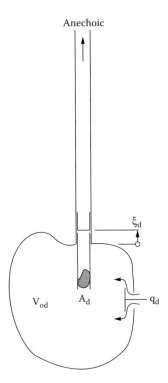

FIGURE 10.5 Simple discharge system with an anechoic pipe.

However, because

$$p = c_{od}\rho_{od}\dot{\xi}_d \tag{10.59}$$

we get

$$\dot{\xi}_d + \frac{c_{od}A_d}{V_{od}}\xi_d = \frac{c_{od}}{\rho_{od}V_{od}}\int_0^t \dot{m}_d dt. \tag{10.60}$$

Computer simulation will show that in this case there will be a pressure increase as the valve discharges, and then a decay of pressure will follow until the next discharge.

10.2.6 Resonator Plus Anechoic Pipe

This system is quite common because it is the simplest muffler design possible when an anechoic pipe is involved. It is sketched in Figure 10.6.

In this case, the equation describing the motion of the mass in the resonator neck is the same as in the case where two resonators were put into series:

$$A_{1d}(p_{1d} - p_{2d}) - L_{1d}A_{1d}\rho_{od}\ddot{\xi}_{1d} - c_{1d}\dot{\xi}_{1d} = 0 \tag{10.61}$$

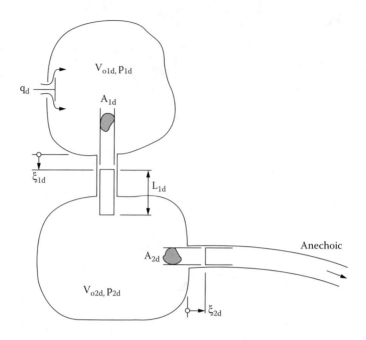

FIGURE 10.6 Two-volume discharge system with anechoic pipe.

while, as before,

$$p_{1d} = \frac{c_{od}^2}{V_{o1d}} \int_0^t \dot{m}_d \, dt - \frac{\rho_{od} c_{od}^2 A_{1d}}{V_{o1d}} \xi_{1d}. \tag{10.62}$$

We get now for p_{2d}

$$p_{2d} = \frac{\rho_{od} c_{od}^2}{V_{o2d}} (A_{1d} \xi_{1d} - A_{2d} \xi_{2d}) \tag{10.63}$$

where

$$p_{2d} = c_{od} \rho_{od} \dot{\xi}_{2d} \tag{10.64}$$

because of the anechoic termination. Combining Equations 10.63 and 10.64 gives

$$c_{od} \rho_{od} \dot{\xi}_{2d} + \frac{\rho_{od} c_{od}^2 A_{2d}}{V_{o2d}} \xi_{2d} - \frac{\rho_{od} c_{od}^2 A_{1d}}{V_{o2d}} \xi_{1d} = 0. \tag{10.65}$$

Equation 10.61 becomes, as before

$$L_{1d}A_{1d}\rho_{od}\ddot{\xi}_{1d} + D_{1d}\dot{\xi}_{1d} + \rho_{od}c_{od}^2 A_{1d}^2 \left(\frac{1}{V_{o1d}} + \frac{1}{V_{o2d}}\right)\xi_{1d}$$
$$- \frac{\rho_{od}c_{od}^2 A_{2d}A_{1d}}{V_{o2d}}\xi_{2d} = \frac{A_{1d}c_{od}^2}{V_{old}}\int_0^t \dot{m}_d \, dt. \qquad (10.66)$$

In matrix form, multiplying Equation 10.65 by A_{2d}, we get

$$\rho_{od}\begin{bmatrix} L_{1d}A_{1d} & 0 \\ 0 & 0 \end{bmatrix}\begin{Bmatrix} \ddot{\xi}_{1d} \\ \ddot{\xi}_{2d} \end{Bmatrix} + \begin{bmatrix} D_{1d} & 0 \\ 0 & A_{2d}c_{od}\rho_{od} \end{bmatrix}\begin{Bmatrix} \dot{\xi}_{1d} \\ \dot{\xi}_{2d} \end{Bmatrix}$$
$$+ \rho_{od}c_{od}^2 \begin{bmatrix} A_{1d}^2\left(\dfrac{1}{V_{o1d}} + \dfrac{1}{V_{o2d}}\right) & -\dfrac{A_{2d}A_{1d}}{V_{o2d}} \\ -\dfrac{A_{1d}A_{2d}}{V_{o2d}} & \dfrac{A_{2d}^2}{V_{o2d}} \end{bmatrix}\begin{Bmatrix} \xi_{1d} \\ \xi_{2d} \end{Bmatrix}$$
$$= \begin{Bmatrix} \dfrac{c_{od}^2 A_{1d}}{V_{old}}\int_0^t \dot{m}_d \, dt \\ 0 \end{Bmatrix}. \qquad (10.67)$$

We see, therefore, that the anechoic termination has a damping effect. This is expected; it is a continuous drain of oscillatory energy because waves entering the anechoic termination are not reflected and their energy is therefore lost to the system.

Note that if we let V_{o2d} approach infinity, we effectively decouple V_{old} and the neck L_{1d} from the anechoic termination, and we have a case where we discharge into a large pressure vessel.

It is therefore convenient to use this system also on the suction side because it allows the description of both high-side and low-side compressors. The only difference is that we change subscripts d to s and replace $\dot{m}_d$ by $\dot{m}_s$.

An example program that incorporates both suction and discharge systems of this kind was first presented by Dhar and Soedel (1978a).

10.2.7 Discharge System for a Two-Cylinder Compressor

Let us now look at the case where a two-cylinder compressor discharges into a common discharge system (see also Soedel et al., 1973). This is shown in Figure 10.7, and a schematic is shown in Figure 10.8. The equations of motion are set up as

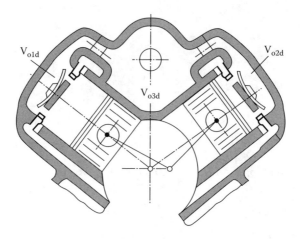

FIGURE 10.7 Discharge system of a two-cylinder compressor. A schematic interpretation is shown in Figure 10.8.

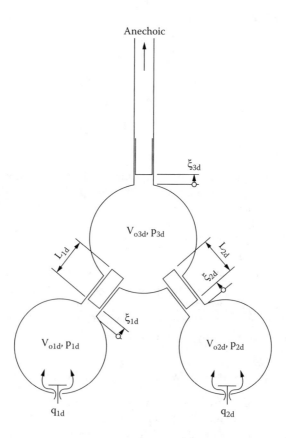

FIGURE 10.8 Schema of a two-cylinder compressor with anechoic termination.

Suction or Discharge System Gas Pulsations and Mufflers

before, considering the forces acting on each gas plug.

$$A_{1d}(p_{1d} - p_{3d}) - L_{1d}A_{1d}\rho_{od}\ddot{\xi}_{1d} - D_{1d}\dot{\xi}_{1d} = 0 \quad (10.68)$$

$$A_{2d}(p_{2d} - p_{3d}) - L_{2d}A_{2d}\rho_{od}\ddot{\xi}_{1d} - D_{2d}\dot{\xi}_{2d} = 0. \quad (10.69)$$

The changes in pressure are given by

$$p_{1d} = \frac{c_{od}^2}{V_{o1d}} \int_0^t \dot{m}_{d1} dt - \frac{\rho_{od}c_{od}^2 A_{1d}}{V_{o1d}} \xi_{1d} \quad (10.70)$$

$$p_{2d} = \frac{c_{od}^2}{V_{o2d}} \int_0^t \dot{m}_{d2} dt - \frac{\rho_{od}c_{od}^2 A_{2d}}{V_{o2d}} \xi_{2d} \quad (10.71)$$

and

$$p_{3d} = \frac{\rho_{od}c_{od}^2}{V_{o3d}} (A_{1d}\xi_{1d} + A_{2d}\xi_{2d} - A_{3d}\xi_{3d}). \quad (10.72)$$

In addition, we have the anechoic termination equation

$$p_{3d} = c_{od}\rho_{od}\dot{\xi}_{3d}. \quad (10.73)$$

Let us now combine the equations. We obtain

$$L_{1d}A_{1d}\rho_{od}\ddot{\xi}_{1d} + D_{1d}\dot{\xi}_{1d} + \rho_{od}c_{od}^2 A_{1d}^2 \left(\frac{1}{V_{o1d}} + \frac{1}{V_{o3d}}\right)\xi_{1d}$$

$$+ \frac{\rho_{od}c_{od}^2 A_{1d}A_{2d}}{V_{o3d}} \xi_{2d} - \frac{\rho_{od}c_{od}^2 A_{1d}A_{3d}}{V_{o3d}} \xi_{3d} \quad (10.74)$$

$$= \frac{A_{1d}c_{od}^2}{V_{o1d}} \int_0^t \dot{m}_{d1} dt$$

$$L_{2d}A_{2d}\rho_{od}\ddot{\xi}_{1d} + D_{2d}\dot{\xi}_{2d} + \rho_{od}c_{od}^2 A_{2d}^2 \left(\frac{1}{V_{o2d}} + \frac{1}{V_{o3d}}\right)\xi_{2d}$$

$$+ \frac{\rho_{od}c_{od}^2 A_{1d}A_{2d}}{V_{o3d}} \xi_{1d} - \frac{\rho_{od}c_{od}^2 A_{2d}A_{3d}}{V_{o3d}} \xi_{3d} \quad (10.75)$$

$$= \frac{A_{2d}c_{od}^2}{V_{o2d}} \int_0^t \dot{m}_{d2} dt.$$

Combining Equations 10.72 and 10.73 we obtain the third equation as

$$A_{3d}c_{od}\rho_{od}\dot{\xi}_{3d} + \frac{\rho_{od}c_{od}^2 A_{3d}^2}{V_{03d}}\xi_{3d} - \frac{\rho_{od}c_{od}^2 A_{1d}A_{3d}}{V_{03d}}\xi_{1d}$$
$$- \frac{\rho_{od}c_{od}^2 A_{2d}A_{3d}}{V_{03d}}\xi_{2d} = 0.$$
(10.76)

In matrix form, we obtain

$$[m]\{\ddot{\xi}_d\} + [D]\{\dot{\xi}_d\} + [K]\{\xi_d\} = \{F\} \qquad (10.77)$$

where

$$[m] = \rho_{od}\begin{bmatrix} L_{1d}A_{1d} & 0 & 0 \\ 0 & L_{2d}A_{2d} & 0 \\ 0 & 0 & 0 \end{bmatrix} \qquad (10.78)$$

$$[D] = \begin{bmatrix} D_{1d} & 0 & 0 \\ 0 & D_{2d} & 0 \\ 0 & 0 & A_{3d}c_{od}\rho_{od} \end{bmatrix} \qquad (10.79)$$

$$[K] = \rho_{od}c_{od}^2 \begin{bmatrix} A_{1d}^2\left(\dfrac{1}{V_{01d}}+\dfrac{1}{V_{03d}}\right) & \dfrac{A_{1d}A_{2d}}{V_{03d}} & -\dfrac{A_{1d}A_{3d}}{V_{03d}} \\ -\dfrac{A_{1d}A_{2d}}{V_{03d}} & A_{2d}^2\left[\dfrac{1}{V_{02d}}+\dfrac{1}{V_{03d}}\right] & -\dfrac{A_{2d}A_{3d}}{V_{03d}} \\ -\dfrac{A_{1d}A_{3d}}{V_{03d}} & -\dfrac{A_{2d}A_{3d}}{V_{03d}} & \dfrac{A_{3d}^2}{V_{03d}} \end{bmatrix} \qquad (10.80)$$

$$\{F\} = c_{od}^2 \begin{Bmatrix} \dfrac{A_{1d}}{V_{01d}}\int_0^t \dot{m}_{d1}dt \\ \dfrac{A_{2d}}{V_{02d}}\int_0^t \dot{m}_{d2}dt \\ 0 \end{Bmatrix} \qquad (10.81)$$

$$\{\xi_d\} = \begin{Bmatrix} \xi_{1d} \\ \xi_{2d} \\ \xi_{3d} \end{Bmatrix}. \qquad (10.82)$$

Suction or Discharge System Gas Pulsations and Mufflers

See also Soedel et al. (1973), where the termination of the condenser was modeled as an equivalent orifice and not an anechoic termination, as in the equations derived here. In retrospect, it is felt that an anechoic termination model would have been a better choice, even though the results were acceptable in their agreement with the measurements. The equations using the anechoic termination as presented here are recommended for use in similar cases. For other two-cylinder compressor modeling see, for example, Elson and Soedel (1974) and Singh and Soedel (1975, 1976, 1978c, 1979).

10.2.8 Discharge System Model for More Than Two Cylinders

The approach is the same as for the two-cylinder case. The equation of motion is in the same form as Equation 10.77, except that the matrix sizes are larger. Note also that it is not necessary that all cylinders discharge into a common plenum, as was the case in the two-cylinder example. Figure 10.9 is a sketch of a four-cylinder compressor example (see Soedel and Baum, 1976). Here we consider only the special case where all volumes, neck lengths, and cross-sections are identical (labeled as V, L, and A) in order to keep the example simple. This is redrawn in schematic form in Figure 10.10.

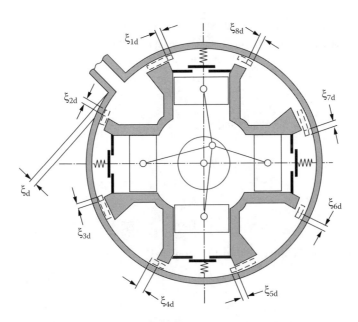

FIGURE 10.9 Four-cylinder compressor. The schematic interpretation is shown in Figure 10.10.

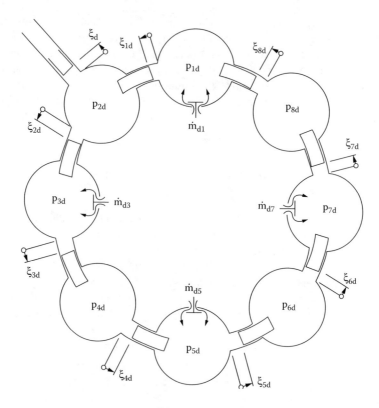

FIGURE 10.10 Schema of the four cylinder model of Figure 10.9. For the sake of simplicity of presentation, all volumes, cross-sectional areas, and effective lengths were taken to be the same, designated as V, A, and L.

Proceeding as for the two-cylinder case, we obtain the following equations of motion:

$$A(p_{1d} - p_{2d}) - LA\rho_{od}\ddot{\xi}_{1d} - D\dot{\xi}_{1d} = 0$$
$$A(p_{2d} - p_{3d}) - LA\rho_{od}\ddot{\xi}_{2d} - D\dot{\xi}_{2d} = 0$$
$$\vdots$$
$$A(p_{8d} - p_{1d}) - LA\rho_{od}\ddot{\xi}_{8d} - D\dot{\xi}_{8d} = 0. \tag{10.83}$$

The equation for the anechoic termination is

$$p_{2d} - c_{od}\rho_{od}\dot{\xi}_d = 0. \tag{10.84}$$

Writing these equations in matrix form gives

$$[m]\{\ddot{\xi}\} + [D]\{\dot{\xi}\} + [G]\{p\} = 0 \tag{10.85}$$

where

$$[m] = \rho_{od} LA \begin{bmatrix} 1 & 0 & 0 & 0 & 0 & 0 & 0 & 0 & 0 \\ 0 & 1 & 0 & 0 & 0 & 0 & 0 & 0 & 0 \\ 0 & 0 & 1 & 0 & 0 & 0 & 0 & 0 & 0 \\ 0 & 0 & 0 & 1 & 0 & 0 & 0 & 0 & 0 \\ 0 & 0 & 0 & 0 & 1 & 0 & 0 & 0 & 0 \\ 0 & 0 & 0 & 0 & 0 & 1 & 0 & 0 & 0 \\ 0 & 0 & 0 & 0 & 0 & 0 & 1 & 0 & 0 \\ 0 & 0 & 0 & 0 & 0 & 0 & 0 & 1 & 0 \\ 0 & 0 & 0 & 0 & 0 & 0 & 0 & 0 & 1 \end{bmatrix} \quad (10.86)$$

$$[D] = D \begin{bmatrix} 1 & 0 & 0 & 0 & 0 & 0 & 0 & 0 & 0 \\ 0 & 1 & 0 & 0 & 0 & 0 & 0 & 0 & 0 \\ 0 & 0 & 1 & 0 & 0 & 0 & 0 & 0 & 0 \\ 0 & 0 & 0 & 1 & 0 & 0 & 0 & 0 & 0 \\ 0 & 0 & 0 & 0 & 1 & 0 & 0 & 0 & 0 \\ 0 & 0 & 0 & 0 & 0 & 1 & 0 & 0 & 0 \\ 0 & 0 & 0 & 0 & 0 & 0 & 1 & 0 & 0 \\ 0 & 0 & 0 & 0 & 0 & 0 & 0 & 1 & 0 \\ 0 & 0 & 0 & 0 & 0 & 0 & 0 & 0 & A\dfrac{c_{od}\rho_{od}}{D} \end{bmatrix} \quad (10.87)$$

$$[G] = A \begin{bmatrix} -1 & 1 & 0 & 0 & 0 & 0 & 0 & 0 & 0 \\ 0 & -1 & 1 & 0 & 0 & 0 & 0 & 0 & 0 \\ 0 & 0 & -1 & 1 & 0 & 0 & 0 & 0 & 0 \\ 0 & 0 & 0 & -1 & 1 & 0 & 0 & 0 & 0 \\ 0 & 0 & 0 & 0 & -1 & 1 & 0 & 0 & 0 \\ 0 & 0 & 0 & 0 & 0 & -1 & 1 & 0 & 0 \\ 0 & 0 & 0 & 0 & 0 & 0 & -1 & 1 & 0 \\ 1 & 0 & 0 & 0 & 0 & 0 & 0 & -1 & 0 \\ 0 & -1 & 0 & 0 & 0 & 0 & 0 & 0 & 0 \end{bmatrix} \quad (10.88)$$

$$\{p\}^T = \{p_{1d}, p_{2d}, p_{3d}, p_{4d}, p_{5d}, p_{6d}, p_{7d}, p_{8d}, 0\} \quad (10.89)$$

$$\{\xi\}^T = \{\xi_{1d}, \xi_{2d}, \xi_{3d}, \xi_{4d}, \xi_{5d}, \xi_{6d}, \xi_{7d}, \xi_{8d}, \xi_d\}. \quad (10.90)$$

The changes in pressure are given by

$$P_{1d} = \frac{c_{od}^2}{V} \int_0^t \dot{m}_{d1} dt - \frac{\rho_{od} c_{od}^2 A}{V} (\xi_{1d} - \xi_{8d}) \tag{10.91}$$

$$P_{2d} = \frac{\rho_{od} c_{od}^2 A}{V} (\xi_{2d} - \xi_{1d}) - \frac{\rho_{od} c_{od}^2 A}{V} \xi_d \tag{10.92}$$

$$P_{3d} = \frac{c_{od}^2}{V} \int_0^t \dot{m}_{d3} dt - \frac{\rho_{od} c_{od}^2 A}{V} (\xi_{3d} - \xi_{2d}) \tag{10.93}$$

$$P_{4d} = \frac{\rho_{od} c_{od}^2 A}{V} (\xi_{4d} - \xi_{3d}) \tag{10.94}$$

$$\vdots$$

$$P_{8d} = \frac{\rho_{od} c_{od}^2 A}{V} (\xi_{8d} - \xi_{7d}) \tag{10.95}$$

or in matrix form,

$$\{p\} = [B] \left\{ \int_0^t \dot{m} \, dt \right\} + [E] \{\xi\} \tag{10.96}$$

where

$$[B] = \frac{c_{od}^2}{V} \begin{bmatrix} 1 & 0 & 0 & 0 & 0 & 0 & 0 & 0 & 0 \\ 0 & 0 & 0 & 0 & 0 & 0 & 0 & 0 & 0 \\ 0 & 0 & 1 & 0 & 0 & 0 & 0 & 0 & 0 \\ 0 & 0 & 0 & 0 & 0 & 0 & 0 & 0 & 0 \\ 0 & 0 & 0 & 0 & 1 & 0 & 0 & 0 & 0 \\ 0 & 0 & 0 & 0 & 0 & 0 & 0 & 0 & 0 \\ 0 & 0 & 0 & 0 & 0 & 0 & 1 & 0 & 0 \\ 0 & 0 & 0 & 0 & 0 & 0 & 0 & 0 & 0 \\ 0 & 0 & 0 & 0 & 0 & 0 & 0 & 0 & 0 \end{bmatrix} \tag{10.97}$$

$$[E] = \frac{\rho_{od} c_{od}^2 A}{V} \begin{bmatrix} -1 & 0 & 0 & 0 & 0 & 0 & 0 & 1 & 0 \\ 1 & -1 & 0 & 0 & 0 & 0 & 0 & 0 & -1 \\ 0 & 1 & -1 & 0 & 0 & 0 & 0 & 0 & 0 \\ 0 & 0 & 1 & -1 & 0 & 0 & 0 & 0 & 0 \\ 0 & 0 & 0 & 1 & -1 & 0 & 0 & 0 & 0 \\ 0 & 0 & 0 & 0 & 1 & -1 & 0 & 0 & 0 \\ 0 & 0 & 0 & 0 & 0 & 1 & -1 & 0 & 0 \\ 1 & 0 & 0 & 0 & 0 & 0 & 1 & -1 & 0 \\ 0 & 0 & 0 & 0 & 0 & 0 & 0 & 0 & 0 \end{bmatrix} \tag{10.98}$$

$$\left\{ \int_0^t \dot{m} \, dt \right\}^T = \left\{ \int_0^t \dot{m}_{d1} \, dt, 0, \int_0^t \dot{m}_{d3} \, dt, 0, \int_0^t \dot{m}_{d5} \, dt, 0, \int_0^t \dot{m}_{d7} \, dt, 0, 0 \right\}^T. \tag{10.99}$$

Combining the two equations gives

$$[m]\{\ddot{\xi}\} + [D]\{\dot{\xi}\} + [K]\{\xi\} = \{F\} \qquad (10.100)$$

where

$$[K] = [G][E] = \frac{\rho_{od} c_{od}^2 A^2}{V} \begin{bmatrix} 2 & -1 & 0 & 0 & 0 & 0 & 0 & -1 & -1 \\ -1 & 2 & -1 & 0 & 0 & 0 & 0 & 0 & 1 \\ 0 & -1 & 2 & -1 & 0 & 0 & 0 & 0 & 0 \\ 0 & 0 & -1 & 2 & -1 & 0 & 0 & 0 & 0 \\ 0 & 0 & 0 & -1 & 2 & -1 & 0 & 0 & 0 \\ 0 & 0 & 0 & 0 & -1 & 2 & -1 & 0 & 0 \\ 0 & 0 & 0 & 0 & 0 & -1 & 2 & -1 & 0 \\ -1 & 0 & 0 & 0 & 0 & 0 & -1 & 2 & 0 \\ -1 & 1 & 0 & 0 & 0 & 0 & 0 & 0 & 1 \end{bmatrix} \qquad (10.101)$$

and where

$$\{F\} = [G][B]\left\{\int_0^t \dot{m}\, dt\right\}. \qquad (10.102)$$

This gives

$$\{F\}^T = \left\{-\int_0^t \dot{m}_{d1} dt, \int_0^t \dot{m}_{d3} dt, -\int_0^t \dot{m}_{d3} dt, \int_0^t \dot{m}_{d5} dt, -\int_0^t \dot{m}_{d5} dt, \int_0^t \dot{m}_{d7} dt, \right.$$
$$\left. -\int_0^t \dot{m}_{d7} dt, \int_0^t \dot{m}_{d1} dt, 0 \right\}^T \qquad (10.103)$$

10.2.9 Steady-State Harmonic Response of a Discharge System with Two Resonators in Series

If the periodic approach is used, the equations of motion can be evaluated in a steady state by considering the source components to be harmonic in time. For example, let us examine an air compressor that discharges into a large tank whose volume can be taken as infinite, or in an approximate sense, a high-side refrigeration compressor that discharges into a large volume formed by the housing and the casing

(Figure 10.4). Setting in Equation 10.56:

$$A_1 \dot{\xi}_{1d} = Q_{1d} \tag{10.104}$$

$$A_2 \dot{\xi}_{2d} = Q_{2d} \tag{10.105}$$

$$\dot{m}_d = \rho_o Q_d e^{j\omega t} \tag{10.106}$$

where Q_{1d} and Q_{2d} are volume flow velocities in the passages and Q_d is the input volume flow velocity from the discharge valve, and differentiating Equation 10.56 with respect to time gives

$$\begin{bmatrix} 1 & 0 \\ 0 & 1 \end{bmatrix} \begin{Bmatrix} \ddot{Q}_{1d} \\ \ddot{Q}_{2d} \end{Bmatrix} + \begin{bmatrix} 2\zeta_{1d}\omega_{11}^2 & 0 \\ 0 & 2\zeta_{2d}\omega_{22}^2 \end{bmatrix} \begin{bmatrix} \dot{Q}_{1d} \\ \dot{Q}_{2d} \end{bmatrix}$$
$$+ \begin{bmatrix} \omega_{11}^2 & -\omega_{12}^2 \\ -\omega_{21}^2 & \omega_{22}^2 \end{bmatrix} \begin{Bmatrix} Q_{1d} \\ Q_{2d} \end{Bmatrix} = \begin{bmatrix} \omega_{01}^2 Q_d e^{j\omega t} \\ 0 \end{bmatrix} \tag{10.107}$$

where

$$\omega_{11}^2 = c_{od}^2 \left(\frac{1}{V_{1d}} + \frac{1}{V_{2d}} \right) \frac{A_{1d}}{L_{1d}} \tag{10.108}$$

$$\omega_{22}^2 = \omega_{21}^2 = c_{od}^2 \frac{A_{2d}}{V_{2d} L_{2d}} \tag{10.109}$$

$$\omega_{12}^2 = c_{od}^2 \frac{A_{1d}}{V_{2d} L_{1d}} \tag{10.110}$$

$$\omega_{01}^2 = c_{od}^2 \frac{A_{1d}}{V_{1d} L_{1d}} \tag{10.111}$$

$$\zeta_{1d} = \frac{D_{1d}}{2\omega_{11}^2 L_{1d} \zeta_{1d}} \tag{10.112}$$

$$\zeta_{2d} = \frac{D_{2d}}{2\omega_{22}^2 L_{2d} \zeta_{2d}}. \tag{10.113}$$

Suction or Discharge System Gas Pulsations and Mufflers

A harmonic volume velocity input will produce harmonic volume velocity outputs, but phase shifted. Thus, we set up the solution as

$$Q_{1d} = \tilde{Q}_{1d} e^{j\omega t} \qquad (10.114)$$

$$Q_{2d} = \tilde{Q}_{2d} e^{j\omega t} \qquad (10.115)$$

where

$$\tilde{Q}_{1d} = |\tilde{Q}_{1d}| e^{-j\phi_1} \qquad (10.116)$$

$$\tilde{Q}_{2d} = |\tilde{Q}_{2d}| e^{-j\phi_2} \qquad (10.117)$$

and where $|\tilde{Q}_{1d}|, |\tilde{Q}_{2d}|$ are the absolute amplitudes of the volume velocity gas oscillations and ϕ_1, ϕ_2 are the phase lag angles.

Equations 10.114 and 10.115 will satisfy Equation 10.107 and we obtain

$$\tilde{Q}_{1d} = \frac{1}{D(\omega)} \left[\left(\omega_{22}^2 - \omega^2 \right) + 2j\omega_{22}\omega \right] Q_d \omega_{01}^2 \qquad (10.118)$$

$$\tilde{Q}_{2d} = \frac{1}{D(\omega)} \omega_{21}^2 \omega_{01}^2 Q_d \qquad (10.119)$$

where

$$D(\omega) = A + jB \qquad (10.120)$$

and where

$$A = \left(\omega_{11}^2 - \omega^2 \right)\left(\omega_{22}^2 - \omega^2 \right) - 4\zeta_{1d}\zeta_{2d}\omega_{11}^2\omega^2 - \omega_{12}^2\omega_{21}^2 \qquad (10.121)$$

$$B = 2\omega \left[\zeta_{2d}\omega_{22}\left(\omega_{11}^2 - \omega^2 \right) + \zeta_{1d}\omega_{11}\left(\omega_{22}^2 - \omega^2 \right) \right]. \qquad (10.122)$$

The magnitudes and phase angles can be determined from this solution in the usual way.

For the typical periodic (but not harmonic) volume flow through the valves of a real compressor, we obtain the harmonic components by a Fourier analysis (Chapter 7), evaluate the above solution for each harmonic component, and sum all the individual results.

10.2.10 Low-Frequency Cutoff Formulas

One of the most difficult problems in muffler design is handling effectiveness at low frequencies. Taking as an example the two-volume muffler case of the previous section (see Soedel and Soedel, 1992), only when

$$\frac{|\tilde{Q}_{2d}|}{Q_d} < 1 \tag{10.123}$$

will the muffler-discharge system combination attenuate gas pulsations. In the low-frequency range, we expect two resonances; thus for an undamped muffler, there will be four points (Figure 10.11) at which

$$\frac{|\tilde{Q}_{2d}|}{Q_d} = 1. \tag{10.124}$$

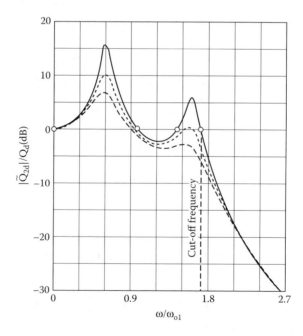

FIGURE 10.11 Defining the cutoff frequency below which the two-degree-of-freedom Helmholtz resonator muffler does not attenuate sound and may even magnify it. Introducing damping may control possible low-frequency magnifications. The damping ratios plotted are $\zeta = 0.05$ (—), 0.10 (---) and 0.15 (----). The cutoff frequency is determined for $\zeta = 0$.

Suction or Discharge System Gas Pulsations and Mufflers

The last one of theses points will determine the cutoff frequency above which Equation 10.123 will be satisfied (until still higher frequencies are reached, at which we may have standing wave resonances). To find this cutoff frequency, we remove damping ($\zeta_{1d} = \zeta_{2d} = 0$) and solve Equation 10.124 for the ω that satisfies this equation. There will be four values. The highest value is the cutoff frequency ω_c. It is, as given in Soedel and Soedel (1992),

$$\omega_c = \omega_{01} \sqrt{1 + \left(\frac{V_1}{V_2}\right) + \left(\frac{A_2}{A_1}\right)\left(\frac{V_1}{V_2}\right)\left(\frac{L_1}{L_2}\right)}. \qquad (10.125)$$

The requirement is that ω_c is lower than the lowest frequency ω of interest. Attenuation (in a broad sense) will occur if

$$\omega_c < \omega. \qquad (10.126)$$

Note that the subscript d for the discharge muffler was dropped in Equation 10.125 because this equation is also valid for suction mufflers.

Because it is often difficult to modify the volumes of a discharge system (including the muffler), it is important that we determine how the ratio of the effective neck length L_1 to the tail pipe length L_2 influences the attenuation. Because the neck length L_1 may not be changeable, we see from Equation 10.125 that the tail pipe length L_2 should be made as long as possible, as long as the Helmholtz theory restrictions are not violated. Also, given a volume V_1, which is the manifold volume close to the valve, one should select the largest muffler volume V_2 (the volume before the tail pipe) possible, again keeping in mind that we should not violate the Helmholtz restrictions, even while from a very approximate design trend viewpoint they can be somewhat relaxed.

10.2.11 OSCILLATION EFFECTS CAUSED BY CYLINDER VOLUME AND VALVE PASSAGE MASSES

In all previous chapters we have assumed that the mass flow rate of gas through the valves is given by a compressor simulation. This simulation takes the elasticity of the cylinder volume into account by way of the thermodynamic model and therefore handles, in an approximate sense, what some people call the *source impedance*.

The effect of the mass of the gas in the valve passages is usually not considered. Trella and Soedel (1971, 1972, 1974) have investigated the conditions and applications for which this is important. The method of investigation resulted essentially in the addition of another resonator, with the cylinder volume as the time-varying resonator volume (considered in the simulation, anyway), and the mass in the valve passages during opening time as a time-varying resonator neck mass, as sketched in Figure 10.12. They showed that for the typical pulsation frequency range of interest (in the range up to 500 Hz), this effect can be neglected for compressors, but if the interest range is extended to 1000 Hz and higher, this effect may have to be taken into consideration.

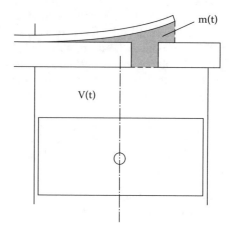

FIGURE 10.12 During discharge, the effective mass in the valve passages acts like the mass of a Helmholtz resonator neck, and the time-dependent cylinder volume is equivalent to the volume.

In this case, the effect is that of a noise source rather than an energy drain. The control volume approach taken by Trella and Soedel requires major modifications in a typical basic compressor simulation program. However, it should be possible to simplify the approach to a point where it can be used with practical valve configurations. A similar approach was taken by Yee and Soedel (1983, 1988) when calculating the re-expansion oscillations in a rotary vane compressor. Figure 10.13 illustrates that the compressed gas in the valve port passage re-expands like a discharging Helmholtz resonator into the trailing volume once the vane has passed the relief slot cutoff. One can approximately view the Helmholtz resonator neck as being time dependent in terms of equivalent length and cross-section.

10.2.12 Discharge System with a Long Pipe Modeled in the Time Domain

The neck of a Helmholtz resonator can only be treated as incompressible if it is short with respect to the shortest wavelength that is of interest. If the neck length is equal to the wavelength of interest, results will be poor. If the neck length is one-eighth of the wavelength of interest, the results can be expected to be very reasonable. Judgment must be applied for lengths in between. Thus, as a compromise it is required that

$$L < \frac{\lambda}{4} \tag{10.127}$$

where λ is the wavelength [m]. Because

$$\lambda = \frac{c_o}{f} \tag{10.128}$$

Suction or Discharge System Gas Pulsations and Mufflers

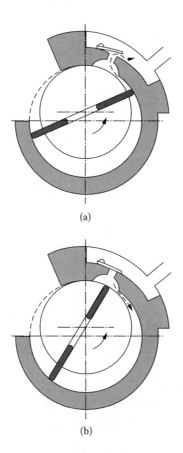

(a)

(b)

FIGURE 10.13 Gas trapped in the discharge port of a rotary vane compressor at discharge pressure re-expands explosively into the trailing volume, which is at lower pressure. (a) Compressed gas is discharged until the vane passes the edge of the relief slot of the discharge port. (b) As the vane passes the relief slot, a Helmholtz resonator neck forms, and an oscillatory discharge into the trailing volume takes place.

where c_o = the speed of sound [m/sec] and f = the highest frequency of interest [Hz], we require that the length L conforms to

$$L < \frac{c_o}{4f}. \tag{10.129}$$

For example, if $c_o = 400$ m/s and L = 0.05 m, we can trust the analytical prediction up to 2000 Hz. In typical compressors, the gas oscillations of importance have strong frequency content in the range of 50 to 800 Hz. This would mean that any one neck length could not exceed 0.125 m, given that f_{max} = 800 Hz. This requirement is often not met, especially in large gas or air compressors.

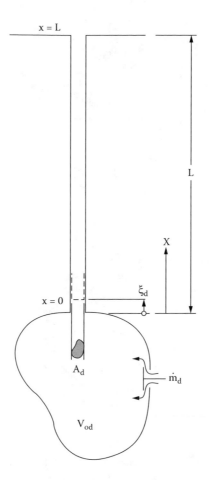

FIGURE 10.14 Long pipe connected to a discharge volume.

Thus, we must treat long necks (pipes) as continuous systems. As will be pointed out in Section 10.3, the wave solutions required are preferably solved in the frequency domain. However, compressor simulations (thermodynamics, valve flutter, and so forth.) are simulated in the time domain. Thus, a hybrid approach involving iterations is employed. But if we prefer to have the entire solution done in the time domain, which is a strength of the Helmholtz approximation, we may wish to express the wave solution for a long neck or pipe in the time domain also.

For the volume–pipe combination of Figure 10.14, where the pipe terminates in a large tank, we may write (this equation is derived in Section 10.3; see also Soedel, 1976b):

$$p_d = \frac{2c_{od}^2 \rho_{od}}{L} \sum_{n=1}^{\infty} \frac{1}{\omega_n \sqrt{1-\zeta_n^2}} \int_0^t \left(\frac{d^2\xi_d}{dt^2}\right)_{t=\tau} e^{-\zeta_n \omega_n (t-\tau)} \sin \omega_n \sqrt{1-\zeta_n^2}\,(t-\tau)d\tau$$

(10.130)

where ζ_n is the damping ratio, and where the natural frequencies of the pipe by itself are

$$\omega_n = (2n-1)\frac{\pi}{2}\frac{c_o}{L}. \tag{10.131}$$

Also, because

$$p_d = \frac{c_{od}^2}{V_{od}}\int_0^t \dot{m}_d \, dt - \frac{\rho_{od}c_{od}^2 A_d}{V_{od}}\xi_d, \tag{10.132}$$

we now have two equations and are able to solve for the two unknowns: p_d and ξ_d.

10.3 THE CONTINUOUS SYSTEM APPROACH APPLIED TO TUBELIKE COMPRESSOR SUCTION AND DISCHARGE MANIFOLDS

10.3.1 THE WAVE EQUATION IN ONE DIMENSION

In the case of a long tubular element, certain assumptions made in the Helmholtz resonator development are not possible. We can no longer assign inertia only or elasticity only to the regions of the system, but must consider both properties simultaneously. However, we are still able to utilize the idea of a bulk modulus.

In the following we will confine ourselves to flow that can be considered one-dimensional. In necks or long pipes, this is easily recognized as a good assumption. Volumes will be viewed as short tubes of constant cross-section. This may cause some difficulty when trying to fit an irregular volume. The key here is that the volume of the equivalent pipe should be equal to the actual volume, while the length of the equivalent pipe should be equal to the length between the entrance and exit of the actual volume.

The governing equation for the gas pulsation in a tube can be derived by reduction from the general three-dimensional wave equation. It can also be derived by linearizing the Navier-Stokes equation. However, to point out the similarities between the continuous theory and the Helmholtz resonator theory, it is derived here from basic principles. Because the governing equation will be a wave equation as used in acoustics, the derivation offers nothing new, but is given for a full understanding of all assumptions.

As a disturbance, such as one caused by a volume velocity input, passes down a tube, the molecules of gas are displaced. Figure 10.15a shows them to be momentarily displaced in the direction of propagation. In this particular derivation, there is no need to specify the shape of the cross-section, but the cross-section must be constant throughout the tube. Note that we assume that the particle displacements are assumed to be approximately planar. This assumes that the pulsation velocities are not laminar because the velocity profile would then be parabolic. The assumption

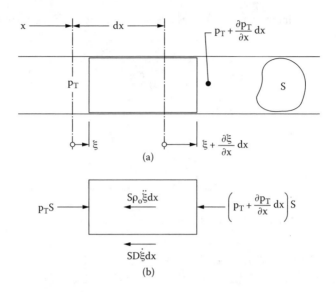

FIGURE 10.15 Element of a continuous gas column.

also restricts the one-dimensional theory to waves that do not have cross-modes, that is, there is no gas pulsation normal to the tube axis. If the largest dimension across the tube cross-sectional area is d, the requirement is, as before in the Helmholtz approach, that

$$d < \frac{c_o}{4f_{max}} \tag{10.133}$$

where c_o = the speed of sound [m/sec], and f_{max} = the highest frequency of interest [Hz].

For example, if c_o = 400 m/sec, and we want to adequately predict what happens in a frequency range of up to 4000 Hz, the largest tube diameter should be less than 2.5 cm. The difference in the Helmholtz theory is that we now have no restriction on the permissible tube length. It is possible to get good agreement even for larger tube diameters in cases where cross-modes are not excited. This depends on various factors and has to be evaluated from case to case.

Let us now examine Figure 10.15a. The original volume is

$$V_o = S\,dx \tag{10.134}$$

where S is the cross-sectional area. The volume of the displaced element is

$$V_o + dV = S\left(dx + \frac{\partial \xi}{\partial x}dx\right). \tag{10.135}$$

Thus, the volume increase is

$$dV = S\frac{\partial \xi}{\partial x}dx. \qquad (10.136)$$

The mass in the original volume is equal to the mass in the increased volume. We may use the bulk modulus formula, if we are willing to use a linearized compression process:

$$p = -K_o \frac{dV}{V_o} \qquad (10.137)$$

where

$$K_o = \rho_o c_o^2. \qquad (10.138)$$

This gives

$$p = -K_o \frac{\partial \xi}{\partial x}. \qquad (10.139)$$

Note that the implications are the same as for the Helmholtz development. Use of the bulk modulus K_o implies that we are operating in the same linear range of compression with an allowable pressure change p of up to ±20% of the mean pressure.

Next, let us examine the free body diagram of the displaced mass element, shown in Figure 10.15b. We assume an equivalent viscous damping model where the force is proportional to volume velocity. Summing all forces gives

$$S(p_o + p) - S\rho_o\ddot{\xi}dx - S\left(p_o + p + \frac{\partial p}{\partial x}dx\right) - r_1 s\dot{\xi}dx = 0 \qquad (10.140)$$

or

$$\rho_o\ddot{\xi} + r_1\dot{\xi} + \frac{\partial p}{\partial x} = 0. \qquad (10.141)$$

Substituting Equation 10.137 gives

$$\ddot{\xi} + \frac{r_1}{\rho_o}\dot{\xi} = c_o^2 \frac{\partial^2 \xi}{\partial x^2} \qquad (10.142)$$

where r_1 = the equivalent viscous damping coefficient [N sec/m⁴], ρ_o = the mean density [N sec²/m⁴], and c_o = the mean speed of sound [m/sec]. The value for r_1 may, in some cases, be calculated using the Helmholtz-Stokes model

$$r_1 = \frac{2\rho_o}{d}\sqrt{2\nu n\Omega} \tag{10.143}$$

where Ω = the rotational speed of the compressor in rad/sec, n = the harmonic number (n = 1, 2, . . .), d = effective tube diameter [m], and ν = the effective kinematic viscosity [m²/sec]. For instance, the kinematic viscosity of air at atmospheric conditions and at 20 °C is $\nu = 15.1(10^{-6})$ [m²/sec]. The density is $\rho_o = 1.207$ [N sec²/m⁴]. For an internal pipe diameter d = 0.01 [m], a compressor speed of $\Omega = 376.8$ [rad/sec] (3600 RPM) and the tenth harmonic, n = 10, we obtain $r_1 = 57.6$ [N sec/m⁴].

Note that the relationship between absolute viscosity μ [N sec/m²] and kinematic viscosity ν is

$$\mu = \rho_o \nu \tag{10.144}$$

in cases where viscosity information is given in this form.

When the system is at a resonance, damping values given by Equation 10.143 were found to be too low by as much as an order of magnitude, even while the general behavior as described by Equation 10.142 seems to be predicted satisfactorily. Therefore, some investigations (see, for example, Singh and Soedel, 1978b) have used

$$r_1 = \zeta \frac{2\rho_o}{d}\sqrt{2\nu n\Omega} \tag{10.145}$$

where ζ is a relatively large correction coefficient that is adjusted for particular manifold geometries by comparing the pressure pulsation simulation with measured values for a prototype design. This probably accounts for turbulence generation.

It is advantageous to write Equation 10.142 in terms of volume velocities q [m³/sec]:

$$q = S\dot{\xi} \tag{10.146}$$

where S is the cross-sectional area of the tube. It becomes, after differentiation with respect to time,

$$\ddot{q} + \frac{r_1}{\rho_o}\dot{q} = c_o^2 \frac{\partial^2 q}{\partial x^2}. \tag{10.147}$$

10.3.2 THE SOLUTION OF THE UNDAMPED WAVE EQUATION

The solution of

$$\frac{\partial^2 \xi}{\partial t^2} = c_o^2 \frac{\partial^2 \xi}{\partial x^2} \qquad (10.148)$$

was given by d'Alembert as early as the 18th century and is

$$\xi = f_1\left(t - \frac{x}{c_o}\right) + f_2\left(t + \frac{x}{c_o}\right) \qquad (10.149)$$

where $f_1(t - \frac{x}{c_o})$ represents a wave of arbitrary shape traveling in the positive x–direction with velocity c_o, and $f_2(t + \frac{x}{c_o})$ represents a wave of arbitrary shape traveling in the negative x–direction with velocity c_o.

One possibility is to utilize this solution directly and follow the individual waves as they travel and reflect. However, the amount of labor would approach that of the method of characteristics in nonlinear gas dynamics, which for the linearized case is entirely equivalent to the d'Alembert's solution.

Because we can argue that the gas pulsations in a compressor running at constant speed are a periodic phenomenon, we can think of the pulsations, no matter of what complicated form, as a Fourier series of harmonic components. This allows us to specialize. We set

$$f_1(\tau) = A_1 e^{j\omega\tau} \qquad (10.150)$$

$$f_2(\tau) = B_1 e^{j\omega\tau} \qquad (10.151)$$

where

$$\omega = n\Omega \qquad (10.152)$$

and where A_1 and A_2 are constants.

D'Alembert's solution becomes

$$\xi(x,t) = A_1 e^{j(\omega t - kx)} + B_1 e^{j(\omega t + kx)} \qquad (10.153)$$

where

$$k = \frac{\omega}{c_o} = \frac{2\pi}{\lambda} = \text{the wave number.} \qquad (10.154)$$

A_1 and B_1 are constants that have to be evaluated from the boundary conditions.

The pressure is

$$p = jk\rho_o c_o^2 \left[A_1 e^{j(\omega t - kx)} - B_1 e^{j(\omega t + kx)} \right]. \tag{10.155}$$

The volume velocity is

$$q(x,t) = S\dot{\xi} \tag{10.156}$$

or

$$q(x,t) = j\omega S \left[A_1 e^{j(\omega t - kx)} + B_1 e^{j(\omega t + kx)} \right]. \tag{10.157}$$

10.3.3 The Solution of the Damped Wave Equation

The solution of

$$\frac{\partial^2 \xi}{\partial t^2} + \gamma_1 \frac{\partial \xi}{\partial t} = c_o^2 \frac{\partial^2 \xi}{\partial x^2} \tag{10.158}$$

where

$$\gamma_1 = \frac{r_1}{\rho_o} \tag{10.159}$$

is, for harmonic waves,

$$\xi(x,t) = A_1 e^{j(\omega t - k_1 x)} + B_1 e^{j(\omega t + k_1 x)} \tag{10.160}$$

where $k_1 = \omega/c_1$ = the modified wave number and c_1 = the modified speed of sound. Substituting the solution into the equation of motion gives

$$-\omega^2 \xi + \gamma_i (j\omega)\xi = c_o^2 (-k_1^2)\xi \tag{10.161}$$

or

$$k_1 = \left(\frac{\omega}{c_o}\right)\sqrt{1 - j\frac{\gamma_1}{\omega}} \tag{10.162}$$

where $c_1 = c_o/\sqrt{1 - j\frac{\gamma_1}{\omega}}$. Because $\frac{\gamma_1}{\omega} \ll 1$, we may expand the square root and get approximately

$$k_1 = \frac{\omega}{c_o}\left(1 - j\frac{\gamma_1}{2\omega}\right) = \frac{\omega}{c_o} - j\frac{\gamma_1}{2c_o} \tag{10.163}$$

and

$$jk_1 = j\frac{\omega}{c_o} + \frac{\gamma_1}{2c_o}. \tag{10.164}$$

Let

$$k = \frac{\omega}{c_o} \tag{10.165}$$

$$a = \frac{\gamma_1}{2c_o}. \tag{10.166}$$

The solution becomes, therefore,

$$\xi(x,t) = A_1 e^{-ax} e^{j(\omega t - kx)} + B_1 e^{+ax} e^{j(\omega t + kx)}. \tag{10.167}$$

The pressure is

$$p = -\rho_o c_o^2 \frac{\partial \xi}{\partial x} \tag{10.168}$$

or

$$p(x,t) = \rho_o c_o^2 (a + jk)[A_1 e^{-ax} e^{j(\omega t - kx)} - B_1 e^{ax} e^{j(\omega t + kx)}]. \tag{10.169}$$

The volume velocity is

$$q(x,t) = j\omega S [A_1 e^{-ax} e^{j(\omega t - kx)} + B_1 e^{+ax} e^{j(\omega t + kx)}]. \tag{10.170}$$

The solutions may also be written

$$p(x,t) = \rho_o c_o^2 \gamma [A_1 e^{-\gamma x} - B_1 e^{+\gamma x}] e^{j\omega t} \tag{10.171}$$

and

$$q(x,t) = j\omega S [A_1 e^{-\gamma x} + B_1 e^{+\gamma x}] e^{j\omega t} \tag{10.172}$$

where

$$\gamma = a + jk \tag{10.173}$$

and where we define

$$P(x) = \rho_o c_o^2 \gamma [A_1 e^{-\gamma x} - B_1 e^{+\gamma x}] \quad (10.174)$$

$$Q(x) = j\omega S [A_1 e^{-\gamma x} + B_1 e^{+\gamma x}]. \quad (10.175)$$

Thus, the input impedance at $x = 0$ is

$$Z_o = \frac{P(0)}{Q(0)}. \quad (10.176)$$

10.3.4 Example: Open Pipe with Volume Velocity Input

Let us examine the case of a pipe where at one end the volume velocity is defined as input, while the other end is open to a large tank. This is the simplest example possible and is shown in Figure 10.16.

FIGURE 10.16 Volume velocity excitation of an open pipe.

Suction or Discharge System Gas Pulsations and Mufflers

The boundary conditions are that at x = 0, Q is specified,

$$Q(0) = Q_o \tag{10.177}$$

and that at x = L, no pressure oscillation of the order that exists in the pipe occurs:

$$P(L) = 0. \tag{10.178}$$

Substituting Equations 10.174 and 10.175 gives

$$j\omega S (A_1 + B_1) = Q_o \tag{10.179}$$

$$\rho_o c_o^2 \gamma (A_1 e^{-\gamma L} - B_1 e^{\gamma L}) = 0. \tag{10.180}$$

From this we obtain

$$A_1 = \frac{Q_o e^{\gamma L}}{2S j\omega \cosh \gamma L} \tag{10.181}$$

$$B_1 = \frac{Q_o e^{-\gamma L}}{2S j\omega \cosh \gamma L} \tag{10.182}$$

because

$$\cosh \gamma L = \frac{e^{\gamma L} + e^{-\gamma L}}{2}. \tag{10.183}$$

This gives

$$P(x) = \frac{\rho_o c_o^2 \gamma Q_o}{2S j\omega \cosh \gamma L} \sinh \gamma (L - x) \tag{10.184}$$

$$Q(x) = \frac{Q_o}{\cosh \gamma L} \cosh \gamma (L - x). \tag{10.185}$$

To obtain the input impedance, we set x = 0 and substitute into Equation 10.176:

$$Z_o = \frac{\rho_o c_o^2 \gamma}{j\omega S} \tanh \gamma L. \tag{10.186}$$

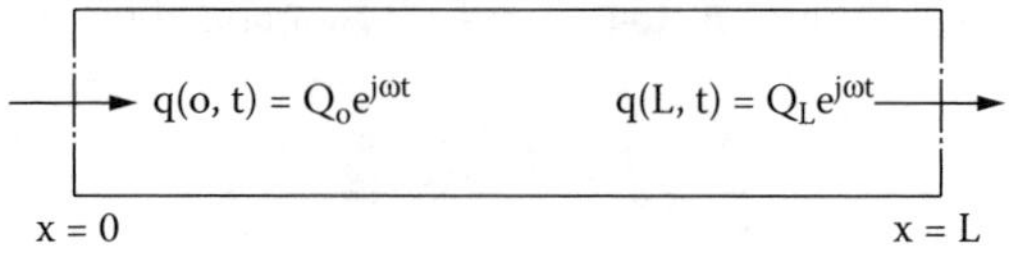

```
                ┌─────────────────────────────────────────────┐
       ────────→│  q(o, t) = Q₀eʲωᵗ         q(L, t) = Q_Leʲωᵗ │────────→
                └─────────────────────────────────────────────┘
                  x = 0                              x = L
```

FIGURE 10.17 The four pole element.

10.3.5 THE FOUR POLE CONCEPT

Let us now look at the case where there are harmonic volume velocities at the same frequency defined at each end of the tube, as shown in Figure 10.17. This is the case that we encounter if another tube is joined at $x = L$. The boundary conditions are

$$Q(0) = Q_o \tag{10.187}$$

$$Q(L) = Q_L. \tag{10.188}$$

The pressures at both ends are unknown. We substitute Equations 10.174 and 10.175 and get

$$j\omega S(A_1 + B_1) = Q_o \tag{10.189}$$

$$j\omega S(A_1 e^{-\gamma L} + B_1 e^{\gamma L}) = Q_L. \tag{10.190}$$

Solving for A_1 and B_1 gives

$$A_1 = \frac{1}{2Sj\omega \sinh \gamma L}(Q_o e^{\gamma L} - Q_L) \tag{10.191}$$

$$B_1 = \frac{1}{2Sj\omega \sinh \gamma L}(Q_L - Q_o e^{-\gamma L}). \tag{10.192}$$

This gives, therefore

$$P(x) = \frac{\rho_o c_o^2 \gamma}{2Sj\omega \sinh \gamma L}(Q_o \cosh \gamma(L-x) - Q_L \cosh \gamma x) \tag{10.193}$$

$$Q(x) = \frac{1}{\sinh \gamma L}(Q_o \sinh \gamma(L-x) + Q_L \sinh \gamma x). \tag{10.194}$$

Evaluating $P(x)$ at $x = 0$ and $x = L$ gives us the boundary values

$$P(0) = P_o = \frac{\rho_o c_o^2 \gamma}{Sj\omega \sinh \gamma L} (Q_o \cosh \gamma L - Q_L) \qquad (10.195)$$

$$P(L) = P_L = \frac{\rho_o c_o^2 \gamma}{Sj\omega \sinh \gamma L} (Q_o - Q_L \cosh \gamma L). \qquad (10.196)$$

These equations can be rearranged into

$$Q_o = Q_L \cosh \gamma L + P_L \frac{j\omega S}{\rho_o c_o^2 \gamma} \sinh \gamma L \qquad (10.197)$$

$$P_o = Q_L \frac{\rho_o c_o^2 \gamma}{j\omega S} \sinh \gamma L + P_L \cosh \gamma L. \qquad (10.198)$$

They relate the conditions at one end of the tube to conditions at the other end. In matrix form, they are written as

$$\begin{Bmatrix} Q_o \\ P_o \end{Bmatrix} = \begin{bmatrix} A & B \\ C & D \end{bmatrix} \begin{Bmatrix} Q_L \\ P_L \end{Bmatrix} \qquad (10.199)$$

where

$$A = \cosh \gamma L = D \qquad (10.200)$$

$$B = \frac{j\omega S}{\rho_o c_o^2 \gamma} \sinh \gamma L \qquad (10.201)$$

$$C = \frac{\rho_o c_o^2 \gamma}{j\omega S} \sinh \gamma L. \qquad (10.202)$$

This is also known as the four pole description of a uniform tube of length L. Note that Q_o and Q_L are positive in the positive x–direction.

In the case where Q_o and P_o are given, we solve for Q_L and P_L:

$$Q_L = \frac{\begin{vmatrix} Q_o & B \\ P_o & D \end{vmatrix}}{D} = \frac{DQ_o - BP_o}{D} \qquad (10.203)$$

$$P_L = \frac{\begin{vmatrix} A & Q_o \\ C & P_o \end{vmatrix}}{D} = \frac{-CQ_o + AP_o}{D} \qquad (10.204)$$

where

$$D = AD - BC = \cosh^2 \gamma L - \sinh^2 \gamma L = 1. \qquad (10.205)$$

Because

$$A = D \qquad (10.206)$$

we obtain

$$\begin{Bmatrix} Q_L \\ P_L \end{Bmatrix} = \begin{bmatrix} A & -B \\ -C & D \end{bmatrix} \begin{Bmatrix} Q_o \\ P_o \end{Bmatrix}. \qquad (10.207)$$

10.3.6 Example: Open Pipe

Let us consider the same example (Figure 10.16) treated before, except that this time we utilize the four pole.

Q_o is given and

$$P_L = 0. \qquad (10.208)$$

Unknowns are P_o and Q_L. Thus

$$\begin{Bmatrix} Q_o \\ P_o \end{Bmatrix} = \begin{bmatrix} A & B \\ C & D \end{bmatrix} \begin{Bmatrix} Q_L \\ 0 \end{Bmatrix} \qquad (10.209)$$

or

$$Q_o = AQ_L \qquad (10.210)$$

$$P_o = CQ_L = \frac{C}{A} Q_o. \qquad (10.211)$$

This gives

$$Q_L = \frac{Q_o}{\cosh \gamma L} \qquad (10.212)$$

and

$$P_o = \frac{\rho_o c_o^2 \gamma Q_o}{j\omega S} \tanh \gamma L. \qquad (10.213)$$

The input point impedance is, therefore,

$$Z_o = \frac{P_o}{Q_o} = \frac{\rho_o c_o^2 \gamma}{j\omega S} \tanh \gamma L. \qquad (10.214)$$

This is, of course, the same result we obtained before in Equation 10.186.

10.3.7 GLOBAL FOUR POLES FROM LOCAL ELEMENT FOUR POLES: TUBES IN SERIES

Let us first consider the problem where we have to generate the four pole for a combination of two tubes, as sketched in Figure 10.18.

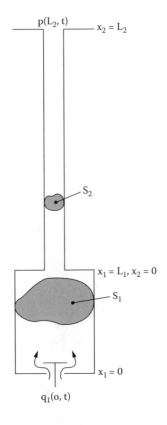

FIGURE 10.18 Combination of two tubes.

The four pole of the first tube is

$$\begin{Bmatrix} Q_{o1} \\ P_{o1} \end{Bmatrix} = \begin{bmatrix} A_1 & B_1 \\ C_1 & D_1 \end{bmatrix} \begin{Bmatrix} Q_{L1} \\ P_{L1} \end{Bmatrix} \quad (10.215)$$

where, if the speed of sound and the damping factor do change from tube to tube,

$$A_1 = \cosh \gamma_1 L_1 = D_1 \quad (10.216)$$

$$B_1 = \frac{j\omega S_1}{\rho_{o1} c_{o1}^2 \gamma_1} \sinh \gamma_1 L_1 \quad (10.217)$$

$$C_1 = \frac{\rho_{o1} c_{o1}^2 \gamma_1}{j\omega S_1} \sinh \gamma_1 L_1. \quad (10.218)$$

The four pole of the second tube is

$$\begin{Bmatrix} Q_{o2} \\ P_{o2} \end{Bmatrix} = \begin{bmatrix} A_2 & B_2 \\ C_2 & D_2 \end{bmatrix} \begin{Bmatrix} Q_{L2} \\ P_{L2} \end{Bmatrix} \quad (10.219)$$

where

$$A_2 = \cosh \gamma_2 L_2 = D_2 \quad (10.220)$$

$$B_2 = \frac{j\omega S_2}{\rho_{o2} c_{o2}^2 \gamma_2} \sinh \gamma_2 L_2 \quad (10.221)$$

$$C_2 = \frac{\rho_{o2} c_{o2}^2 \gamma_2}{j\omega S_2} \sinh \gamma_2 L_2. \quad (10.222)$$

At the boundary between the two tubes

$$\begin{Bmatrix} Q_{L1} \\ P_{L1} \end{Bmatrix} = \begin{Bmatrix} Q_{o2} \\ P_{o2} \end{Bmatrix}. \quad (10.223)$$

Thus, we get

$$\begin{Bmatrix} Q_{o1} \\ P_{o1} \end{Bmatrix} = \begin{bmatrix} A_1 & B_1 \\ C_1 & D_1 \end{bmatrix} \begin{bmatrix} A_2 & B_2 \\ C_2 & D_2 \end{bmatrix} \begin{Bmatrix} Q_{L2} \\ P_{L2} \end{Bmatrix} \quad (10.224)$$

Suction or Discharge System Gas Pulsations and Mufflers

or

$$\begin{Bmatrix} Q_{o1} \\ P_{o1} \end{Bmatrix} = \begin{bmatrix} A_T & B_T \\ C_T & D_T \end{bmatrix} \begin{Bmatrix} Q_{L2} \\ P_{L2} \end{Bmatrix} \quad (10.225)$$

where

$$A_T = A_1 A_2 + B_1 C_2 \quad (10.226)$$

$$B_T = A_1 B_2 + B_1 D_2 \quad (10.227)$$

$$C_T = C_1 A_2 + D_1 C_2 \quad (10.228)$$

$$D_T = C_1 B_2 + D_1 D_2. \quad (10.229)$$

Because in our example case, Q_{o1} is given and $P_{L2} = 0$, we get

$$\begin{Bmatrix} Q_{o1} \\ P_{o1} \end{Bmatrix} = \begin{bmatrix} A_T & B_T \\ C_T & D_T \end{bmatrix} \begin{Bmatrix} Q_{L2} \\ 0 \end{Bmatrix} \quad (10.230)$$

or

$$Q_{L2} = \frac{1}{A_T} Q_{o1} \quad (10.231)$$

and

$$P_{o1} = C_T Q_{L2} = \frac{C_T}{A_T} Q_{o1}. \quad (10.232)$$

Thus, our input point impedance is

$$Z_o = \frac{P_{o1}}{Q_{o1}} = \frac{C_T}{A_T}. \quad (10.233)$$

Note that once we know what P_{o1} is, we can go into the first sub–four pole and find P_{L1} and Q_{L1}

$$\begin{Bmatrix} Q_{L1} \\ P_{L1} \end{Bmatrix} = \begin{bmatrix} A_1 & -B_1 \\ -C_1 & D_1 \end{bmatrix} \begin{Bmatrix} Q_{o1} \\ P_{o1} \end{Bmatrix}. \quad (10.234)$$

Thus, we now know $P_{o1}, P_{L1} = P_{o2}, Q_{L1} = Q_{o2}, Q_{L2}$. Because Q_{o1} is the given input and $P_{L2} = 0$, we know the pressures and volume velocities at every piping joint. Often this is sufficient. If it is necessary to know pressures and velocities inside the pipe sections, we can use Equations 10.193 and 10.194. We obtain, for the first tube,

$$P_{x1} = \frac{\rho_{o1} c_{o1}^2 \gamma_1}{j\omega S_1 \sinh \gamma_1 L_1} [Q_{o1} \cosh \gamma_1 (L_1 - x_1) - Q_{L1} \cosh \gamma_1 x_1] \qquad (10.235)$$

$$Q_{x1} = \frac{1}{\sinh \gamma_1 L_1} [Q_{o1} \sinh \gamma_1 (L_1 - x_1) + Q_{L1} \sinh \gamma_1 x_1]. \qquad (10.236)$$

For the second tube we get

$$P_{x2} = \frac{\rho_{o2} c_{o2}^2 \gamma_2}{j\omega S_2 \sinh \gamma_2 L_2} [Q_{o2} \cosh \gamma_2 (L_2 - x_2) - Q_{L2} \cosh \gamma_2 x_2] \qquad (10.237)$$

$$Q_{x2} = \frac{1}{\sinh \gamma_2 L_2} [Q_{o2} \sinh \gamma_2 (L_2 - x_2) + Q_{L2} \sinh \gamma_2 x_2]. \qquad (10.238)$$

The principle of connecting tubes in series is by now obvious. If we consider n tubes as shown in Figure 10.19 we get

$$\begin{Bmatrix} Q_{o1} \\ P_{o1} \end{Bmatrix} = \begin{bmatrix} A_1 & B_1 \\ C_1 & D_1 \end{bmatrix} \begin{bmatrix} A_2 & B_2 \\ C_2 & D_2 \end{bmatrix} \begin{bmatrix} A_3 & B_3 \\ C_3 & D_3 \end{bmatrix} \cdots \begin{bmatrix} A_n & B_n \\ C_n & D_n \end{bmatrix} \begin{Bmatrix} Q_{Ln} \\ P_{Ln} \end{Bmatrix}. \qquad (10.239)$$

10.3.8 Branched Tubes

Let us start with the simple example shown in Figure 10.20. The four pole relationships are

$$\begin{Bmatrix} Q_{o1} \\ P_{o1} \end{Bmatrix} = \begin{bmatrix} A_1 & B_1 \\ C_1 & D_1 \end{bmatrix} \begin{Bmatrix} Q_{L1} \\ P_{L1} \end{Bmatrix} \qquad (10.240)$$

$$\begin{Bmatrix} Q_{o2} \\ P_{o2} \end{Bmatrix} = \begin{bmatrix} A_2 & B_2 \\ C_2 & D_2 \end{bmatrix} \begin{Bmatrix} Q_{L2} \\ P_{L2} \end{Bmatrix} \qquad (10.241)$$

$$\begin{Bmatrix} Q_{o3} \\ P_{o3} \end{Bmatrix} = \begin{bmatrix} A_3 & B_3 \\ C_3 & D_3 \end{bmatrix} \begin{Bmatrix} Q_{L3} \\ P_{L3} \end{Bmatrix}. \qquad (10.242)$$

Suction or Discharge System Gas Pulsations and Mufflers

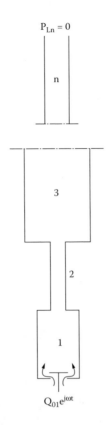

FIGURE 10.19 Many tubes in series.

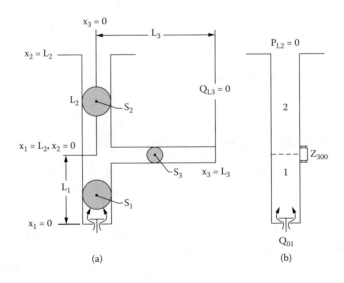

FIGURE 10.20 (a) Branched tubes example, (b) equivalent system.

At the branch junction, we have

$$Q_{L1} = Q_{o2} + Q_{o3} \tag{10.243}$$

$$P_{L1} = P_{o2} = P_{o3}. \tag{10.244}$$

Let the point impedance at $x_3 = 0$ for the side branch tube be Z_{300}:

$$Z_{300} = \frac{P_{o3}}{Q_{o3}} \tag{10.245}$$

From the branch junction conditions we therefore get

$$Q_{L1} = Q_{o2} + \frac{P_{o2}}{Z_{300}} \tag{10.246}$$

$$P_{L1} = P_{o2}. \tag{10.247}$$

In matrix form, this gives

$$\begin{Bmatrix} Q_{L1} \\ P_{L1} \end{Bmatrix} = \begin{bmatrix} 1 & \frac{1}{Z_{300}} \\ 0 & 1 \end{bmatrix} \begin{Bmatrix} Q_{o2} \\ P_{o2} \end{Bmatrix}. \tag{10.248}$$

Thus, combining all equations, we get

$$\begin{Bmatrix} Q_{o1} \\ P_{o1} \end{Bmatrix} = \begin{bmatrix} A_1 & B_1 \\ C_1 & D_1 \end{bmatrix} \begin{bmatrix} 1 & \frac{1}{Z_{300}} \\ 0 & 1 \end{bmatrix} \begin{bmatrix} A_2 & B_2 \\ C_2 & D_2 \end{bmatrix} \begin{Bmatrix} Q_{L2} \\ P_{L2} \end{Bmatrix} \tag{10.249}$$

where $P_{L2} = 0$ and where Q_{o1} is the input. The two unknowns are P_{o1} and Q_{L2}. We get the impedance Z_{300} from

$$Z_{300} = \frac{P_{o3}}{Q_{o3}} = \frac{C_3 Q_{L3} + D_3 P_{L3}}{A_3 Q_{L3} + B_3 P_{L3}}. \tag{10.250}$$

In our case

$$Q_{L3} = 0. \tag{10.251}$$

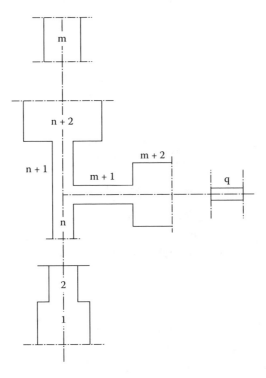

FIGURE 10.21 Generalized branched tubes.

Thus

$$Z_{300} = \frac{D_3}{B_3}. \tag{10.252}$$

Let us now generalize the example to that shown in Figure 10.21. We may write immediately

$$\begin{Bmatrix} Q_{o1} \\ P_{o1} \end{Bmatrix} = \begin{bmatrix} A_1 & B_1 \\ C_1 & D_1 \end{bmatrix} \begin{bmatrix} A_2 & B_2 \\ C_2 & D_2 \end{bmatrix} \cdots \begin{bmatrix} A_n & B_n \\ C_n & D_n \end{bmatrix}$$

$$\begin{bmatrix} 1 & \frac{1}{Z_{(m+1)00}} \\ 0 & 1 \end{bmatrix} \begin{bmatrix} A_{n+1} & B_{n+1} \\ C_{n+1} & D_{n+1} \end{bmatrix} \begin{bmatrix} A_{n+2} & B_{n+2} \\ C_{n+2} & D_{n+2} \end{bmatrix} \cdots \begin{bmatrix} A_m & B_m \\ C_m & D_m \end{bmatrix} \begin{Bmatrix} Q_{Lm} \\ P_{Lm} \end{Bmatrix} \tag{10.253}$$

where

$$Z_{(m+1)oo} = \frac{P_{o(m+1)}}{Q_{o(m+1)}} \tag{10.254}$$

and where $P_{o(m+1)}$ and $Q_{o(m+1)}$ are obtained from

$$\begin{Bmatrix} Q_{o(m+1)} \\ P_{o(m+1)} \end{Bmatrix} = \begin{bmatrix} A_{m+1} & B_{m+1} \\ C_{m+1} & D_{m+1} \end{bmatrix} \begin{bmatrix} A_{m+2} & B_{m+2} \\ C_{m+2} & D_{m+2} \end{bmatrix} \cdots \begin{bmatrix} A_q & B_q \\ C_q & D_q \end{bmatrix} \begin{Bmatrix} Q_{Lq} \\ P_{Lq} \end{Bmatrix}. \tag{10.255}$$

10.3.9 Anechoic Termination to a Muffler

As we have seen in the discussion of the Helmholtz resonator approach, condensers and evaporators are often modeled as anechoic pipes. For such a pipe, the volume velocity at its entrance is related to the entrance pressure by

$$q(o,t) = \frac{S}{c_o \rho_o} p(o,t). \tag{10.256}$$

As an example, let us treat the case shown in Figure 10.22. The four pole describing tube 1 is given by

$$\begin{Bmatrix} Q_{o1} \\ P_{o1} \end{Bmatrix} = \begin{bmatrix} A_1 & B_1 \\ C_1 & D_1 \end{bmatrix} \begin{Bmatrix} Q_{L1} \\ P_{L1} \end{Bmatrix}. \tag{10.257}$$

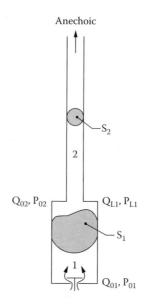

FIGURE 10.22 Tube with anechoic termination.

Because

$$\begin{Bmatrix} Q_{L1} \\ P_{L1} \end{Bmatrix} = \begin{Bmatrix} Q_{o2} \\ P_{o2} \end{Bmatrix} \tag{10.258}$$

and because of the anechoic termination

$$Q_{o2} = \frac{S_2}{c_{o2}\rho_{o2}} P_{o2}, \tag{10.259}$$

which means that the impedance of the anechoic termination is

$$Z_{2oo} = \frac{P_{o2}}{Q_{o2}} = \frac{c_{o2}\rho_{o2}}{S_2} \tag{10.260}$$

we get

$$\begin{Bmatrix} Q_{L1} \\ P_{L1} \end{Bmatrix} = \begin{Bmatrix} \dfrac{S_2}{c_{o2}\rho_{o2}} \\ 1 \end{Bmatrix} P_{L1}. \tag{10.261}$$

Thus

$$\begin{Bmatrix} Q_{o1} \\ P_{o1} \end{Bmatrix} = \begin{bmatrix} A_1 & B_1 \\ C_1 & D_1 \end{bmatrix} \begin{Bmatrix} \dfrac{S_2}{c_{o2}\rho_{o2}} \\ 1 \end{Bmatrix} P_{L1}. \tag{10.262}$$

Because Q_{o1} is given, the two unknowns are P_{o1} and P_{L1}. They are now found to be

$$P_{L1} = \frac{Q_{o1}}{A_1 \dfrac{S_2}{c_{o2}\rho_{o2}} + B_1} \tag{10.263}$$

$$P_{o1} = P_{L1} \left[C_1 \frac{S_2}{c_{o2}\rho_{o2}} + D_1 \right] = Q_{o1} \frac{C_1 \dfrac{S_2}{c_{o2}\rho_{o2}} + D_1}{A_1 \dfrac{S_2}{c_{o2}\rho_{o2}} + B_1}. \tag{10.264}$$

Let us look at the case where damping is zero

$$\gamma_1 = jk_1 = j\frac{\omega_1}{c_{ol}}. \tag{10.265}$$

This gives

$$A_1 = D_1 = \cosh jk_1 L_1 = \cos k_1 L_1 \tag{10.266}$$

$$B_1 = \frac{S_1}{\rho_{ol} c_{ol}} \sinh jk_1 L_1 = \frac{jS_1}{\rho_{ol} c_{ol}} \sin k_1 L_1 \tag{10.267}$$

$$C_1 = \frac{\rho_{ol} c_{ol}}{S_1} \sinh jk_1 L_1 = j\frac{\rho_{ol} c_{ol}}{S_1} \sin k_1 L_1. \tag{10.268}$$

And, therefore, for $\rho_{o1} = \rho_{o2} = \rho_o$ and $c_{o1} = c_{o2} = c_o$

$$P_{ol} = Q_{ol} \frac{c_o \rho_o}{S_1} \frac{\frac{S_1}{S_2} \cos k_1 L_1 + j \sin k_1 L_1}{\cos k_1 L_1 + j\frac{S_1}{S_2} \sin k_1 L_1}. \tag{10.269}$$

Let us now generalize this result to the general system shown in Figure 10.23. Because

$$\begin{Bmatrix} Q_{o(n-1)} \\ P_{o(n-1)} \end{Bmatrix} = \begin{bmatrix} A_{n-1} & B_{n-1} \\ C_{n-1} & D_{n-1} \end{bmatrix} \begin{Bmatrix} \frac{S_n}{c_{on} \rho_{on}} \\ 1 \end{Bmatrix} P_{L(n-1)} \tag{10.270}$$

we get

$$\begin{Bmatrix} Q_{ol} \\ P_{ol} \end{Bmatrix} = \begin{bmatrix} A_1 & B_1 \\ C_1 & D_1 \end{bmatrix} \begin{bmatrix} A_2 & B_2 \\ C_2 & D_2 \end{bmatrix} \cdots \begin{bmatrix} A_{n-1} & B_{n-1} \\ C_{n-1} & D_{n-1} \end{bmatrix} \begin{Bmatrix} \frac{S_n}{c_{on} \rho_{on}} \\ 1 \end{Bmatrix} P_{L(n-1)}. \tag{10.271}$$

Again, because Q_{ol} is given, we can solve P_{ol} and $P_{L(n-1)}$. Next, the values at the other joining points can be calculated.

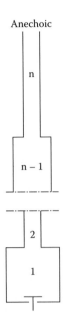

FIGURE 10.23 System with anechoic termination.

At times, for experimental purposes, an anechoic termination is introduced deliberately, as, for example, in an experimental discharge system with an excessively long pipe (see Elson and Soedel, 1974).

10.3.10 Termination Defined by an Impedance

Often, a termination is defined by an experimentally determined impedance (for example, see Singh and Soedel, 1978a). In the case shown in Figure 10.24a, we have

$$Q_{o2} = Z_{oo2} P_{o2}. \tag{10.272}$$

However, because $Q_{o2} = Q_{L1}$ and $P_{o2} = P_{L1}$ we get

$$\begin{Bmatrix} Q_{L1} \\ P_{L1} \end{Bmatrix} = \begin{Bmatrix} Z_{oo2} \\ 1 \end{Bmatrix} P_{o2} \tag{10.273}$$

$$\begin{Bmatrix} Q_{o1} \\ P_{o1} \end{Bmatrix} = \begin{bmatrix} A_1 & B_1 \\ C_1 & D_1 \end{bmatrix} \begin{Bmatrix} Z_{oo2} \\ 1 \end{Bmatrix} P_{o2}. \tag{10.274}$$

Note that if we have an anechoic pipe, $Z_{oo2} = S_2 / c_{o2} \rho_{o2}$, our equation reduces to that for the anechoic termination case.

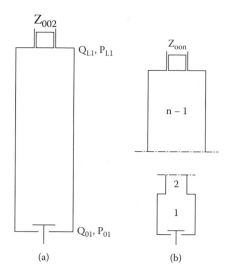

FIGURE 10.24 Termination defined by an impedance: (a) single tube, (b) multiple tubes in series.

The general case corresponding to Figure 10.24 (b) is

$$\begin{Bmatrix} Q_{o1} \\ P_{o1} \end{Bmatrix} = \begin{bmatrix} A_1 & B_1 \\ C_1 & D_1 \end{bmatrix} \begin{bmatrix} A_2 & B_2 \\ C_2 & D_2 \end{bmatrix} \cdots \begin{bmatrix} A_{n-1} & B_{n-1} \\ C_{n-1} & D_{n-1} \end{bmatrix} \begin{Bmatrix} Z_{oon} \\ 1 \end{Bmatrix} P_{L(n-1)}. \quad (10.275)$$

10.3.11 MULTICYLINDER COMPRESSORS

As an example, let us investigate the case shown in Figure 10.21. Let us assume that inputs are the following at tube 1 of the form

$$q_1(o,t) = Q_{o1} e^{j\omega t} \quad (10.276)$$

and at tube m they are

$$q_m(L,t) = Q_{Lm} e^{j(\omega t - \phi)}. \quad (10.277)$$

Note that the time base for the Fourier analysis of the compressor inputs has to be the same, even if the cylinders are out of phase, because all pistons are actuated by the same shaft. For the special case of identical cylinders where the inputs are identical, but only shifted by a time lag T_1 as shown in Figure 10.25, we may write

$$q_m(L,t) = q_1(o, t - T_1). \quad (10.278)$$

Suction or Discharge System Gas Pulsations and Mufflers

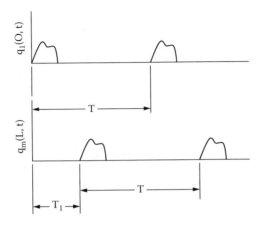

FIGURE 10.25 Shifted volume velocity inputs for a two-cylinder compressor.

Thus, every component in the Fourier series is shifted by T_1:

$$q_m(L,t) = a_o + \sum_{n=1}^{\infty} a_n \cos\omega(t-T_1) + b_n \sin\omega(t-T_1) \qquad (10.279)$$

but the coefficients of the Fourier series remain the same.

Thus, our inputs are of the form

$$q_1(o,t) = Q_{o1} e^{j\omega t} \qquad (10.280)$$

$$q_m(L,t) = Q_{o1} e^{j\omega(t-T_1)} = Q_{o1} e^{-j\omega T_1} e^{j\omega t}. \qquad (10.281)$$

Therefore

$$Q_{Lm} = Q_{o1} e^{-j\omega T_1} \qquad (10.282)$$

Let us now proceed with our example. The fundamental relationship is, as before

$$\begin{Bmatrix} Q_{o1} \\ P_{o1} \end{Bmatrix} = \begin{bmatrix} A_1 & B_1 \\ C_1 & D_1 \end{bmatrix} \begin{bmatrix} A_2 & B_2 \\ C_2 & D_2 \end{bmatrix} \cdots \begin{bmatrix} A_n & B_n \\ C_n & D_n \end{bmatrix} \begin{bmatrix} 1 & \dfrac{1}{Z_{(m+1)oo}} \\ 0 & 1 \end{bmatrix} \qquad (10.283)$$

$$\begin{bmatrix} A_{n+1} & B_{n+1} \\ C_{n+1} & D_{n+1} \end{bmatrix} \begin{bmatrix} A_{n+2} & B_{n+2} \\ C_{n+2} & D_{n+2} \end{bmatrix} \cdots \begin{bmatrix} A_m & B_m \\ C_m & D_m \end{bmatrix} \begin{Bmatrix} Q_{Lm} \\ P_{Lm} \end{Bmatrix}. \qquad (10.284)$$

The only difference now is that instead of a boundary condition, we have a defined input at the other end. Therefore, Q_{o1} and Q_{Lm} are known; the unknowns are P_{o1} and P_{Lm}, which can now be obtained from the equation.

This approach can easily be extended to compressors with more than two cylinders, and has been applied to compressors by Singh and Soedel (1979).

10.3.12 Gas Pulsations in Intercoolers

In multistage air compressors, for the example of a two-stage compressor, the discharge gas of stage 1 is passed through an intercooler to the suction intake of stage 2. Discharge and intake are in phase. Therefore, if we consider the discharge from the first stage, to be given by (at location $x = 0$, of the first intercooler element, the entrance)

$$q_{o1}(t) = Q_{o1} e^{j\omega t} \tag{10.285}$$

and the exit to the second stage at $x = L_n$ of the last intercooler element, to be given by

$$q_{Ln}(t) = Q_{Ln} e^{j\omega t} \tag{10.286}$$

we obtain Equation 10.239, except that in this case P_{o1} and P_{ln} are unknown while Q_{o1} and Q_{Ln} are given. The various four poles of the intercooler elements will have different sound speeds. Also, the magnitudes of Q_{Ln} will be less than for Q_{o1} (the mass flow rate will be the same). This difference can either be determined by a total simulation, or judiciously be assumed utilizing a cyclic thermodynamic model which gives the swept volumes of the two compressor stages.

10.3.13 Derivation of the Anechoic Termination Relationship

Here, the relationship in Equation 10.57 will be derived, which relates the velocity at the entrance of an anechoic pipe to the pressure at the entrance. This is the same expression as Equation 10.256, except that there the pressure is related to the volume velocity.

10.3.13.1 Propagation of a Unit Volume Velocity Impulse in an Anechoic Pipe

An anechoic pipe is, in the applied mechanics sense, a semi-infinite pipe. The acoustic displacements are given by the well-known one-dimensional wave equation

$$\frac{\partial^2 \xi}{\partial t^2} = c^2 \frac{\partial^2 \xi}{\partial x^2} \tag{10.287}$$

Suction or Discharge System Gas Pulsations and Mufflers

with the initial conditions

$$p(x,0) = 0 \qquad (10.288)$$

$$\frac{\partial \xi}{\partial t}(x,0) = 0 \qquad (10.289)$$

and boundary conditions

$$\frac{\partial \xi}{\partial t}(0,t) = \frac{1}{A}\delta(t-\tau) \qquad (10.290)$$

$$p(\infty,t) = 0. \qquad (10.291)$$

The first condition defines the unit volume velocity impulse, applied at $t = \tau$ and located mathematically by $\delta(t-\tau)$, the Dirac-delta function. The second condition defines the anechoic termination. The cross-sectional area of the pipe is A in this example. The acoustic pressure is related to the acoustic displacement by Equation 10.139:

$$p(x,t) = -\rho c^2 \frac{\partial \xi}{\partial x}(x,t) \qquad (10.292)$$

Differentiating Equation 10.282 with respect to x and substituting Equation 10.292 gives

$$\frac{\partial^2 p}{\partial t^2} = c^2 \frac{\partial^2 p}{\partial x^2}. \qquad (10.293)$$

Let us apply the Laplace transform to Equation 10.293, with respect to time. We get

$$\frac{s^2}{c^2} p(x,s) = \frac{\partial^2 p(x,s)}{\partial x^2}. \qquad (10.294)$$

The solution of this equation is

$$p(x,s) = \overline{A}\, e^{-\frac{x}{c}s} + \overline{B}\, e^{\frac{x}{c}s}. \qquad (10.295)$$

Because of the boundary condition in Equation 10.291,

$$\overline{B} = 0 \qquad (10.296)$$

and thus

$$p(x,s) = \bar{A}\, e^{-\frac{x}{c}s}. \qquad (10.297)$$

To evaluate $\bar{A}$, we make use of the relationship

$$\frac{\partial p}{\partial x} = -\rho \frac{\partial^2 \xi}{\partial t^2} \qquad (10.298)$$

which we obtain from Equations 10.287 and 10.292. Taking the Laplace transformation of Equation 10.298 gives

$$\frac{\partial p(x,s)}{\partial x} = -s^2 \rho\, \xi(x,s), \qquad (10.299)$$

and substituting Equation 10.297 results in

$$-\frac{\bar{A}s}{c} e^{-\frac{x}{c}s} = -s^2 \rho\, \xi(x,s). \qquad (10.300)$$

At $x = 0$, we get

$$\bar{A} = sc\rho\, \xi(0,s). \qquad (10.301)$$

By transforming the boundary condition in Equation 10.290, we obtain

$$\xi(0,s) = \frac{1}{A}\frac{e^{-\tau s}}{s} \qquad (10.302)$$

and thus

$$\bar{A} = \frac{c\rho}{A} e^{-\tau s}. \qquad (10.303)$$

Therefore, Equation 10.295 becomes

$$p(x,s) = \frac{c\rho}{A} e^{-\left(\frac{x}{c}+\tau\right)s}. \qquad (10.304)$$

Suction or Discharge System Gas Pulsations and Mufflers

Inverting this expression gives

$$p(x,t) = \frac{c\rho}{A}\delta\left(t - \frac{x}{c} - \tau\right). \tag{10.305}$$

We recognize this expression as a dynamic Green's function and write it as

$$G(x,t;o,\tau) = \frac{c\rho}{A}\delta\left(t - \frac{x}{c} - \tau\right) \tag{10.306}$$

where $G(x,t;o,\tau)$ is the acoustic pressure at pipe location x and at time t, due to a unit volume velocity impulse at $x = 0$ (pipe entrance) and time τ.

10.3.13.2 Response of Anechoic Pipe to General Volume Velocities at the Pipe Entrance

Instantaneous pressure, in terms of the dynamic Green's function and the instantaneous volume velocity at the pipe entrance $q(o,t)$, is given by the convolution integral

$$p(x,t) = \int_o^t q(o,\tau)\,G(x,t;o,\tau)\,d\tau. \tag{10.307}$$

Substituting Equation 10.305 gives

$$p(x,t) = \frac{c\rho}{A} q\left(o, t - \frac{x}{c}\right). \tag{10.308}$$

This means, for instance, that the pressure at a time t, and a pipe position x, is equal to $\frac{c\rho}{A}$ multiplied by the instantaneous volume velocity $q(o,t)$ at time $t = t_1 - \frac{x_1}{c}$. Thus, for instance, for the case of a rectangular volume velocity pulse, the propagating pressure is also a rectangular pulse, propagating with velocity c.

Of special interest is the case at $x = 0$:

$$p(o,t) = \frac{c\rho}{A} q(o,t) \tag{10.309}$$

For instance, if $p(o,t)$ is calculated first, we get

$$q(o,t) = \frac{A}{c\rho} p(o,t) \tag{10.310}$$

and

$$p(x,t) = p\left(0, t - \frac{x}{c}\right). \qquad (10.311)$$

Of primary interest for the Helmholtz resonator model application is the acoustic displacement at the anechoic pipe entrance. Because

$$\frac{d\xi}{dt}(0,t) = \frac{1}{A}q(0,t), \qquad (10.312)$$

we obtain, utilizing Equation 10.310:

$$\frac{d\xi}{dt}(0,t) = \frac{1}{\rho c}p(0,t), \qquad (10.313)$$

which is Equation 10.57. It is understood that the speed of sound c is the average speed of sound c_o, and the mass density ρ is the average mass density ρ_o.

In terms of volume velocity, this can also be written in the form of Equation 10.256:

$$q(0,t) = \frac{A}{c\rho}p(0,t). \qquad (10.314)$$

Where the cross-section of the pipe is A or S (S was not used here to avoid confusion with the Laplace transformation variable s). For harmonically varying pressure and volume velocities, this becomes

$$Q(0,t) = \frac{A}{c\rho}P(0,t) \qquad (10.315)$$

where we have set $q = Qe^{j\omega t}$ and $p = Pe^{j\omega t}$.

10.3.14 Time Response of a Finite Gas Column

In order to preserve the advantage of the Helmholtz resonator model which allows simulations directly in the time domain (also desirable when modeling thermodynamics and heat transfer), we have to overcome the wavelength limitation, which restricts the allowable length of the manifold passages and muffler pipes. This will also allow the treatment of long, finite pipes in compressor systems. Here we will relate the pressure response of a finite gas column of uniform cross-section to a volume flow oscillation at its entrance.

Suction or Discharge System Gas Pulsations and Mufflers

As an example, let us consider a pipe of uniform cross-section. The termination is a large collection volume or tank, as may occur in an air compressor installation. At the entrance of the pipe, we wish to know the pressure as a function of the mass flow at the pipe entrance.

The equation of motion (Equation 10.142) for gas particles for uniform or piecewise uniform pipes is for zero damping:

$$\frac{\partial^2 \xi}{\partial x^2} - \frac{1}{c^2}\frac{\partial^2 \xi}{\partial t^2} = 0 . \tag{10.316}$$

If we designate $x = 0$ as the entrance to the pipe system and $x = L$ as the exit into the large collection volume, we have as boundary conditions

$$\frac{\partial \xi}{\partial x}(L,t) = 0 \quad \text{and} \tag{10.317}$$

$$\rho_o A\, \xi(o,t) = \int_o^t \dot{M}\, dt \tag{10.318}$$

where $\dot{M}$ is the mass flow rate; for example, from a discharge valve. To solve this problem, we must make the following substitution:

$$\xi(x,t) = \xi(o,t) + \eta(x,t). \tag{10.319}$$

Then

$$\frac{\partial^2 \eta}{\partial x^2} - \frac{1}{c^2}\frac{\partial^2 \eta}{\partial t^2} = \frac{1}{c^2 \rho_o A}\frac{d\dot{M}}{dt}. \tag{10.320}$$

The boundary conditions become

$$\frac{\partial \eta}{\partial x}(L,t) = 0 \tag{10.321}$$

$$\eta(o,t) = 0. \tag{10.322}$$

Therefore, we have to solve Equation 10.320, with Equations 10.321 and 10.322 as boundary conditions. The particle displacement solution is then given by Equation 10.319. The pressure solution is given by

$$p(x,t) = -c^2 \rho_o \frac{\partial \xi(x,t)}{\partial x}. \tag{10.323}$$

Note that if the collection tank is not large, the boundary condition in Equation 10.321 modifies to

$$\frac{\partial \eta}{\partial x}(L,t) - \frac{A}{V}\eta(L,t) = 0 \tag{10.324}$$

where A is the cross-section of the pipe and V is the volume of the receiving tank.

10.3.14.1 Eigenvalues of the Gas Column

Solving the homogeneous part of Equation 10.320, we recognize solutions of the form

$$\eta(x,t) = \bar{\eta}(x)e^{i\omega t}. \tag{10.325}$$

Substituting this into

$$\frac{\partial^2 \eta}{\partial x^2} - \frac{1}{c^2}\frac{\partial^2 \eta}{\partial t^2} = 0 \tag{10.326}$$

gives

$$\frac{d^2\bar{\eta}}{dt^2} + \left(\frac{\omega}{c}\right)^2 \bar{\eta} = 0. \tag{10.327}$$

The solution is

$$\bar{\eta}(x) = A\sin\frac{\omega}{c}x + B\cos\frac{\omega}{c}x. \tag{10.328}$$

Equations 10.321 and 10.322 become

$$\frac{\partial \bar{\eta}}{\partial x}(L) = 0 \tag{10.329}$$

$$\bar{\eta}(o) = 0. \tag{10.330}$$

Substituting Equation 10.328 into Equation 10.330 gives

$$B = 0. \tag{10.331}$$

Substituting Equation 10.328 into Equation 10.329 gives

$$\cos \frac{\omega}{c} L = 0. \qquad (10.332)$$

This is satisfied whenever

$$\frac{\omega L}{c} = \frac{\pi}{2}, \frac{3\pi}{2}, \frac{5\pi}{2}, \ldots \qquad (10.333)$$

or

$$\omega_n = (2n-1)\frac{\pi}{2}\frac{c}{L} \qquad (n = 1, 2, \ldots). \qquad (10.334)$$

The associated eigenfunction (mode shape) is, from Equation 10.328

$$\bar{\eta}_n(x) = \sin \frac{\omega_n}{c} x. \qquad (10.335)$$

10.3.14.2 Solution by Modal Expansion

We are now able to solve Equation 10.320 by modal expansion. Let

$$\eta(x,t) = \sum_{n=1}^{\infty} q_n(t) \bar{\eta}_n(x). \qquad (10.336)$$

Substituting this into Equation 10.320 and making use of Equation 10.327 gives

$$-\sum_{}^{\infty} \left(\frac{d^2 q_n}{dt^2} + \omega_n^2 q_n \right) \bar{\eta}_n(x) = \frac{1}{\rho_o A} \frac{d\dot{M}}{dt}. \qquad (10.337)$$

Because of the orthogonality of the eigenfunctions

$$\int_{x=0}^{L} \bar{\eta}_n(x) \bar{\eta}_m(x)\, ds = \begin{cases} 0, & n \neq m \\ \int_0^L \bar{\eta}_n^2(x)dx, & n = m \end{cases} \qquad (10.338)$$

we may multiply Equation 10.337 by $\bar{\eta}_m(x)$ and integrate. This allows us to remove the summation and Equation 10.337 becomes

$$\frac{d^2 q_n}{dt^2} + \omega_n^2 q_n = Q_n(t) \qquad (10.339)$$

where

$$Q_n(t) = -\frac{1}{\rho_o A N_n}\frac{dM}{dt}\int_o^L \bar{\eta}_n(x)\,dx \qquad (10.340)$$

$$N_n = \int_o^L \bar{\eta}_n^2(x)\,dx. \qquad (10.341)$$

For zero initial conditions, the solution of Equation 10.339 is

$$q_n(t) = \frac{1}{\omega_n}\int_o^t Q_n(\tau)\sin\omega_n(t-\tau)\,dt. \qquad (10.342)$$

Because the impulse response of Equation 10.339 is

$$g(t) = \frac{1}{\omega_n}\sin\omega_n t, \qquad (10.343)$$

Equation 10.336 becomes

$$\eta(x,t) = -\sum_{n=1}^{\infty}\bar{\eta}_n(x)\int_o^L \bar{\eta}_n(x)\,dx\,\frac{1}{\omega_n N_n}$$

$$\int_o^t \left(\frac{dM}{dt}\right)_{t=\tau}\sin\omega_n(t-\tau)\,d\tau. \qquad (10.344)$$

Equation 10.323 becomes, utilizing Equations 10.319 and 10.344,

$$p(x,t) = c^2\rho_o\sum_{n=1}^{\infty}\frac{\partial\bar{\eta}_n(x)}{\partial x}\int_o^L \bar{\eta}_n(x)\,dx\,\frac{1}{\omega_n N_n}$$

$$\int_o^t \left(\frac{dM}{dt}\right)_{t=\tau}\sin\omega_n(t-\tau)\,d\tau. \qquad (10.345)$$

This equation relates the input mass flow rate $\dot{M}$ at the pipe entrance to the pressure at any position x along the pipe.

Note that because

$$\frac{d\dot{M}}{dt} = \frac{1}{\rho_o A} \frac{d\xi(o,t)}{dt}, \quad (10.346)$$

Equation 10.346 can also be interpreted as relating pressure in the pipe to input displacement. Let us now substitute Equation 10.335. Equation 10.346 becomes

$$p(x,t) = \frac{2c^2\rho_o}{L} \sum_{n=1}^{\infty} \frac{\cos \frac{\omega_n}{c} x}{\omega_n}$$
$$\int_o^t \left(\frac{d\dot{M}}{dt}\right)_{t=\tau} \sin \omega_n (t-\tau) d\tau. \quad (10.347)$$

At the entrance of the pipe, the pressure is

$$p(o,t) = \frac{2c^2\rho_o}{L} \sum_{n=1}^{\infty} \frac{1}{\omega_n}$$
$$\int_o^t \left(\frac{d\dot{M}}{dt}\right)_{t=\tau} \sin \omega_n (t-\tau) d\tau. \quad (10.348)$$

Equations 10.347 and 10.348 can now be combined with time domain Helmholtz resonator models or can be used in their own right.

If we need to consider damping, it is best to work in terms of a modal damping coefficient ζ_n. This coefficient varies with the mode number n and becomes larger as n increases. This means that higher frequency modes are damped more. The best way to find ζ_n is by measurement or by assuming values based on experience. Considering modal damping, Equation 10.339 becomes

$$\frac{d^2 q_n}{dt^2} + 2\zeta_n \omega_n \frac{dq_n}{dt} + \omega_n^2 q_n = Q_n(t). \quad (10.349)$$

The impulse response is

$$g(t) = \frac{1}{\omega_n \sqrt{1-\zeta_n^2}} e^{-\zeta_n \omega_n t} \sin \sqrt{1-\zeta^2} \omega_n t \quad (10.350)$$

and thus

$$p(x,t) = \frac{2c^2\rho_o}{L} \sum_{n=1}^{\infty} \frac{\cos\frac{\omega_n}{c}x}{\omega_n\sqrt{1-\zeta_n^2}} \quad (10.351)$$

$$\int_o^t \left(\frac{dM}{dt}\right)_{t=\tau} e^{-\zeta_n\omega_n(t-\tau)} \sin\sqrt{1-\zeta_n^2}\,(t-\tau)d\tau.$$

The integral can easily be evaluated for any standard form of mass flow rate input. For arbitrary inputs in a simulation, the solution must be set up as a numerical integration (see also Soedel 1974a and Kim and Soedel, 1992b, 1994).

10.4 TYPICAL BEHAVIOR OF SIMPLE COMPRESSOR SUCTION OR DISCHARGE MUFFLERS

10.4.1 Defining Transmission Loss

In compressor simulation models, but also in engineering practice, inputs to suction and discharge systems, including mufflers, are defined (and thought of) as volume velocity inputs. The desired outputs are also volume velocities because they are the sources for acoustic field calculations, be they for the ambient field in the case of open suction manifolds of air compressors, or the sound field in the volume formed between a hermetic shell and the compressor casing. Rarely is it useful to think in terms of pressure sources or pressure outputs. The main reason that information in the muffler literature is often given in terms of pressures is because this is the most easily measured. Transmission loss is often defined in terms of a pressure ratio. In this book, transmission loss is defined in terms of a volume velocity ratio. Linearly, it is

$$\eta = \frac{Q_{in} - Q_{out}}{Q_{in}} \quad (10.352)$$

and logarithmically, in dB, it is

$$TL = 20\log_{10}\eta = 20\log_{10}\frac{Q_{in}}{Q_{out}}. \quad (10.353)$$

Q_{in} is the amplitude of a harmonic volume velocity input and Q_{out} is the amplitude of a harmonic volume velocity output, both of which are a function of frequency. Phasing is, from a practical viewpoint, less important and will not be reported here. (It is more important from a mathematical viewpoint if various mufflers or muffler elements are to be combined into a larger muffler system, but then it can be easily generated.) When Q_{out} is much larger than Q_{in}, TL shows a minimum (TL will be negative for small damping), which indicates a resonance of the muffler (the muffler may actually amplify noise in this frequency band).

If the muffler is characterized by an overall four pole such that

$$\begin{Bmatrix} Q_{in} \\ P_{in} \end{Bmatrix} = \begin{bmatrix} A & B \\ C & D \end{bmatrix} \begin{Bmatrix} Q_{out} \\ P_{out} \end{Bmatrix}, \qquad (10.354)$$

where $P_{out} = 0$ (the usual assumption unless we have an anechoic termination). From Equation 10.354 we obtain

$$\frac{Q_{out}}{Q_{in}} = \frac{1}{A} \qquad (10.355)$$

or

$$TL = 20\log_{10} A. \qquad (10.356)$$

A resonance of the muffler system occurs whenever (for zero damping)

$$A = 0 \qquad (10.357)$$

at which point

$$TL = -\infty. \qquad (10.358)$$

10.4.2 BACKPRESSURE

The pressure magnitude P_{in} at the location of the muffler where the input volume velocity magnitude Q_{in} is defined, characterizes the backpressure behavior of the muffler. Therefore, the ratios P_{in}/Q_{in} will be plotted in the following as a function of frequency. However, this ratio alone will not tell us how much the muffler will interfere with the opening of the valve; phasing information is also necessary, and will be given in the following. Mathematically, the ratio of the pressure at the input (the backpressure) to the volume velocity at the output is for $P_{out} = 0$, from Equation 10.354,

$$\frac{P_{in}}{Q_{out}} = C. \qquad (10.359)$$

And, because of Equation 10.355, the ratio of the backpressure to the input volume velocity is

$$\frac{P_{in}}{Q_{in}} = \frac{C}{A}. \qquad (10.360)$$

10.4.3 REACTION MUFFLER ELEMENT ARRANGED IN LINE

In the following, examples of typical (see Kim and Soedel, 1992 and Soedel, 1992) arrangements of muffler components are shown, as one may find them in small, fractional-horsepower, refrigeration compressors. Dimensions are defined in the respective figures, where for these specific examples $L_1 = 20$ mm, $D_1 = 16$ mm, $D_2 = 4$ mm, and the speed of sound is $c = 140$ m/sec, unless otherwise indicated. For larger compressors, these designs can be scaled up, keeping in mind that diameters have to be small enough for cross-modes not to be excited, otherwise the theory used here is invalid (for scaling rules, see Laville and Soedel, 1978). But even with cross-modes present, conclusions reached by examining these results will still be of help to designers.

10.4.3.1 Volume-Tailpipe Muffler

This is the simplest arrangement. The mathematical model of the system four pole is

$$\begin{bmatrix} A & B \\ C & D \end{bmatrix} = \begin{bmatrix} A_1 & B_1 \\ C_1 & D_1 \end{bmatrix} \begin{bmatrix} A_2 & B_2 \\ C_2 & D_2 \end{bmatrix} \quad (10.361)$$

where the subscripts 1 and 2 refer to the two elements that form this type of muffler (volume = 1, tailpipe = 2).

Figure 10.26 shows a typical transmission ratio in terms of the ratio of the volume velocities at the tailpipe exit to the harmonic input velocity at the valve. This silencer has an approximate low-frequency cutoff of

$$f_c = \sqrt{2}\, f_1 \quad (10.362)$$

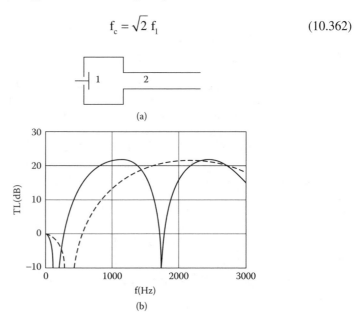

FIGURE 10.26 (a) Typical expansion chamber–tailpipe design, (b) associated transmission loss:—, $L_2 = 2L_1$; ---, $L_2 = 0.5\, L_1$.

where

$$f_1 = \frac{c}{2\pi}\sqrt{\frac{A_2}{V_1 L_2}}. \tag{10.363}$$

The silencer is unable to attenuate pulsations below its cutoff frequency. This illustrates why it is so difficult to attenuate low-frequency pulses. In order to lower the cutoff frequency, it is desirable to have a large length L_2, and a large volume V_1. The transfer pipe cross-section A_2 is usually selected based on general flow criteria, for example, low Mach number, and is assumed to be a given quantity. It is interesting to see that as L_2 is made larger, from $0.5\,L_1$ to $2\,L_1$, the first resonance frequency of the tailpipe gas column, approximately at

$$f = \frac{c}{2L_2} \tag{10.364}$$

shifts into the frequency region of interest. In the vicinity of this frequency, the muffler will not attenuate; as a matter of fact, it may amplify. Only when we truly go to $L_2 = \infty$ will we have both a zero cutoff frequency and an uninterrupted region of attenuation.

The following concept, therefore, emerges. While a long tailpipe has the advantage that the muffler becomes effective at a lower cutoff frequency, it has the disadvantage that it will shift tailpipe resonances into the region of interest. Therefore, the proper tailpipe length has to be adapted to the physical requirements of the design.

A tailpipe is absolutely necessary when the low frequency cutoff is of concern. If, for some reason, the tailpipe length cannot be altered, then the expansion volume has to be increased to attain a lower cutoff frequency. In many practical mufflers, the low frequency cutoff determines the overall size (volume) of the design.

Figure 10.27 shows the backpressure characteristics for the case in Figure 10.26. The backpressure magnitude reaches a peak whenever there is a muffler resonance. Below the first resonance, the pressure is leading the volume velocity input by $\pi/2$ at the first resonance, the backpressure harmonic is in phase with the volume velocity, and above the first resonance it is lagging the volume velocity.

This can be approximately illustrated by solving Equation 10.18 for a harmonic volume velocity input $(Q_{in} = Q_d)$:

$$Q_d = Q_o e^{j\omega t} \tag{10.365}$$

recognizing that $(Q_{out} = Q_L)$

$$Q_L = A_d \dot{\xi}_d. \tag{10.366}$$

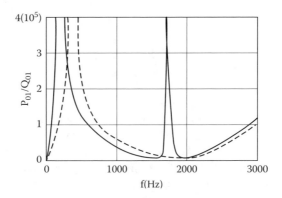

FIGURE 10.27 Backpressure magnitudes at the valve. Key as in Figure 10.26.

This equation describes the Helmholtz simplification, while Figure 10.26 is based on the continuous approach, but it will illustrate the phasing. We get

$$Q_L = \frac{\omega_{nd}^2}{(\omega_{nd}^2 - \omega^2)} Q_o. \tag{10.367}$$

From Equation 10.17, when $\omega < \omega_{nd}$, we obtain,

$$p_d = P_o e^{j(\omega t + \frac{\pi}{2})} \tag{10.368}$$

where

$$P_o = \frac{K_{od}}{V_{od}} \frac{\omega}{(\omega_{nd}^2 - \omega^2)} Q_o \tag{10.369}$$

and when $\omega > \omega_{nd}$,

$$p_d = P_o e^{j(\omega t - \frac{\pi}{2})} \tag{10.370}$$

where

$$P_o = \frac{K_{od}}{V_{od}} \frac{\omega}{(\omega^2 - \omega_{nd}^2)} Q_o. \tag{10.371}$$

Unfortunately, the interpretation is not simple because it depends on (a) knowledge of the harmonics, which are important to the flow characteristics of the valve (positive backpressure, if present during valve opening, will impede valve performance), and (b) knowledge of how the important harmonic waves should ideally be phased with respect to the valve opening. A total simulation is necessary.

However, the safest course seems to be to (a) avoid the first resonance, because it will create unreasonably large gas pulsations, even while in certain circumstances the phasing may be such that we obtain very small pulsation losses, and (b) select a design that gives the smallest P_o/Q_o ratio at the valve opening frequency (assuming this to be the most important criterion for satisfactory flow performance). Therefore, in the particular example (assuming 3600 RPM or $\omega = 377 \text{rad/s}$), it appears that $L_2 = 2L_1 = 40$ mm is the best choice. What the graph does not show is that making the expansion volume as large as possible is an additional requirement for the lowest P_o/Q_o ratio.

Another way of looking at this is to assume that the interpretation depends on the frequency that is the pumping frequency of the valve. In a single-cylinder compressor of 3600 RPM, this frequency is 60 Hz ($\omega = 2\pi(60)$ rad/s). The wave amplitude of the fundamental volume flow velocity component is approximately centered at the midpoint of the valve opening. What the results tell us is that one should not select the valve opening period to coincide with the first resonance of the manifold and muffler system. Rather, being above or below this resonance tends to average the effect of the valve backpressure to approximately zero because the pressure is $\pi/2$ leading or lagging.

There is still another interpretation that has to do with waves traveling back and forth through the system (the well-known ramming effect argument), but this becomes too complicated if we deviate very much from a simple pipe. Also, simulations seem to indicate that the best one can achieve with ramming is to create a situation where there is zero backpressure buildup. A truly negative backpressure when the discharge valve opens, which seems to be the theoretical preference, seems impossible to achieve.

10.4.3.2 Scaling

It can be shown that if we scale all cross-sectional areas equally, but we keep the length of each tubular element the same, the transmission loss curve is not changed. It is true that oscillation velocities in the various cross-sections reduce with increasing diameters, but in wave guide acoustics, it is the volume velocity that is important, not the linear velocity (see also Laville and Soedel, 1978).

On the other hand, it can be shown that the backpressure is greatly reduced as we scale to twice the original cross-sections, holding to previous statements that the backpressure criterion dictates the size of the muffler.

A beneficial effect in terms of transmission loss is only obtained if we increase the volume diameter, but keep the tailpipe diameter the same, as shown in Figure 10.28. This lowers the cutoff frequency and increases the transmission loss in general. Because the tailpipe length was not changed, the first tailpipe resonance (standing wave) is not changed, and the frequency band of attenuation is preserved. It should, however, be noted that we cannot increase the diameter (cross-sectional area) of the volume ad infinitum because the plane wave theory used to obtain these curves will break down due to the excitation of the cross-modes.

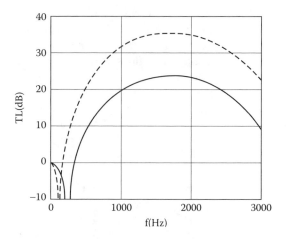

FIGURE 10.28 Influence of changing the first cross-section:—, A_1; ---, $2A_1$, for $L_2 = 2L_1$ in both cases.

Another instructive case is shown in Figure 10.29, where the diameters were kept constant and only the length dimensions were scaled. Increasing the length dimensions will reduce the cutoff frequency, as expected, but will shift tailpipe resonances into the region of interest, thereby disrupting the attenuation characteristics. Therefore, unless the cutoff frequency is of prime concern, there is no particular advantage in scaling up the length of a muffler from a transmission loss viewpoint; however, it can be shown that the backpressure characteristic will improve.

It should also be noted that damping smoothes out the attenuation "holes" due to tailpipe resonances. It will prevent noise amplification at these resonances by removing the transmission loss singularities.

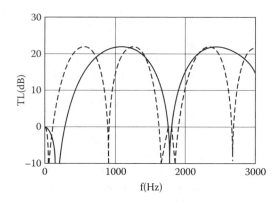

FIGURE 10.29 Doubling all length dimensions:—, $L_2 = 2L_1$; ---, $L_2 = 4L_1$.

Suction or Discharge System Gas Pulsations and Mufflers

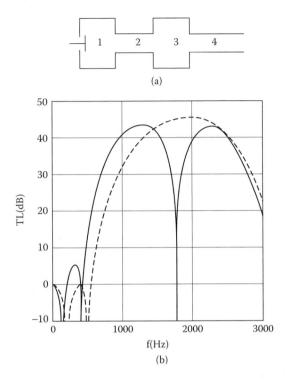

FIGURE 10.30 (a) Typical case of two identical expansion volumes in series; (b) associated transmission loss:—, $L_4 = 2L_1$;---, $L_4 = 0.5L_1$; otherwise $L_1, L_2 = L_1$, $L_3 = L_1$, $A_1 = A_1, A_2, A_3$, $A_4 = A_2$.

10.4.3.3 Two-Volume Muffler with Inertia Tube and Tailpipe

Figure 10.30 illustrates the typical behavior of a two-volume muffler (two volume-tailpipe mufflers placed in series). The mathematical model is

$$\begin{bmatrix} A & B \\ C & D \end{bmatrix} = \begin{bmatrix} A_1 & B_1 \\ C_1 & D_1 \end{bmatrix} \begin{bmatrix} A_2 & B_2 \\ C_2 & D_2 \end{bmatrix} \begin{bmatrix} A_3 & B_3 \\ C_3 & D_3 \end{bmatrix} \begin{bmatrix} A_4 & B_4 \\ C_4 & D_4 \end{bmatrix} \quad (10.372)$$

where the subscripts refer to the following: first volume = 1, inertia tube = 2, second volume = 3, tailpipe = 4. The pipe between the two volumes is termed the inertia tube because its function is to block gas pulsations by the mass of the gas contained in it. This design can also be viewed mathematically as

$$\begin{bmatrix} A & B \\ C & D \end{bmatrix} = \begin{bmatrix} A_I & B_I \\ C_I & D_I \end{bmatrix} \begin{bmatrix} A_{II} & B_{II} \\ C_{II} & D_{II} \end{bmatrix} \quad (10.373)$$

where the four pole of the first volume-tailpipe muffler is

$$\begin{bmatrix} A_I & B_I \\ C_I & D_I \end{bmatrix} = \begin{bmatrix} A_1 & B_1 \\ C_1 & D_1 \end{bmatrix} \begin{bmatrix} A_2 & B_2 \\ C_2 & D_2 \end{bmatrix} \tag{10.374}$$

and the four pole of the second volume-tailpipe muffler is

$$\begin{bmatrix} A_{II} & B_{II} \\ C_{II} & D_{II} \end{bmatrix} = \begin{bmatrix} A_3 & B_3 \\ C_3 & D_3 \end{bmatrix} \begin{bmatrix} A_4 & B_4 \\ C_4 & D_4 \end{bmatrix}. \tag{10.375}$$

This design is relatively common. If we compare the $L_2 = L_1$ case here with the $L_2 = L_1$ single-volume case in Figure 10.26, we see that the maximum transmission loss magnitude in decibels has doubled, but that on the other hand, the cutoff frequency has increased. Because of the practical importance of the two-volume muffler, an approximate formula for the cutoff frequency is useful (see Soedel and Soedel, 1992) which is

$$f_c = f_{01} \sqrt{1 + \left(\frac{V_1}{V_2}\right) + \left(\frac{A_2}{A_1}\right)\left(\frac{V_1}{V_2}\right)\left(\frac{L_1}{L_2}\right)} \tag{10.376}$$

where

$$f_{01} = \frac{c}{2\pi} \sqrt{\left(\frac{A_1}{V_1 L_1}\right)} \tag{10.377}$$

and where V_1, V_2 are the two volumes, A_1, A_2 the two cross-sections, and L_1, L_2 are the two lengths.

It is shown in Figure 10.30, that an increase in the tailpipe length L_2 causes the cutoff frequency to be lowered, but if L_2 is too large, the first tailpipe resonance shifts into the frequency band of interest. When comparing Figure 10.30 with Figure 10.26, it is seen that the tailpipe resonance is hardly affected by the fact that we now have two instead of one volume. These pipe resonances can, as a rule, always be estimated considering the free-free pipe case.

The potential problems with backpressure have increased since the cutoff frequency of the two-muffler case, as compared to the cutoff frequency of the single-muffler case (in spite of the fact that we have doubled the total muffler volume), is higher. It may be necessary to increase the volume V_1 somewhat to compensate.

10.4.3.4 Triple-Volume Muffler

Placing three volume-tailpipe mufflers in series continues the trend established for the double volume muffler. The mathematical model is

$$\begin{bmatrix} A & B \\ C & D \end{bmatrix} = \begin{bmatrix} A_I & B_I \\ C_I & D_I \end{bmatrix} \begin{bmatrix} A_{II} & B_{II} \\ C_{II} & D_{II} \end{bmatrix} \begin{bmatrix} A_{III} & B_{III} \\ C_{III} & D_{III} \end{bmatrix} \quad (10.378)$$

where

$$\begin{bmatrix} A_I & B_I \\ C_I & D_I \end{bmatrix} = \begin{bmatrix} A_1 & B_1 \\ C_1 & D_1 \end{bmatrix} \begin{bmatrix} A_2 & B_2 \\ C_2 & D_2 \end{bmatrix} \quad (10.379)$$

$$\begin{bmatrix} A_{II} & B_{II} \\ C_{II} & D_{II} \end{bmatrix} = \begin{bmatrix} A_3 & B_3 \\ C_3 & D_3 \end{bmatrix} \begin{bmatrix} A_4 & B_4 \\ C_4 & D_4 \end{bmatrix} \quad (10.380)$$

$$\begin{bmatrix} A_{III} & B_{III} \\ C_{III} & D_{III} \end{bmatrix} = \begin{bmatrix} A_5 & B_5 \\ C_5 & D_5 \end{bmatrix} \begin{bmatrix} A_6 & B_6 \\ C_6 & D_6 \end{bmatrix} \quad (10.381)$$

and where the subscripts refer to, from the left: first volume = 1, first inertia tube = 2, second volume = 3, second inertia tube = 4, third volume = 5, tailpipe = 6.

The triple-volume muffler has a total volume that is three times as large as the single-volume muffler. For this expenditure in space and cost, we obtain the advantage of potentially tripling the transmission loss in decibels if we can place the lobe maximum at the frequencies of interest. We lose, potentially, low-frequency effectiveness (in the example, the low-frequency cutoff is now at about 650 Hz, while for the one-volume muffler it was at approximately 400 Hz) and the frequency bands for which the muffler is effective become narrower.

10.4.4 SIDE-BRANCH ATTENUATORS

The principle of a single-branch attenuator is that it takes oscillatory energy from the main branch and converts it into oscillations of the side-branch where it will be dissipated into heat.

10.4.4.1 Helmholtz Resonator Side-Branch Attenuator

In this kind of design, the fundamental natural frequency of the side-branch resonator is tuned to the approximate center of the frequency band that is to be attenuated.
The mathematical model is

$$\begin{bmatrix} A & B \\ C & D \end{bmatrix} = \begin{bmatrix} 1 & 1/Z_{300} \\ 0 & 1 \end{bmatrix} \begin{bmatrix} A_2 & B_2 \\ C_2 & D_2 \end{bmatrix} \quad (10.382)$$

where

$$Z_{300} = \frac{P_{03}}{Q_{03}} \qquad (10.383)$$

$$\begin{Bmatrix} Q_{03} \\ P_{03} \end{Bmatrix} = \begin{bmatrix} A_3 & B_3 \\ C_3 & D_3 \end{bmatrix} \begin{bmatrix} A_4 & B_4 \\ C_4 & D_4 \end{bmatrix} \begin{Bmatrix} 0 \\ P_{L4} \end{Bmatrix}. \qquad (10.384)$$

The elements are numbered as follows: tailpipe = 2, side-branch resonator neck = 3, resonator volume = 4.

In Figure 10.31, the tuning frequency of the side-branch Helmholtz resonator is approximately 1000 Hz ($V_4 = 6.24(10^{-7})$ m³, $L_3 = 0.5$ $L_1 = 0.01$ m, $A = 1.256(10^{-5})$ m², c = 140 m/sec.. The sharp transmission loss peak becomes rounded off as damping is introduced (not shown). The width of the attenuated frequency band is a function of the rest of the system; in this case simply a pipe of length L_2. The

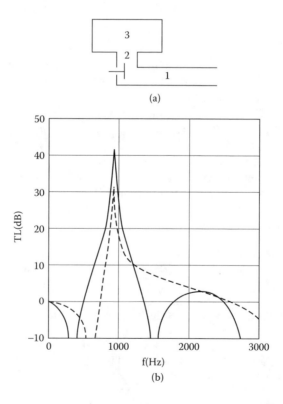

FIGURE 10.31 (a) Typical side-branch resonator located close to the valve. (b) Associated transmission loss:—, $L_2 = 2L_1$; ---, $L_2 = 0.5L_1$. Otherwise: $L_2 = 0.5L_1$, $L_3 = L_1$; also, $A_1 = A_2$.

attenuated frequency region to the left of the tuning frequency diminishes with decreasing L_2, while it widens to the right.

As L_2 increases from $0.5 L_1$, the process reverses, and for $L_2 = L_1$, the first tailpipe resonance shifts into the region of interest. Note that this resonance (approximately 2500 Hz) is appreciably shifted from the open-open gas column frequency because the pipe of length L_2 is not separated from the side-branch attenuator by a volume.

10.4.4.2 Influence of Attachment Location

A commonly asked question concerns whether a side-branch attenuator should best be attached to a volume or a pipe section. Figures 10.32a and 10.32b illustrate the two cases under consideration.

The mathematical model is, for the case in Figure 10.32a,

$$\begin{bmatrix} A & B \\ C & D \end{bmatrix} = \begin{bmatrix} A_1 & B_1 \\ C_1 & D_1 \end{bmatrix} \begin{bmatrix} 1 & 1/Z_{400} \\ 0 & 1 \end{bmatrix} \begin{bmatrix} A_2 & B_2 \\ C_2 & D_2 \end{bmatrix} \begin{bmatrix} A_3 & B_3 \\ C_3 & D_3 \end{bmatrix} \quad (10.385)$$

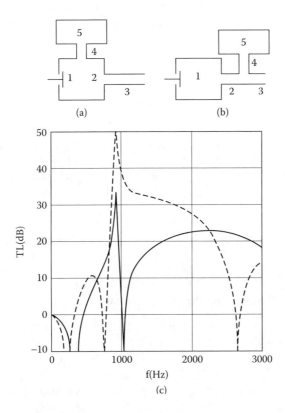

FIGURE 10.32 Typical side-branch resonator located (a) off the expansion volume, (b) off the tail pipe. (c) Associated transmission loss:—, for case (a); ---, for case (b).

and for the case in 10.32b,

$$\begin{bmatrix} A & B \\ C & D \end{bmatrix} = \begin{bmatrix} A_1 & B_1 \\ C_1 & D_1 \end{bmatrix}\begin{bmatrix} A_2 & B_2 \\ C_2 & D_2 \end{bmatrix}\begin{bmatrix} 1 & 1/Z_{400} \\ 0 & 1 \end{bmatrix}\begin{bmatrix} A_3 & B_3 \\ C_3 & D_3 \end{bmatrix} \quad (10.386)$$

where, for both cases

$$Z_{400} = \frac{P_{04}}{Q_{04}} \quad (10.387)$$

$$\begin{Bmatrix} Q_{04} \\ P_{04} \end{Bmatrix} = \begin{bmatrix} A_4 & B_4 \\ C_4 & D_4 \end{bmatrix}\begin{bmatrix} A_5 & B_5 \\ C_5 & D_5 \end{bmatrix}\begin{Bmatrix} 0 \\ P_{L5} \end{Bmatrix}. \quad (10.388)$$

The elements for through flow are labeled 1, 2, 3; the side-branch resonator tube = 4 and the volume = 5.

The answer to the question is that an attachment of the side attenuation to a pipe section is probably preferable. This is illustrated in Figure 10.32c, where it is shown that a higher transmission loss can be achieved in a frequency band to the right of the tuning frequency when the attenuator is attached to a tube section. The tuning frequency, as calculated in the design stage by only considering the Helmholtz resonator frequency

$$f = \frac{c}{2\pi}\sqrt{\frac{A_4}{V_5 L_4}} \quad (10.389)$$

where A_4 and L_4 are tube cross-section and length of the resonator neck (this A_4 has nothing to do with A_4 in the matrix mathematical model) and V_5 is the resonator volume, is not exactly the final tuning frequency because the rest of the system influences it. For example, in Figure 10.32c, it is approximately 944 Hz, but it is 1000 Hz when estimated by Equation 10.389. Therefore, some trial-and-error development will always be necessary.

The backpressure behavior is becoming more complicated. It can be shown that an additional backpressure peak is generated, whose location seems to be at the tuning frequency. This seems to indicate that one should not tune a side-branch resonator to the fundamental pumping frequency.

10.4.4.3 Gas Column Side-Branch Attenuator

In principle, the shape of the side-branch attenuator is only important in the sense that Helmholtz resonator shapes can be tuned to very low frequencies, while a pipe resonator has to be made fairly long in order to be able to be tuned to the same frequency. In the example in Figure 10.33, the side-branch pipe resonator was tuned nominally to 1400 Hz. Note that a side-branch tube of this type is often referred to as a quarter wave tube, but this does not necessarily mean that the quarter-wave formula ($f = 4c/(2L_1) = 1400$ Hz, in this case) will describe the actual tuning frequency. It can be used only in a very approximate sense because of the influence of the rest of

Suction or Discharge System Gas Pulsations and Mufflers

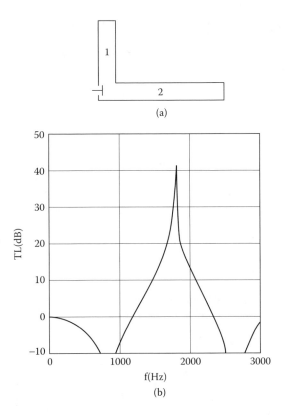

FIGURE 10.33 (a) Typical side-branch resonator in the form of a tube (quarter wave tube). (b) Associated transmission loss ($L_2 = 2L_1$).

the system. The actual tuning frequency turned out to be about 1750 Hz. Meaningful approximations can sometimes be made only by experts. The exact analytical solutions or simulation programs should be used by less experienced designers.

The mathematical model is

$$\begin{bmatrix} A & B \\ C & D \end{bmatrix} = \begin{bmatrix} 1 & 1/Z_{300} \\ 0 & 1 \end{bmatrix} \begin{bmatrix} A_2 & B_2 \\ C_2 & D_2 \end{bmatrix} \qquad (10.390)$$

where

$$Z_{300} = P_{03}/Q_{03} \qquad (10.391)$$

$$\begin{Bmatrix} Q_{03} \\ P_{03} \end{Bmatrix} = \begin{bmatrix} A_3 & B_3 \\ C_3 & D_3 \end{bmatrix} \begin{Bmatrix} 0 \\ P_{L3} \end{Bmatrix}. \qquad (10.392)$$

The side-branch attenuator is a continuous system with many resonance frequencies of its own.

The backpressure developed by this example can be shown to be very large. This case was only selected for its educational value. Tubular side-branch resonators on tube perform from a backpressure viewpoint, better when expansion volumes are present.

10.4.5 Tube Penetrating Volumes

Taking as an example the two-volume muffler of Figure 10.30, there is no fundamental reason why tubes cannot be placed inside the muffler volumes, as shown, for example, in Figure 10.35, if one remembers that the volumes should be increased by the amount that the volumes are diminished by the presence of the tubes. The low-frequency cutoff will be more or less the same, as long as the tube lengths are the same. The only difference is that side-branch pockets are created, which may be beneficial in the higher frequency range (quarter- and half-wave elimination).

10.4.5.1 Single-Volume Muffler with Penetrating Tailpipe

Figure 10.34 shows the case where the tailpipe penetration is kept constant at $0.5\ L_1$, but the total length of the tailpipe is changed.

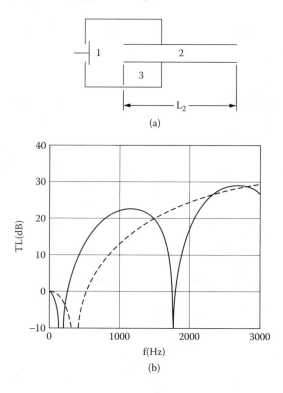

FIGURE 10.34 (a) Typical design when the tailpipe extends into the expansion volume. (b) Associated transmission loss:—, $L_2 = L_1$; ---, $L_2 = 0.5 L_1$. Otherwise, $L_3 = L_1$.

Suction or Discharge System Gas Pulsations and Mufflers

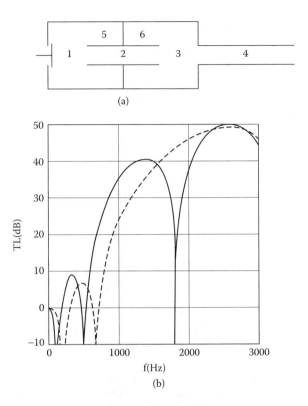

FIGURE 10.35 (a) Typical two expansion volume muffler with penetrating inertia tube connection. (b) Associated transmission loss:—, $L_2 = 2L_1$; ---, $L_2 = 0.5L_1$. Otherwise, $L_1 = L_5 = L_6 = L_3$ and $L_2 = L_5 + L_6$.

The mathematical model is

$$\begin{bmatrix} A & B \\ C & D \end{bmatrix} = \begin{bmatrix} A_1 & B_1 \\ C_1 & D_1 \end{bmatrix} \begin{bmatrix} 1 & 1/Z_{300} \\ 0 & 1 \end{bmatrix} \begin{bmatrix} A_2 & B_2 \\ C_2 & D_2 \end{bmatrix} \quad (10.393)$$

$$Z_{300} = P_{03}/Q_{03} \quad (10.394)$$

$$\begin{Bmatrix} Q_{03} \\ P_{03} \end{Bmatrix} = \begin{bmatrix} A_3 & B_3 \\ C_3 & D_3 \end{bmatrix} \begin{Bmatrix} 0 \\ P_{L3} \end{Bmatrix}. \quad (10.395)$$

We see, essentially, that the results are similar to Figure 10.26. The low-frequency cutoff is somewhat higher because of the decreased expansion volume.

10.4.5.2 Two-Volume Muffler with Penetrating Inertia Tube

The configuration is shown in Figure 10.35a. The mathematical model is

$$\begin{bmatrix} A & B \\ C & D \end{bmatrix} = \begin{bmatrix} A_1 & B_1 \\ C_1 & D_1 \end{bmatrix} \begin{bmatrix} 1 & \frac{1}{Z_{500}} \\ 0 & 1 \end{bmatrix} \begin{bmatrix} A_2 & B_2 \\ C_2 & D_2 \end{bmatrix} \begin{bmatrix} 1 & \frac{1}{Z_{600}} \\ 0 & 1 \end{bmatrix} \begin{bmatrix} A_3 & B_3 \\ C_3 & D_3 \end{bmatrix} \begin{bmatrix} A_4 & B_4 \\ C_4 & D_4 \end{bmatrix}$$

(10.396)

$$Z_{500} = \frac{P_{05}}{Q_{05}}, \quad Z_{600} = \frac{P_{06}}{Q_{06}} \quad (10.397)$$

$$\begin{Bmatrix} Q_{05} \\ P_{05} \end{Bmatrix} = \begin{bmatrix} A_5 & B_5 \\ C_5 & D_5 \end{bmatrix} \begin{Bmatrix} 0 \\ P_{L5} \end{Bmatrix} \quad (10.398)$$

$$\begin{Bmatrix} Q_{06} \\ P_{06} \end{Bmatrix} = \begin{bmatrix} A_6 & B_6 \\ C_6 & D_6 \end{bmatrix} \begin{Bmatrix} 0 \\ P_{L6} \end{Bmatrix}. \quad (10.399)$$

The subscripts are selected as follows: the first part of first expansion volume = 1, inertia tube = 2, second part of second expansion volume = 3, tailpipe = 4, the second part of the first expansion volume (first side branch) = 5, and the first part of the second expansion volume (second side branch) = 6.

The results in Figure 10.35b show the influence of tailpipe length and are almost identical to the two-volume muffler results in Figure 10.30. The main difference is that the low-frequency cutoff is higher because the two volumes were not increased by the space lost due to the presence of the tubes. The estimation formula (Equation 10.376) for the effective low-frequency cutoff remains valid, but with the volumes adjusted appropriately.

It can be shown that the backpressure peaks have shifted with the increased low-frequency cutoff, but have not changed in character.

If the tailpipe length is kept constant and the inertia tube length is varied, it can be shown (see Kim and Soedel, 1992 and Soedel, 1992) that the low-frequency cutoff is higher for a shorter tube and lower for a longer tube, because of the change in effective mass of the system. When the inertia tube length is increased more and more, the side-branch effect and the tube resonance both shift into the frequency region of interest.

The obvious question is whether it is of advantage to take one volume-tailpipe muffler and divide the expansion chamber into two chambers with a connecting inertia pipe. While the length of the inertia tube obviously matters and may influence final design conclusions, it seems, in general, that the low frequency cutoff is increased (which is a disadvantage), but that in higher frequency bands the transmission loss is significantly increased (an advantage).

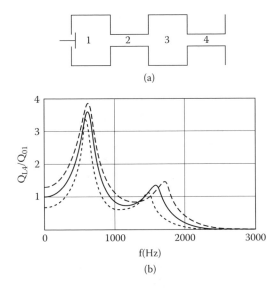

FIGURE 10.36 (a) Typical two expansion volume muffler subjected to different temperatures. (b) Typical transmission loss:—, $T_1 = T_2 = T_3 = T_4$; - - -, $T_3 = 1.2\, T_2 = 1.44 T_1$; ---, $T_3 = 0.8 T_2 = 0.64 T_1$. Otherwise, $L_1 = L_2 = L_3 = 30$ mm, $A_1 = A_3 = 707$ mm², $A_3 = A_4 = 20$ mm²

10.4.6 Influence of Temperature

For the system shown in Figure 10.36a, the influence of heating and cooling on the transmission ratio Q_{L4}/Q_{01} is illustrated. The mathematical model is (based on the simplified Helmholtz resonator model in Soedel and Soedel, 1992)

$$\begin{Bmatrix} Q_{01} \\ P_{01} \end{Bmatrix} = \begin{bmatrix} A & B \\ C & D \end{bmatrix} \begin{Bmatrix} Q_{L4} \\ 0 \end{Bmatrix} \qquad (10.404)$$

where

$$\begin{bmatrix} A & B \\ C & D \end{bmatrix} = \begin{bmatrix} A_1 & B_1 \\ C_1 & D_1 \end{bmatrix} \begin{bmatrix} A_2 & B_2 \\ C_2 & D_2 \end{bmatrix} \begin{bmatrix} A_3 & B_3 \\ C_3 & D_3 \end{bmatrix} \begin{bmatrix} A_4 & B_4 \\ C_4 & D_4 \end{bmatrix} \qquad (10.405)$$

and where the speeds of sound are different:

$$c_i^2 = kRT_i. \qquad (10.406)$$

The temperature of the first expansion volume is T_1, of the second T_3, and the exit volume, approximated as infinite, has a temperature T_5. The inertial tubes in between are assumed to have temperatures that are averages of the adjoining volumes. Not unlike a typical compressor, the temperature of the entering flow is assumed to be influenced only by the compression process (in the case of a discharge system), and therefore T_1 is held constant. If we assume that $T_5 = T_3 = T_1$, then we obtain the solid line response curve as shown in Figure 10.36b. In the low-frequency range shown, there are two resonances. If the gas is heated up as it passes through the muffler, for example, in a high-side compressor or due to high motor and friction losses, so that $T_5 = 1.2, T_3 = 1.44 T_1$ (the temperature differences are exaggerated to make the point clear), both response peaks are shifted to the right (the second one more dramatically than the first) due to the resonance frequencies being higher because the average speed of sound is higher. Also, the response magnitudes increase because the exit volume velocity is larger due to the increased gas temperature at the exit. The cutoff frequency is higher. If we cool the gas, as shown by the $T_5 = 0.8, T_3 = 0.64 T_1$ curve (again, the temperature differences are exaggerated), the trend reverses as expected; response magnitudes decrease, response resonances shift to the left, and the effective cutoff frequency is lower.

10.4.7 Muffler Synthesis Example

In the following, an example sequence of designing a muffler for attenuating pulsations in the 3000 Hz band is sketched. This case study is an example only and other possible choices could have been made.

As the first step, we identify the desired low-frequency cutoff from basic thermodynamic calculations or from measured data (gas pulsations or even noise data). Let us say 500 Hz is an acceptable cutoff. The transmission loss of our first design based on a volume-tailpipe combination (Figure 10.37a) is shown in Figure 10.38a and let us pretend that it is too low in the 3000 Hz region.

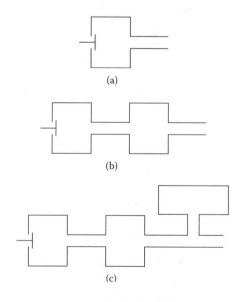

FIGURE 10.37 Development of a compressor suction muffler with special attention to the transmission loss at 3000 Hz: (a) single volume-tailpipe muffler, (b) two volume-tailpipe combinations in series, (c) a side-branch resonator is added.

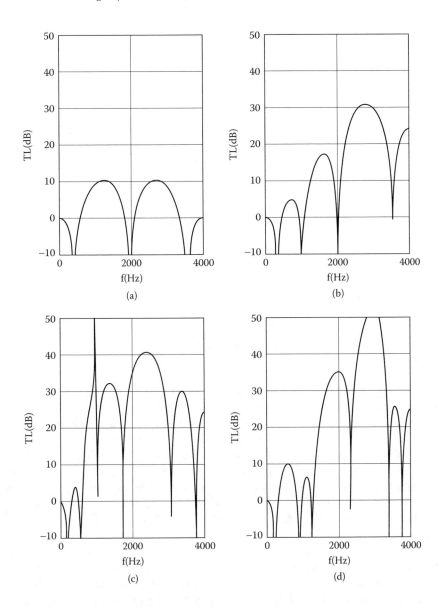

FIGURE 10.38 Typical transmission loss curves that correspond to the evolution of the muffler design shown in Figure 10.37: (a) volume-tailpipe design, (b) two-volume muffler, (c) side-branch resonator tuned nominally to about 900 Hz, (d) side-branch resonator tuned nominally to 3000 Hz.

Because space constraints do not permit us to increase the diameters of the volumes in this case study, after some trial and error we add a second volume as shown in Figure 10.37b (assuming that length is unrestricted). The transmission loss curve is now much improved, as shown in Figure 10.38b. In the 3000 Hz region we have improved from 10 dB to 30 dB. On the negative side, we have introduced a new

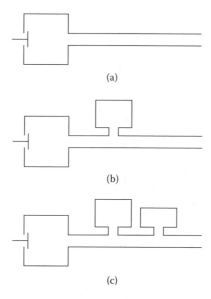

FIGURE 10.39 (a) Initial volume-tailpipe design, (b) a side-branch resonator is added, which is tuned nominally to 1100 Hz, (c) a second side-branch resonator is added, which is tuned nominally to 2500 Hz.

low-frequency cutoff at 1000 Hz, because of the two-degrees-of-freedom system base configuration.

To improve the fact that at about 1000 Hz we have no attenuation and possibly an amplification (if sufficient damping cannot be introduced), we design a side-branch Helmholtz resonator (Figure 10.37c) with a turning frequency of about 900 Hz to compensate for the rest of the system. This results in Figure 10.38c, where the sharp dip at 1000 Hz can easily be eliminated by very reasonable damping (not shown). In spite of some new resonances appearing, one would judge this to be a better design.

As an alternative, if a low-frequency cutoff of 1000 Hz is acceptable, we can use the side-branch resonator to further enhance the transmission loss in the 3000 Hz region, if we so desire. Therefore, tuning the side-branch resonator to approximately 3000 Hz results in Figure 10.38d, where the theoretical transmission loss in this region is much larger. Negative aspects of this design are that we now have an effective low-frequency cutoff of about 1200 Hz, and that there are more resonance frequencies.

After the design has been agreed upon from a sound attenuation viewpoint, the backpressure has to be calculated, and the design must be checked out using a compressor gas pulsation simulation, or preferably an experimental investigation.

10.4.8 Multiple Side Branch Attenuators

It is possible to use two or more side-branch resonators to attenuate several frequency bands at once? Some trial and error, by simulation and experiment, is involved because the attenuators may influence each other.

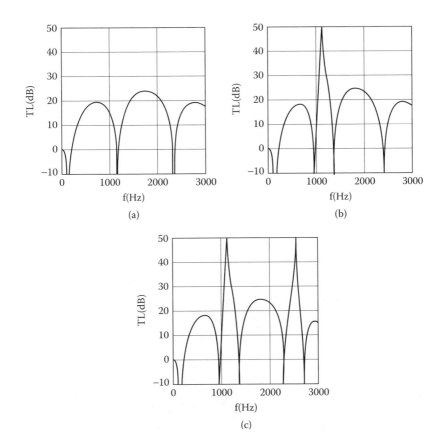

FIGURE 10.40 Typical transmission loss curves associated with cases (a) to (c) of Figure 10.39.

In Figure 10.39a, an expansion volume-tailpipe muffler example is shown. The tailpipe was made relatively large (3 L_1) for demonstration purposes. The transmission loss in Figure 10.40 (a) shows the cutoff frequency and two gas column resonances in the tailpipe in the frequency range shown. Attaching a side-branch resonator, as shown in Figure 10.40b, nominally tuned by the simple standard Helmholtz formula to 1100 Hz, shows the improved transmission loss diagram of Figure 10.40b in the 1100 Hz region. Introducing a second side-branch resonator (Figure 10.40c) tuned to 2500 Hz results in the transmission loss curve of Figure 10.40 (c). It can be shown (Kim and Soedel, 1992b) that in Figure 10.40c, the influence of the spacing between the two side-branch Helmholtz resonators is negligible in the lower frequency ranges.

It should be noted that as we introduce side-branch resonators, we may pay a price in terms of a changed cutoff frequency, which may result in poor attenuation or even magnification of the lower-frequency pulses. Of course, in such a case, countermeasures can again be taken.

11 Multidimensional Compressor Sound

Because the gas cavity formed between the compressor housing (shell) and the compressor casing, for example, cannot be approached using the one-dimensional wave equation, a two- or three-dimensional wave equation is needed. Also, if we wish to study the acoustic field that is radiated from the compressor housing (shell), we need to approach it three dimensionally.

11.1 THREE-DIMENSIONAL ACOUSTIC WAVE EQUATION

Rather than deriving this equation from a control volume using Newton's second law, it will be obtained by reduction from the three-dimensional equations of motion for a solid (see Section 7.1), starting with the Cartesian coordinate form. One can also derive it, by reduction, from the Navier-Stokes equation, of course.

First, gases can sustain only fairly negligible shear stresses (it is a function of viscosity). Therefore,

$$\sigma_{xy} = \sigma_{xz} = \sigma_{yz} = 0. \tag{11.1}$$

Next, the normal stresses on an infinitesimal element are simply the acoustic pressure (we mentally subtract or add the mean pressure if needed; the resulting equations are for acoustic pressure fluctuations only):

$$\sigma_{xx} = \sigma_{yy} = \sigma_{zz} = -p. \tag{11.2}$$

We obtain, for $q_x = q_y = q_z = 0$.

$$-\frac{\partial p}{\partial x} - \rho \ddot{u}_x = 0 \tag{11.3}$$

$$-\frac{\partial p}{\partial y} - \rho \ddot{u}_y = 0 \tag{11.4}$$

$$-\frac{\partial p}{\partial z} - \rho \ddot{u}_z = 0. \tag{11.5}$$

This can also be written

$$-\frac{\partial^2 p}{\partial x^2} - \rho \frac{\partial^2}{\partial t^2}\left(\frac{\partial u_x}{\partial x}\right) = 0 \tag{11.6}$$

$$-\frac{\partial^2 p}{\partial y^2} - \rho \frac{\partial^2}{\partial t^2}\left(\frac{\partial u_y}{\partial y}\right) = 0 \tag{11.7}$$

$$-\frac{\partial^2 p}{\partial y^2} - \rho \frac{\partial^2}{\partial t^2}\left(\frac{\partial u_z}{\partial z}\right) = 0. \tag{11.8}$$

Adding the three equations gives

$$-\nabla^2 p - \rho \frac{\partial^2}{\partial t^2}\left(\frac{\partial u_x}{\partial x} + \frac{\partial u_y}{\partial y} + \frac{\partial u_z}{\partial z}\right) = 0 \tag{11.9}$$

where

$$\nabla^2(\cdot) = \frac{\partial^2(\cdot)}{\partial x^2} + \frac{\partial^2(\cdot)}{\partial y^2} + \frac{\partial^2(\cdot)}{\partial z^2}. \tag{11.10}$$

From the strain–stress relationships for the solid, which can also be written as

$$\sigma_{11} = 2G\varepsilon_{11} + \lambda(\varepsilon_{11} + \varepsilon_{22} + \varepsilon_{33}) \tag{11.11}$$

$$\sigma_{22} = 2G\varepsilon_{22} + \lambda(\varepsilon_{11} + \varepsilon_{22} + \varepsilon_{33}) \tag{11.12}$$

$$\sigma_{33} = 2G\varepsilon_{33} + \lambda(\varepsilon_{11} + \varepsilon_{22} + \varepsilon_{33}), \tag{11.13}$$

we obtain

$$-p = \lambda\left(\frac{\partial u_x}{\partial x} + \frac{\partial u_y}{\partial y} + \frac{\partial u_z}{\partial z}\right). \tag{11.14}$$

Thus, the equation of motion becomes

$$\nabla^2 p - \frac{\rho}{\lambda}\frac{\partial^2 p}{\partial t^2} = 0. \tag{11.15}$$

Multidimensional Compressor Sound

The definition of the bulk modulus, λ, for a solid does not quite apply here. After all, what would Poisson's ratio for a gas be? Thus, we replace λ by K, where

$$K = \rho c^2, \tag{11.16}$$

which was derived in the previous chapter when the one-dimensional wave equation was discussed. Note that ρ is the mass density of the gas and c is the speed of sound. Average values are used.

$$\nabla^2 p - \frac{1}{c^2}\frac{\partial^2 p}{\partial t^2} = 0 \tag{11.17}$$

If we would have used general curvilinear coordinates in the derivation, the same expression would have resulted, with

$$\nabla^2 = \frac{1}{A_1 A_2 A_3}\left[\frac{\partial}{\partial \alpha_1}\left(\frac{A_2 A_3}{A_1}\frac{\partial}{\partial \alpha_1}\right) + \frac{\partial}{\partial \alpha_2}\left(\frac{A_3 A_1}{A_2}\frac{\partial}{\partial \alpha_2}\right) + \frac{\partial}{\partial \alpha_3}\left(\frac{A_1 A_2}{A_3}\frac{\partial}{\partial \alpha_3}\right)\right]. \tag{11.18}$$

Therefore, we can convert the wave equation easily to cylindrical or spherical coordinates, when required.

In displacement form, the wave equation becomes

$$\nabla^2 u_x - \frac{1}{c^2}\ddot{u}_x = 0 \tag{11.19}$$

$$\nabla^2 u_y - \frac{1}{c^2}\ddot{u}_y = 0 \tag{11.20}$$

$$\nabla^2 u_z - \frac{1}{c^2}\ddot{u}_z = 0. \tag{11.21}$$

If we choose to work in terms of particle velocities, we can either use

$$\nabla^2 v_x - \frac{1}{c^2}\ddot{v}_x = 0 \tag{11.22}$$

$$\nabla^2 v_y - \frac{1}{c^2}\ddot{v}_y = 0 \tag{11.23}$$

$$\nabla^2 v_z - \frac{1}{c^2}\ddot{v}_z = 0 \tag{11.24}$$

or because it is often convenient to introduce a potential function ϕ such that

$$v_x = -\frac{\partial \phi}{\partial x}, \quad v_y = -\frac{\partial \phi}{\partial y}, \quad v_z = -\frac{\partial \phi}{\partial z}, \tag{11.25}$$

we may use also the form

$$\nabla^2 \phi - \frac{1}{c^2} \frac{\partial^2 \phi}{\partial t^2} = 0. \tag{11.26}$$

11.2 SOUND RADIATION FROM THE COMPRESSOR HOUSING (OR CASING)

While there is no closed-form analytical solution for the noise radiation of a real compressor housing, or casing in the case of compressors that are not hermetically sealed, it is instructive to consider an extremely simplified model involving the noise radiation of a cylinder of a diameter equivalent to the compressor housing. This will allow us to illustrate the major parameter of noise radiation and to make a few points. But by way of introduction, we start with a review of the simplest possible radiation source, a monopole. For example, see Reynolds (1981).

11.2.1 MONOPOLE SOURCE

Imagine a small sphere of radius a, pulsing with a normal velocity at all points of

$$v = v_a e^{j\omega t}. \tag{11.27}$$

The source strength Q is

$$Q(t) = 4\pi a^2 v_a e^{j\omega t}, \tag{11.28}$$

which is the product of the surface area of the sphere and the surface velocity.

If the sound produced by this sphere is free to radiate in all directions, we expect that the wave equation in spherical coordinates,

$$A_1 = a, \quad A_2 = a \sin\phi, \quad \alpha_1 = \phi, \quad \alpha_2 = \theta \tag{11.29}$$

is independent of θ and ϕ. In this case, in the potential function form, the equation of motion reduces to

$$\frac{\partial^2 \phi}{\partial t^2} - \frac{c^2}{r^2} \frac{\partial}{\partial r}\left[r^2 \frac{\partial \phi}{\partial r}\right] = 0. \tag{11.30}$$

For harmonic pulsations,

$$\phi(r,t) = \Phi(r)e^{j\omega t}. \tag{11.31}$$

This gives

$$\frac{d^2(r\Phi)}{dr^2} + k^2 r\Phi = 0 \tag{11.32}$$

where $k = \omega/c$. The solution is

$$\Phi(r) = \frac{A}{r}e^{-jkr} + \frac{B}{r}e^{jkr} \tag{11.33}$$

or

$$\phi(r,t) = \frac{A}{r}e^{j(\omega t - kr)} + \frac{B}{r}e^{j(\omega t + kr)}. \tag{11.34}$$

$B = 0$ because waves only travel outward from the monopole. Because from the definition of the potential functions,

$$\frac{\partial \phi(a,t)}{\partial r} = v_a e^{j\omega t}, \tag{11.35}$$

we obtain

$$v_a e^{j\omega t} = -A\left(\frac{1+jka}{a^2}\right)e^{j(\omega t - ka)}. \tag{11.36}$$

This allows us to solve for A, and substituting it into the expression for $\Phi(r,t)$; we obtain (also substituting the source strength)

$$\phi(r,t) = -\frac{Q(t)}{4\pi}\left(\frac{1}{1+jka}\right)\frac{1}{r}e^{-jk(r-a)}. \tag{11.37}$$

The acoustic pressure is, therefore,

$$p = -\rho\frac{\partial \phi}{\partial t} = a^2 v_a \left(\frac{jkz_0}{1+jka}\right)\frac{1}{r}e^{j[\omega t - k(r-a)]} \tag{11.38}$$

where $z_0 = \rho c$ is the so-called *characteristic acoustic impedance*.

Thus, the acoustic pressure amplitude is proportional to the surface velocity amplitude v_a. Imagining the vibrating compressor housing to be a collection of tightly packed monopoles, it is intuitively apparent that the sound pressure p generated by the vibrating housing will be proportional to its average vibration velocity in some way, give or take some geometrically caused complexities.

11.2.2 Radiation from a Circular Cylindrical Housing

Because of the acoustic boundary conditions posed by the end caps, there are no simple, closed-form analytical solutions for something resembling a realistic compressor housing. If we do not wish to resort to a boundary element simulation, the best we can do is to consider the cylindrical shell to be extended axially into infinity at both ends (see Reynolds, 1981), or to consider a finite-length cylindrical shell that vibrates in an acoustic space limited by parallel planes at the end cap locations and having sliding-clamped boundary conditions as illustrated in Figure 11.1 (Kim and Soedel, 1992a). While these cases both ignore the end cap geometry and radiation, they give us valuable information about the way such shells radiate noise and the importance of certain parameters. In the following case involving two parallel baffles, we study this limited radiation field.

For example, the transverse displacement component of the circular cylindrical shell is, when vibrating in its m,n mode (this is one of the cases for which an exact, closed-form solution can be obtained):

$$u_3(x,\theta,t) = A_{mn} \cos\frac{m\pi x}{L} \cos n\theta \, e^{j\omega t} \qquad (11.39)$$

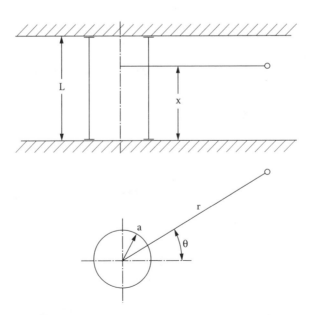

FIGURE 11.1 Simplified model of the sound radiation from an equivalent circular cylindrical compressor housing. The sound field is confined between two parallel baffles.

where A_{mn} is the vibration amplitude, determined from steady-state harmonic response calculations. The transverse velocity is therefore

$$\dot{u}_3(x,\theta,t) = A_{mn} j\omega \cos\frac{m\pi x}{L}\cos n\theta e^{j\omega t} \qquad (11.40)$$

where the velocity amplitude is $A_{mn}\omega$.

The boundary conditions of the acoustic space are

$$v_r(r=a,x,\theta,t) = \dot{u}_3(x,\theta,t) \qquad (11.41)$$

$$v_r(r=\infty,x,\theta,t) = 0 \qquad (11.42)$$

$$v_x(r,x=0,\theta,t) = 0 \qquad (11.43)$$

$$v_x(r,x=L,\theta,t) = 0. \qquad (11.44)$$

The equation of motion, in cylindrical coordinates where $A_1 = 1$, $A_2 = r$, $A_3 = 1$, $\alpha_1 = r, \alpha_2 = \theta, \alpha_3 = x$, is

$$\frac{1}{r}\left[\frac{\partial}{\partial r}\left(r\frac{\partial\phi}{\partial r}\right) + \frac{\partial}{\partial\theta}\left(\frac{1}{r}\frac{\partial\phi}{\partial\theta}\right) + \frac{\partial}{\partial x}\left(r\frac{\partial\phi}{\partial x}\right)\right] - \frac{1}{c^2}\frac{\partial^2\phi}{\partial t^2} = 0. \qquad (11.45)$$

The solution

$$\phi(r,\theta,x,t) = R(r)\cos\frac{m\pi x}{L}\cos n\theta e^{j\omega t} \qquad (11.46)$$

satisfies the boundary conditions at $x = 0, L$, and the continuity condition in the θ – directions.

It satisfies the partial differential equation in the sense that an ordinary differential equation in R is obtained:

$$r^2\frac{d^2R}{dr^2} + r\frac{dR}{dr} + \left[\left(k^2 - \frac{m^2\pi^2}{L^2}\right)r^2 - n^2\right]R = 0 \qquad (11.47)$$

where $k = \omega/c$. Note that a second possible solution exists where $\cos n\theta$ is replaced by $\sin n\theta$, but it will not furnish us with more insight. Whenever $k > m\pi/L$, the solution is

$$R(r) = A_n J_n(\kappa r) + B_n Y_n(\kappa r) \qquad (11.48)$$

where

$$\kappa = k^2 - \frac{m^2\pi^2}{L^2} \qquad (11.49)$$

and J_n and Y_n are Bessel functions of the first and second kind.

A second type of solution exists if $k < m\pi/L$, but it is of little practical significance.
Because of the boundary conditions in the r–direction, it is convenient to write the solution of R in terms of Hankel functions:

$$R(r) = G_n H_n^{(1)}(\kappa r) + D_n H_n^{(2)}(\kappa r) \tag{11.50}$$

where

$$H_n^{(1)}(\kappa r) = J_n(\kappa r) + j Y_n(\kappa r) \tag{11.51}$$

$$H_n^{(2)}(\kappa r) = J_n(\kappa r) - j Y_n(\kappa r). \tag{11.52}$$

Because $H_n^{(1)}(\kappa r)$ will not approach zero as $r \to \infty$, it must be that $G_n = 0$. Therefore,

$$R(r) = D_n H_n^{(2)}(\kappa r). \tag{11.53}$$

The first boundary condition can be written

$$v_r(r=a, x, \theta t) = -\left.\frac{\partial \phi}{\partial r}\right|_{r=a} = j A_{mn} \omega \cos\frac{m\pi x}{L} \cos n\theta e^{j\omega t} \tag{11.54}$$

$$\frac{dR}{dr}(r=a) = -j A_{mn}\omega. \tag{11.55}$$

Substituting for R gives

$$D_n = -\frac{j A_{mn}\omega}{H_n^{(2)'}(\kappa a)} \tag{11.56}$$

where

$$H_n^{(2)'}(\kappa a) = \left.\frac{dH_n^{(2)}(\kappa r)}{dr}\right|_{r=a}. \tag{11.57}$$

Therefore,

$$\phi(r,\theta,x,t) = -j A_{mn}\omega \frac{H_n^{(2)}(\kappa r)}{H_n^{(2)'}(\kappa a)} \cos\frac{m\pi x}{L} \cos n\theta e^{j\omega t}. \tag{11.58}$$

The acoustic radiation pressure is

$$p = \rho \frac{\partial \phi}{\partial t} \tag{11.59}$$

or

$$p = A_{mn}\rho\omega^2 \frac{H_n^{(2)}(\kappa r)}{H_n^{(2)'}(\kappa a)} \cos\frac{m\pi x}{L} \cos n\theta e^{j\omega t}. \tag{11.60}$$

As for the monopole, we see that the pressure is proportional to the amplitude of vibration displacement, or the amplitude of vibration velocity, $A_{mn}\omega$. The sound pressure distribution follows the mode shape in character. The θ-dependency is of some interest for the diagnostics of noise. If the n = 1 mode dominates (a rigid body motion of the housing, usually due to crank-motor shaft imbalance), two lobes of sound pressure level magnitude distribution will typically be measured as shown in Figure 11.2. An n = 2 mode will produce 4 lobes, or in general, a mode of mode number n will produce 2n lobes of sound pressure level.

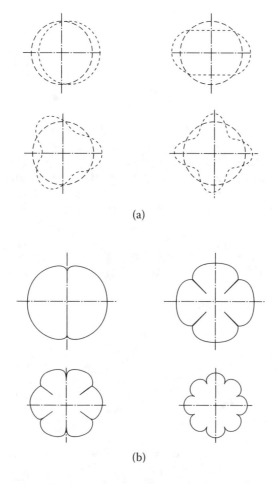

(a)

(b)

FIGURE 11.2 Top view of the compressor housing of Figure 11.1: (a) shell vibration modes, (b) associated lines of equal sound pressure level.

If the shell vibrates not at a resonance where a particular mode dominates, all natural modes will be present in the response, as determined by a shell response solution. The acoustic pressure is then the superposition of all these mode contributions:

$$p = \sum_{n=0}^{\infty} \sum_{m=1}^{\infty} A_{mn} \rho \omega^2 \frac{H_n^{(2)}(\kappa r)}{H_n^{(2)'}(\kappa a)} \cos \frac{m\pi x}{L} \cos n\theta e^{j\omega t}. \qquad (11.61)$$

plus any contributions from possible $\sin n\theta$ type modes.

In such a case, lobes usually cannot be detected, unless a particular mode dominates the vibration.

For additional work on sound radiation see, for example, Biscaldi et al. (1998), Masters et al. (1992), DaSilva et al. (2004), Kawai et al. (1988), Singh and Soedel (1985), Smith et al. (1992), and Trella and Soedel (1971).

11.2.3 Practical Consideration

Physically, the decibel scale describes very well the way we hear because the ear acts as a logarithmic sensor. On this scale, a 3 dB sound reduction can be perceived as an improvement, while less than 3 dB is not always sufficient to be classified as an obvious improvement.

A common problem is that customers and management often do not understand how difficult it is to reduce the sound pressure level of a compressor at a certain frequency by, say, 5 dB. They may say: "to reduce the noise level from 85 to 80 dB is such a small amount, why don't you put some damping material on the housing and have it done by tomorrow?" Let us take a close look at this.

The sound pressure level SPL in dB is

$$\text{SPL} = 20 \log \left(\frac{p}{p_{\text{ref}}} \right) \qquad (11.62)$$

where $p_{\text{ref}} = 0.0002$ microbar. Thus

$$(\text{SPL})_1 = 20 \log \left(\frac{p_1}{p_{\text{ref}}} \right) \qquad (11.63)$$

where $(\text{SPL})_1 = 85$ dB, and

$$(\text{SPL})_2 = 20 \log \left(\frac{p_2}{p_{\text{ref}}} \right) \qquad (11.64)$$

where $(\text{SPL})_2 = 80$ dB.

Subtracting the two equations gives

$$(SPL)_1 - (SPL)_2 = 20 \log\left(\frac{p_1}{p_2}\right) = 5 \text{ dB}. \tag{11.65}$$

Thus

$$\frac{p_1}{p_2} = \log^{-1}(0.25) = 1.78 \cong \frac{V_{a1}}{V_{a2}}. \tag{11.66}$$

This means that we must be able to decrease the average surface velocity of the housing to 56% of the original value to obtain a 5 dB improvement. Because at a particular frequency ω [rad/s], the harmonic response vibration displacement amplitude u_a is related to velocity by

$$u_a = \frac{V_1}{\omega}, \tag{11.67}$$

it means also, in different words, that we must reduce the vibration amplitude to 56% of the original value. As any vibration specialist knows, this can be a difficult order.

For example, if a resonance peak is due to a natural mode of the housing being in resonance, with a reasonably small amount of damping, this mode will dominate the response peak in the vicinity of the resonance and the transverse vibration solution will be of the following approximate form (see Chapter 6)

$$u_3(x,\theta,t) \cong \eta_k U_{3k} \tag{11.68}$$

where U_{3k} is the natural mode and

$$\eta_k = \Lambda_k e^{j(\omega t - \phi_k)} \tag{11.69}$$

and where

$$\Lambda_k \cong \frac{F_k^*}{2\zeta_k \omega^2} \tag{11.70}$$

where $\omega = \omega_k$, the natural frequency of this mode. F_k^* is a constant. This means that the response amplitude at resonance is approximately proportional to the inverse of the modal damping coefficient ζ_k. Therefore, a required decrease of response amplitude to 56% of its original value requires, if we try to accomplish this by damping alone, an increase of the modal damping ratio to 178% of its original value.

This illustrates another problem. If damping was, to begin with, very small (for an untreated steel hosing it might be, say, $\zeta_k = 0.05$), it is relatively easy to reach $(1.78)(0.05) \cong 0.09$. But if the housing already has some damping associated with it, say, $\zeta_k = 0.15$, due to the interaction with interior attachments; for example, it may become virtually impossible to raise the model damping coefficient by an additional factor of 1.78 to 0.27. Measures other than damping will have to be taken. Also, the peak may not be dominated by one mode any longer, making the goal too difficult to reach.

This carries over, philosophically, to other noise control work. As a rule of thumb, if we start with a very noisy compressor, we can probably achieve an overall reduction of (let us be modest) 3 dBA, in a matter of days. To lower the noise by another 3 dBA will require a time expenditure (because we have to be more clever and do more work) on the order of weeks. The third 3 dBA reduction may take another month, and so on. Obviously, it is easier to reduce the sound of a very noisy compressor by 5 dBA than of a compressor, which is already relatively quiet.

11.3 GAS PULSATIONS IN THE CAVITY BETWEEN THE COMPRESSOR CASING AND HOUSING

The volume formed between the compressor casing and the housing can often be approximated as an annular cylinder, with the thickness dimension of the cylinder being smaller than the other dimensions (see Johnson and Hamilton, 1972). For very approximate estimates of certain natural gas frequencies and modes, one can also, at first, neglect the space formed between the top of the casing and the top of the housing. On the other end of the cylinder, the oil sump in a refrigeration compressor forms the barrier (see also Lai et al., 1996, and Lai and Soedel, 1996a–c, 1997, 1998).

For sheetlike gas volumes, Equations 11.17 applies, with Equation 11.18 reduced to

$$\nabla^2 = \frac{1}{A_1 A_2}\left[\frac{\partial}{\partial \alpha_1}\left(\frac{A_2}{A_1}\frac{\partial}{\partial \alpha_1}\right) + \frac{\partial}{\partial \alpha_2}\left(\frac{A_1}{A_2}\frac{\partial}{\partial \alpha_2}\right)\right] \quad (11.71)$$

where α_1, α_2 are curvilinear coordinates and A_1, A_2 are the associated Lamé parameters.

11.3.1 NATURAL FREQUENCIES AND MODES OF GAS IN AN ANNULAR CYLINDER

In this case (see Figure 11.3), utilizing cylindrical coordinates, $\alpha_1 = \theta$, $\alpha_2 = z$, $A_1 = a$, $A_2 = 1$, and $\frac{\partial(\cdot)}{\partial r} = 0$, Equation 11.71 becomes

$$\frac{1}{a^2}\frac{\partial^2 p}{\partial \theta^2} + \frac{\partial^2 p}{\partial z^2} = \frac{1}{c^2}\frac{\partial^2 p}{\partial t^2} \quad (11.72)$$

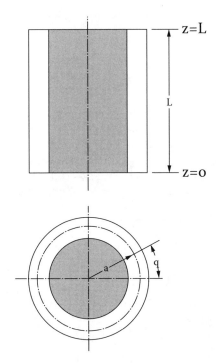

FIGURE 11.3 Volume between compressor casing and the compressor housing approximated as an equivalent annular cylindrical cavity.

with boundary conditions (zero normal acoustic velocity)

$$\frac{\partial p}{\partial z} = 0 \qquad (11.73)$$

at both $z = 0$ and approximately $z = L$ (the $z = L$ condition will be revisited). The boundary condition in the θ–direction is replaced by the condition of continuity

$$p(\theta = 0, z, t) = p(\theta = 2\pi, z, t). \qquad (11.74)$$

Equations 11.72 to 11.74 are satisfied by

$$p_{mn1}(\theta, z, t) = \cos(n\theta)\cos\left(\frac{m\pi z}{L}\right)e^{j\omega t} \qquad (11.75)$$

and

$$p_{mn2}(\theta, z, t) = \sin(n\theta)\cos\left(\frac{m\pi z}{L}\right)e^{j\omega t} \qquad (11.76)$$

where n = 0, 1, 2, ... and m = 0, 1, 2, ... In general, this can be written

$$p_{mni}(\theta, z, t) = \cos n(\theta - \phi_i) \cos\left(\frac{m\pi z}{L}\right) e^{j\omega t}, \quad i = 1, 2 \quad (11.77)$$

where $\phi_1 = 0$ and $\phi_2 = \frac{\pi}{2n}$. This illustrates why, for an axisymmetric gas volume, natural modes have no preferential direction in the θ –direction. It can be shown (Lai and Soedel, 1996 a–c) that the gas pulsation response in such a volume cannot be reduced by changing the suction manifold intake location θ. The gas modes will shift with the intake location shift and always orient themselves such that they are most easily excited. (On the other hand, shifting the z-location of the intake to a gas mode node will be a successful noise abatement measure.)

Substituting Equation 11.77 in Equation 11.72 gives us the natural frequencies of gas oscillation in rad/sec:

$$\omega_{mn} = c\sqrt{\left(\frac{n}{a}\right)^2 + \left(\frac{m\pi}{L}\right)^2} \quad (11.78)$$

where a is the average radius of the gas cylinder and L is the height; c is the speed of sound of the gas. The frequency in Hertz is $f_{mn} = \frac{\omega_{mn}}{2\pi}$.

The natural gas modes are, from Equation 11.77,

$$P_{mni}(\theta, z) = \cos n(\theta - \phi_i) \cos\left(\frac{m\pi z}{L}\right) \quad (11.79)$$

and are sketched in terms of their node lines at r = a in Figure 11.4.

For irregular casing and housing shapes, Equations 11.78 and 11.79 give estimates that are typically not too far from reality for m = 0, 1, 2 and n = 0, 1, 2. This is of value when trying to identify sound spectrum peaks in the frequency region where such modes are typically strongly excited (for fractional-horsepower compressors up to, say, 700 Hz).

A gas pulsation that frequently occurs in a dominant way in compressors involves the m = 0 and n = 1 mode. It is sometimes called the *sloshing mode* because the gas sloshes back and forth in the circumferential direction. Its natural frequency is, from Equation 11.78,

$$\omega_{01} = \frac{c}{a} \quad \text{or} \quad f_{01} = \frac{c}{2\pi a}. \quad (11.80)$$

This particular equation can also be derived by picturing a ring of gas of radius a. Because $c = f\lambda$, where the wavelength λ is $\lambda = 2\pi a$, we obtain $f = c/(2\pi a)$.

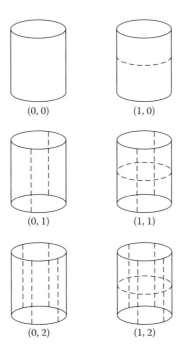

FIGURE 11.4 Natural pressure modes (m, n) of the gas in an annular, circular cylindrical cavity in terms of node lines on the reference surface. Note that (m, n) = (0, 0) represents a body mode of zero natural frequency.

11.3.2 Natural Frequencies and Modes of Gas in a Circular Disk Volume

This case will be important for developing an understanding of how the gas pulsations in the approximately disklike volume between the top of the casing and the top of the housing influence the overall pulsations (Figure 11.5); see also Lai and Soedel (1996d).

For this case, we select polar coordinates, so that $\alpha_1 = r$, $\alpha_2 = \theta$, $A_1 = 1$, $A_2 = r$, and $\frac{\partial(\cdot)}{\partial z} = 0$. Equation 11.71 becomes

$$\frac{1}{r}\frac{\partial}{\partial r}\left(r\frac{\partial p}{\partial r}\right) + \frac{1}{r^2}\frac{\partial^2 p}{\partial \theta^2} = \frac{1}{c^2}\frac{\partial^2 p}{\partial t^2} \tag{11.81}$$

with a boundary condition, at $r = a$, of zero normal acoustic velocity, or

$$\frac{\partial p}{\partial r} = 0. \tag{11.82}$$

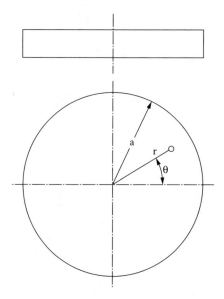

FIGURE 11.5 The volume between the top of the casing and the top of the housing approximated as an equivalent circular disklike cavity.

The boundary condition in the θ–direction is replaced by the condition of continuity

$$p(r, \theta = 0, t) = p(r, \theta = 2\pi, t), \qquad (11.83)$$

and the condition at $r = 0$ is that the mode amplitude cannot be infinite.

We expect a solution of the form

$$p(r, \theta, t) = R(r) \cos n(\theta - \phi) e^{j\omega t}. \qquad (11.84)$$

Substituting it in Equation 11.81 gives

$$r^2 \frac{d^2 R}{dr^2} + r \frac{dR}{dr} + (k^2 r^2 - n^2) = 0 \qquad (11.85)$$

where

$$k = \frac{\omega}{c}. \qquad (11.86)$$

The solution to this Bessel equation is:

$$R(r) = A_n J_n(kr) + B_n Y_n(kr) \qquad (11.87)$$

where J_n and Y_n are Bessel functions of the first and second kind, of order n. Because $Y_n(kr)$ is singular at $r = 0$, it must be that $B_n = 0$. Thus,

$$R(r) = A_n J_n(kr). \tag{11.88}$$

Applying the conditions in Equation 11.82 gives, at $r = a$:

$$\left. \frac{dJ_n(kr)}{dr} \right|_{r=a} = 0. \tag{11.89}$$

The roots $(ka)_{mn}$ of this equation give the natural frequencies in rad/sec:

$$\omega_{mn} = (ka)_{mn} \left(\frac{c}{a} \right) = \frac{c}{a} (ka)_{mn} \tag{11.90}$$

or $f_{mn} = \frac{c}{2\pi a} (ka)_{mn}$ in Hz.

The natural modes are given by (setting $A_n = 1$):

$$P_{mn_i}(r, \theta) = J_n(k_{mn} r) \cos n(\theta - \phi_i), \quad i = 1, 2, \tag{11.91}$$

where $k_{mn} = (ka)_{mn}/a$, and $\phi_1 = 0$, $\phi_2 = \pi/2n$, and they are sketched in Figure 11.6.

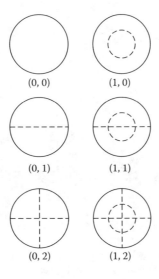

FIGURE 11.6 Natural pressure modes (m, n) of the gas in the equivalent circular disklike cavity of Figure 11.5.

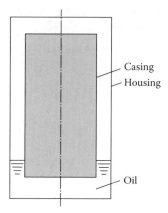

FIGURE 11.7 A more realistic model of the gas volume between the compressor casing and the compressor housing. An equivalent annular, circular cylindrical cavity is joined to an equivalent circular disk-like cavity.

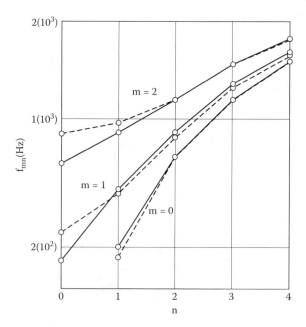

FIGURE 11.8 Typical natural frequencies of the total cavity between the compressor casing and the compressor housing (—), when the model of Figure 11.7 is used and the results if the equivalent annular, circular cylindrical cavity model of Figure 11.3 is used (---). As can be seen, the more simplified model is adequate for frequency estimates.

11.3.3 Natural Frequencies and Modes of Gas in a Volume Consisting of an Annular Cylinder and a Circular Disk

While the annular cylinder model of compressor gas pulsations in the cavity between the housing and the casing is sufficient for a basic understanding, it is instructive to combine the annular cylinder with the circular disk to study how the presence of the latter modifies the natural frequencies and modes of the former. The geometry of the gas cavity is defined in Figure 11.7. The two gas cavities are combined using so-called *line impedances*, as described in Lai and Soedel (1996d). This is similar to the line receptance method alluded to in the chapters on housing vibrations.

Natural frequencies for a representative case are shown in Figure 11.8 and natural modes in terms of *node lines* in Figure 11.9. It can be seen that from a natural frequency viewpoint, the correction of the annular cylindrical cavity model due to the presence of the circular disk cavity is negligible, except for the $n = 0$ set of modes. This is reflected in the shift of the node line locations for the $n = 0$ modes.

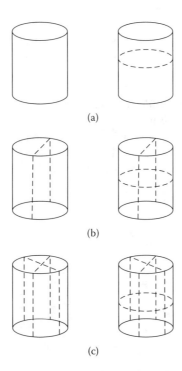

FIGURE 11.9 Typical natural modes of the total cavity approximated by the model of Figure 11.7: (a) the first two $n = 0$ modes, (b) the first two $n = 1$ modes, (c) the first two $n = 2$ modes. Note that the (m, n) classification used for an annular cylinder is not applicable anymore, even while the natural modes are similar.

Physically, the n = 0 modes are not sloshing modes in the sense of an acoustic gas motion in a circumferential direction. They represent oscillations in the axial direction, where one would expect the disk cavity to be of more influence.

11.3.4 Rocking Vibration of Housing Shell

In the annular cavity between the housing shell and the compressor casing, the pressure distribution for the sloshing mode (n = 1, m = 0) is, according to Equation 11.79

$$P_{oli}(\theta,z) = \cos(\theta - \phi_i). \tag{11.92}$$

This sloshing mode occurs, typically, at a relatively low natural frequency (well below the flexural vibration natural frequencies of the housing shell), and will, therefore, not be in resonance with the flexural natural frequencies of the shell. But it can be in resonance with the n = 1 type rigid body mode of the shell as a rigid body moving on its vibration isolation springs with respect to the cabinet frame to which it is attached. While such a resonance is rare because the isolation springs are typically such that the rigid body mode natural frequency of the housing lies well below the sloshing mode natural frequency of the gas in the annular cavity, the simple fact that the sloshing mode rocks the shell back and forth with a nonnegligible force resultant is enough to sometimes produce a relatively high response peak in the acoustic spectrum, which is associated with an n = 1 sound field.

Obviously, if this occurs, stiffening the compressor shell will not reduce this peak unless the stiffening adds so much mass to the housing that its vibration amplitude is reduced by the inertia effect alone. The best course of action seems to be to reduce the excitation of the n = 1 gas mode by a suitable suction muffler. Of course, should the n = 1 gas mode be in resonance with the n = 1 rigid body mode of the housing, the stiffnesses of the isolation springs would also need to be changed.

12 Remarks on Sound and Vibration Measurements and Source Identification

This text focuses mainly on the simplified physical description and explanation of vibration and sound mechanisms in compressors, notably refrigeration and air conditioning compressors, and does not intend to be a text on measurement technology. Therefore, the following is an overview of what should or could be measured, and how sound and vibration sources may be identified utilizing measurements and simplified calculations. For details on the required instrumentation and procedures, the literature on this subject should be consulted, for example, Hamilton (1988), Baars et al. (1998), Bucciarelli et al. (1994), Buligan et al. (2002), Cerrato-Jay and Lowery (2002), Cyklis (2000, 2002), Dreiman et al. (2000), Elson and Soedel (1972 b), Fleming and Brown (1982), Forbes and Mitchell (1974), Fu (2004), Funer and Tauchmann (1963), Gatley and Cohen (1970), Gavric and Badie-Cassagnet (2000), Gavric and Darpas (2002), Gluck, Ukrainetz, and Cohen (1964), Herfat (2002), Hwang et al. (2000), Ih and Jang (1998), Laursen (1990), Lee et al. (1998), Lee et al. (2000), Lee and Kim (1988), Lowery (1998), Lowery and Cohen (1962, 1963, 1971), Ma et al. (2002), Motegi and Nahashima (1996), Shiva-Prasad and Wollatt (2000), Singh and Soedel (1978 a), Ventimiglia et al. (2002), Yoshimura et al. (1992), Yoshimura et al. (1994), Yun (1996).

12.1 ROOMS FOR MEASURING SOUND

For directivity measurements, a fully anechoic room is recommended. Sound-absorbing wedges are supposed to create an open space environment, ideally without reflection from the room-enclosing structures. A fully anechoic room has sound-absorbing wedges not only on the walls and ceiling, but also on the floor. The compressor to be measured is typically suspended on very soft springs. Devices that simulate the refrigerating cycle load are typically placed outside the room, and the suction and discharge gas is fed to the compressor via pipes. For air compressors, the receiving pressure tank is also typically placed outside the room. The floor is sometimes covered with a removable grating so one can approach the compressor. Because the anechoic arrangement is never perfect (reflections do occur, especially for low frequencies), one should place the compressor at least one-half of a wavelength of the lowest frequency of interest from any room wall: $d = \frac{\lambda}{2}$, where the wavelength is $\lambda = \frac{c}{f}$ and where d is the distance from any wall, f is the lowest frequency of interest (say, 300 Hz), and c is the speed of sound of the air at room temperature (say, 300 m/s). Thus, the minimum distance d is 0.5 m.

Whether near-field or far-field measurements are taken depends on the identification approach that is used. In the far field, doubling the distance from the compressor source diminishes the sound pressure level (SPL) by 6 dB (the inverse square law). Satisfying this is a measure of the quality of the room. The far field begins approximately when the microphone is one wavelength of the lowest frequency of interest from the source (about 1 m when using the numbers of our small compressor example), and also one characteristic source dimension from the nearest surface of the source.

The advantage of anechoic room measurements is that a directivity pattern can be obtained (see also the discussion on sound radiated from the housing).

The second type of room is the reverberation room. It is designed (rigid walls) to reflect nearly all of the radiated sound. In principle, this should allow us to obtain the same sound power level anywhere in the room. Therefore, a single measurement of the compressors will be sufficient. Compressors can be theoretically compared to each other with a single sound power level number, while a comparison of different compressors in an anechoic room setting requires an integration of all the sound pressure levels on a sphere with the compressor at its center.

A problem with reverberation rooms is that they have natural frequencies and modes. Exciting these with the sound output of the compressor may selectively amplify the sound at certain resonances, depending also on the microphone location. Making the walls of a reverberation room nonparallel, as is sometimes advocated, will not help much because a room of any shape will still have natural frequencies and modes, the latter now being somewhat less regular in geometry than for a rectangular room, but resonances still arise. The recommended room designs are either rooms where the walls periodically change orientation (very rare due to cost) or large, solid vanes rotate in the room, so that the natural modes continue to change during a measurement and resonances are averaged out, so to speak. Another way of dealing with possible reverberation room resonances is to average the sound power level reading by rotating the microphone on a boom while the reading is taken.

The third type of room (the semi-anechoic room) is often found in industry because of the convenience of the floor, on which the compressor and instrumentation can be supported. The floor is a hard, reflecting surface, with the walls and ceiling covered with sound-absorbing wedges. It simulates a half space, with reflections only from the floor. This type of room is satisfactory for comparison purposes and some directivity measurements, but one must beware of floor reflections, which may invalidate the comparisons if the compressors to be compared are not located at the same average height above the floor. Directional readings for diagnostic purposes may also be invalidated. The reason is that the floor reflections may partially cancel the direct sound rays to the microphone.

This can be illustrated as follows. Figure 12.1a shows the compressor, extremely simplified as a monopole source, in an anechoic room. The microphone receives only a direct ray of sound and we can be confident about the measurement. Next, let us look at the situation in a semi-anechoic room. The compressor source is located at a certain height above the hard floor, as shown in Figure 12.1b. The microphone is located at the same relative distance and direction with respect to the compressors as it was in the anechoic room, but it now receives two acoustic rays: one direct ray, as in the anechoic room measurement, and one reflected ray, where we have used,

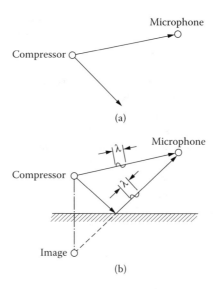

FIGURE 12.1 Sound rays from a small compressor, modeled as a monopole source, to a microphone: (a) in a fully anechoic room, (b) in a semi-anechoic room with a hard floor.

for the purposes of illustration, the method of images. It can be shown that the second ray approaches the microphone as if it is coming from a second monopole, which is the mirror reflection of the original monopole source, the hard floor being the mirror. The natural path of the second ray is longer than the path of the first ray. This means that a pressure wave of length λ (the associated frequency is $f = c/\lambda$, where c is the speed of sound in air) from the image may arrive later than the same pressure wave when it approaches the microphone directly. Therefore, these two pressure waves will either reinforce each other, and the microphone may measure as much as twice the actual sound pressure level at a particular frequency, or they will cancel each other, which means that the microphone may measure as little as zero at the frequency associated with a canceling wavelength λ.

This means that if we wish to compare two compressors, but locate them at different average heights, our comparison will be faulty. It might be better to place the compressor directly on the hard floor because the monopole and its reflection approach coincidence and wave cancellation would not be a problem. The sound pressure levels measured by the microphone would always be twice as high as those measured in a fully anechoic room. Unfortunately, this case is somewhat impractical because compressors are of finite dimensions, and the limit of a floor monopole can only be approached.

12.2 REPEATABILITY OF MEASUREMENTS

Care must be taken that the compressor has properly warmed up to a steady-state condition before sound measurements, or internal pressure measurements, or valve motion measurements are taken. For example, Johnson and Hamilton (1972) showed that a gas pulsation sloshing mode in the cavity formed by the compressor casing and the compressor housing shifted its natural frequency sufficiently during the warm-up period, that sound readings taken before warm-up was completed provided invalid sound data (see also Yun (1996)).

Another problem that sometimes occurs involves disconnecting and then reconnecting a refrigeration compressor to the pipes that connect the suction and discharge gas to the refrigeration cycle simulator (load stand), which can result in reading variations. The reason for this may be that the connection influences the stiffness of the compressor housing, and it may also influence the creation of structural vibrations of the pipes. It must be remembered that for the typically small damping present in the structural parts, resonances come in very sharp in the frequency spectrum and the slightest tuning or mistuning may make a considerable difference in the sound measurement.

Based on the same argument—that slight tuning or mistuning of natural frequencies can at times cause resonances to appear or disappear—one should also not be surprised to find, at times, considerable variations in the sound characteristics of compressors coming from the same assembly line. For example, for hermetically sealed, fractional-horsepower compressors, the housing shell is typically deep drawn from sheet metal in two parts: the top of the housing and the bottom of the housing. After the compressor casing and the electromotor are inserted in the bottom, and electrical connections and structural connections are made, the top half of the housing is welded to the bottom. Despite the appearance of uniformity, there may be enough temperature difference during welding, or enough of a welding speed difference, that thermally caused residual stresses are locked into the housing that are different enough from compressor to compressor that they will cause slight variations in natural frequencies of the housing shell. And slight variations of natural frequencies mean that the harmonics of the operational speed may or may not produce certain resonances. It should be noted that in many other industries, the undesirable effect of residual stress variations due to welding can be removed by annealing (heating the structural material above its recrystallization point), but annealing the housing is obviously not possible with a compressor inside.

Therefore, it is advisable to take measurements enough times, and also on enough compressors of the same type, so that a statistical average can be formulated. This is, of course, very time consuming. Therefore, when doing sound reduction work, it may be sufficient to do it on one or two compressors only until a solution seems to have been found, and only then test it on a statistically significant number of compressors.

Large sound reductions are obviously less sensitive to statistical variations than small sound reductions and are easily recognized. But because sound reduction work typically proceeds in small steps (one works on the largest peak in the response spectrum first, and progresses from there), promising sound reduction solutions may at first only produce a few dB, well within the statistical variation of measurements, but if pursued may result in large improvements. The problem is that such solutions may not be recognized as valuable because the first result does not show a significant enough improvement, being well inside of the expected experimental variations, or worse, no improvement because disconnecting and disassembling the compressor, making the improvement, and going through a warm-up cycle may create enough variation to mask the improvement completely. Good experimental procedures are therefore a must in order to minimize variations as much as possible. It is also advisable to conduct sound and vibration sensitivity studies. Sound abatement measures, which by themselves show high sensitivity to procedural variations, should perhaps not be pursued, or should be improved so that their effects are more broadband in nature.

An example of this is trying to detune a certain compressor housing natural frequency from any of the excitation harmonics by slight structural modifications. If this natural frequency is highly sensitive to assembly operations, the next compressor may show the resonance again. A damping treatment, being more broadband and insensitive to structural residual stresses, would be more advisable (if this is a possible solution; see previous discussion of damping).

12.3 IDENTIFICATION OF SOUND AND VIBRATION MECHANISMS

12.3.1 EXTERIOR AND INTERIOR MEASUREMENTS AND RECOMMENDED CALCULATIONS

Taking again the hermetically sealed refrigeration or air conditioning compressor as an example (but much of the discussion also applies to air and gas compressors), the first step might be to take sound pressure level or sound power level spectra, and examine the sound peaks. Third-octave spectra are perhaps preferable at first because narrowband spectra have so many spikes that the proverb about "not seeing the forest for the trees" applies. With the aid of the simplified models discussed in previous chapters, one should be able to identify the frequency regions in which the housing shell natural frequencies form clusters. For fractional-horsepower compressors, they often occur in the 900 to 1300 Hz region, and there is often a second cluster in the 2500 to 3500 Hz region. It may also be advisable to measure the natural frequencies of the housing alone by shaker or by the impact method.

The vibration accelerations on various strategic locations of the housing shell could also be measured while the compressor is operating. Next, one should calculate, if the spectrum shows a candidate peak in the 200 to 900 Hz range, the natural frequencies of the first few gas modes in the cavity formed between the compressor housing shell and the compressor casing to verify the possibility that they are involved. Eventually these natural frequencies and modes should, of course, also be verified by direct pressure measurements in the gas cavity.

Next, if one suspects that a particular peak in, say, the 400 to 1600 Hz region could be an indication of a fluttering valve, one should calculate the first one or two natural frequencies of the suction and discharge valve reeds or plates. This can be done, if applicable, by considering the valve reeds as beams, or by finite element analysis. Or one could measure the natural frequencies of the valves using electromagnetic excitation. The best course of action is, of course, a direct, *in situ* measurement of the valve motions in the compressor to verify that there is flutter. This is often difficult in fractional-horsepower compressors because one often has no room for the installation of a proximity transducer or for strain gauges (pioneered by Lowery and Cohen, 1962). One may have to remove parts of a valve stop (motion limiter) or a valve seat. But if this is done properly, the measurements can still ascertain if there is valve flutter or not, and to what extent.

It is, of course, most useful to measure cylinder pressures as a function of time. Not only will the pressure trace tell if there is valve flutter (the pressure will be oscillatory during the opening times of the suction or discharge valves if there is flutter), but folded into a pressure-volume diagram, thermodynamic losses due to the valve restrictions will be obtainable at a glance.

Acceleration levels of the discharge pipe (shock loop) inside the housing, in the case of a low-side compressor, or the suction pipe inside the housing in the case of a high-side compressor, should be measured while the compressor is operating. Large vibration levels would indicate a strong vibration transmission to the housing shell. It is also advisable to measure the natural frequencies of these discharge or suction pipes, or utilize a finite element approach to calculate them. Because of the slenderness and relatively large flexibilities of these pipes, it is expected that there will be many natural frequencies in the frequency range of interest.

Knowledge of the internal pressure oscillations of the suction and discharge manifolds, mufflers and, piping are very valuable because they provide information on the effectiveness of mufflers and gas pulsation–caused excitation mechanisms.

The natural frequencies of the compressor casing should be measured because their associated natural modes are excited by the time-varying cylinder pressures and also possibly by valve impact.

Measurements of the natural frequencies of the vibration isolation springs may give insight with respect to the frequency bands in which vibration isolation takes place and those in which it probably does not.

12.3.2 Modifications to the Compressor for Diagnostic Purposes

It is advisable to verify sound and vibration mechanism identifications obtained by a combination of calculations and external measurements by modifying parts of the compressor. For example, if a particular peak in a sound spectrum is identified as being due to valve flutter, the valve could be modified to have a different natural frequency, or a motion limiter could be introduced, or an already existing motion limiter could be replaced by one with a different stop height. Not only will this verify that the sound-producing mechanism was correctly identified, but it will also perhaps point the direction to an effective sound abatement measure.

A frequently asked question is how much the gas path contributes to the overall sound pressure level. Taking as an example a low-side refrigeration compressor, gas pulsations generated by the suction valve will excite the gas modes in the space between the compressor housing shell and the casing. This gas vibration will in turn excite the housing shell, which radiates the sound, which is externally measured. On the other hand, there is also a mechanical path where casing vibrations are transmitted to the housing shell via the discharge tube (shock loop), perhaps by way of discharge tube resonances. Mechanical vibrations of the casing can also be transmitted to the housing shell through the vibration isolation springs. Another possible mechanical path is through the oil sump. An experiment that may clarify the contribution of the gas path is to modify the design in such a way that the suction tube is directly connected to the suction manifold, preferably by a very flexible connection to eliminate the creation of a new mechanical transmission path, thus bypassing the cavity formed between the housing shell and the compressor casing completely. This may not be a desirable practical solution, but these external sound measurements will indicate how much has been gained by eliminating gas pulsations inside the housing shell completely. If there is no appreciable change in the sound spectrum, it can be concluded that the mechanical path mechanisms dominate, and full attention has to be directed toward them. On the other hand, if there is a measurable sound reduction, it will establish a benchmark of how much we can possibly achieve in designing a perfect suction muffler.

It has to be noted that the foregoing experiment has only clarified the role of the pulsating suction gas inside the low-side compressor housing in exciting shell vibrations and thus producing sound. It has not clarified the role of gas pulsations in the discharge pipe, because as discussed earlier, pulsating gas in a bent pipe will produce vibrations due to the Borden tube effect, which are then mechanically transmitted to the housing shell. If this mechanism is suspected, one could try to reduce the discharge gas pulsations that reach the very flexible discharge tube by taking extreme measures in suction muffler design, which may consist of oversized muffler volumes and several low-pass filter mufflers in series, even if this might degrade the performance of the compressor. Again, this would not produce a practical solution, but it would verify the hypothesized sound-generating mechanism. Of course, interior gas pulsation measurements in the discharge tube should also be taken to ascertain the effectiveness of the experimental muffler.

The basic philosophy is that one should try to create large changes in the external sound spectra measurements, even if the modifications that are made to achieve them may be entirely unrealistic from a practical viewpoint. Once a mechanism of sound generation or transmission is identified, work on more realistic solutions can commence.

13 Miscellaneous Sound and Vibration Sources or Effects

13.1 ELECTROMOTORS

Because electromotors are such an integral part of hermetic refrigeration and air conditioning compressors, they are briefly discussed here. On the other hand, power sources for large compressors, be they external electromotors or diesel engines for large air and gas compressors, are not discussed because of the limited scope of this book.

Electromagnetic forces cause vibrations of the electromotor structural elements of the rotor and stator, which are then typically transmitted into the compressor casing–electromotor assembly. They also cause small shaft speed variations that are superimposed on the shaft speed variations caused by the compressor kinematics and gas compression. These electromagnetically originated vibrations are then transmitted mechanically through the vibration isolation springs or through the piping to the housing shell.

In narrowband sound or vibration spectra, their influence tends to show up as spikes, which tend to be multiples of the electric line frequency (60 Hz in the United States and 50 Hz on the European continent), for example. Thus if we observe, for the former, response peaks at 60, 120, 180, 240, ... Hz, they are due to the electromagnetic effects. Nonelectromagnetic effects occur at multiples of the compressor running speed, which is less then 3600 rotations per minute for a two-pole induction motor because of the torque load. For example, if it is 3540 rotations per minute, which corresponds to 59 Hz, we observe peaks in the spectra of 59, 118, 177, 236, ... Hz.

There are, of course, certain frequencies at which the electromagnetic effect is stronger than at others. Hamilton (1988) gives them for induction motors as

$$f_{em} = \frac{R}{P} f_s \pm mf \qquad (13.1)$$

where f_{em} = the frequencies due to the electromagnetic effect, R = the number of rotor slots, P = the number of poles, f_s = the shaft speed in [Hz], and f = the line frequency in [Hz], m = 0, ±2, ±4, ±6, ...

For identification purposes, a compressor should be run at two different line frequencies using a frequency converter. Changes in the sound and vibration spectra are then most likely due to the electromagnetic effects.

It has been reported that eccentricity of the rotor relative to the stator, which would make the gap between rotor and stator nonuniform, will accentuate the electromagnetically generated vibrations and should be avoided.

In certain designs, the cooler suction gas is passed through the motor, notably the gap between the stator and rotor. This cools the electromotor and increases its efficiency. The interaction of the gas with the rotor slots and stator produces aerodynamically generated gas pulsations, probably mostly of a turbulent nature. It has been shown that their effect can be observed in measured sound spectra.

The fundamental frequency of aerodynamically generated sound is

$$f_a = R f_s \tag{13.2}$$

where R = the number of rotor slots and f_s = the shaft speed in Hz. If the rotor slots are closed up such that there are no discontinuities in the circumferential direction, aerodynamic sound will be reduced (Hamilton, 1988).

Finally, there are also mechanical effects, which may be due to roller or ball bearings and due to resonances of the shaft assembly, which consists of the rotor and the shaft connecting the rotor to the compressor kinematic mechanism. These resonances can be due to natural frequencies associated either with bending or with torsion.

For more information, see Hamilton (1988), Medira et al. (2004), Mochizuki et al. (1988), Morimoto et al. (2004), and Roys and Soedel (1989).

13.2 LUBRICATION OIL

In refrigerating and air conditioning compressors, the lubrication oil is located at the bottom of the housing shell from where it is fed by an oil pumping mechanism or by splash lubrication to the moving parts of the compressor. Because of the affinity of many refrigerants to oil, oil can be distributed to the piston and cylinders by the refrigerants, even to the valves. In the latter case, oil provides at times a beneficial cushioning effect that lowers impact stresses. Another design is that oil is forced to the moving parts by the discharge pressure in high-side compressors, for example, in rotary vane compressors.

But no matter how the oil is transported, there is typically an oil sump at the bottom of the housing shell. Because of the low compressibility of oil, it can transmit vibrations from the compressor casing or isolation springs to the housing shell. Also, if the oil transport happens via splash lubrication, where the crankshaft of a reciprocating piston compressor dips into the oil sump, turbulence and thus vibrations of the oil are generated and transmitted.

It has been reported that by changing from splash lubrication to an oil pump design, compressor sound pressures were reduced. The same occurred when avoiding contact of the compressor casing with the oil sump.

It has also been reported that at the start-up of a low-side compressor, when the suction pressure reduces and refrigerant bubbles out of the oil-refrigerant solution, sound pressure levels often seem to reduce significantly, but only temporarily until the bubble activity has ceased. It has been speculated that the beneficial effect is due to the soap bubble–like foam, which may fill the space between the housing and the compressor casing. Another hypothesis is that the bubbles in the oil itself make the oil-bubble mixture more compressible, allowing the oil sump to act as a dynamic vibration absorber that is attached to the housing shell.

In support of the dynamic vibration absorber explanation, I am reminded of a lecture given at Purdue University some time ago by P. Baade (1970). I have used one experiment he demonstrated with great effect in my own lectures. The experiment proceeds as follows. One holds up an empty wine glass (the housing shell) and "pings" (impacts) it with a pencil. Everyone is asked to keep the generated sound in mind. Then one adds water (equivalent to oil without bubbles) to the glass until it is half full. Again, one impacts the edge of the glass with a pencil. The sound is more or less as loud as before, only somewhat different in pitch. Finally, one dissolves a tablet of bicarbonate of soda in the water and generates bubbles (equivalent to bubbles in oil). Impacting the edge of the glass with a pencil produces a dull and low sound only which disappears quickly. In this experiment, there is no foam, only bubbles in the water itself, yet a sound reduction is observed.

References

1. Adams, G.P. and Soedel, W. (1992). Remarks on Oscillating Bearing Loads in Twin Screw Compressors. *Proceedings of the 1992 International Compressor Engineering Conference*, 439–448. West Lafayette, IN, Purdue University.
2. Adams, G.P. and Soedel, W. (1994). Dynamic Simulation of Rotor Contact Forces in Twin Screw Compressors. *Proceedings of the 1994 International Compressor Engineering Conference*, 73–78. West Lafayette, IN, Purdue University.
3. Adams, J.A., Hamilton, J.F., and Soedel, W. (1974). The Prediction of Dynamic Strain in Ring Type Compressor Valves Using Experimentally Determined Strain Modes. *Proceedings of the 1974 Compressor Technology Conference*, 303–311. West Lafayette, IN, Purdue University.
4. Akashi, H., Yagi, A., Sugimoto, S., and Yoshimura, T. (2000). Influence of Pressure Wave in a Suction Path on Performances in Reciprocating Compressor. *Proceedings of the 2000 International Compressor Engineering Conference*, 611–618. West Lafayette, IN, Purdue University.
5. Akella, S., Anantapantula, V.S, and Venkateswarlu, K. (1998). Hermetic Compressor Muffler Design: Tuning of Mufflers for Noise Reduction. *Proceedings of the 1998 International Compressor Engineering Conference*, 349–354. West Lafayette, IN, Purdue University.
6. Asami, K., Ishijima, K., and Tanaka, H. (1982). Improvements of Noise and Efficiency of Rolling Piston Type Refrigeration Compressor for Household Refrigerator and Freezer. *Proceedings of the 1982 Compressor Technology Conference*, 268–274. West Lafayette, IN, Purdue University.
7. ASHRAE (American Society of Heating, Refrigerating and Air Conditioning Engineers) (1981). *Handbook of Fundamentals*. Atlanta, GA, American Society of Heating, Refrigerating and Air Conditioning Engineers.
8. ASME (American Society of Mechanical Engineering) (1959). Fluid Meters — Their Theory and Application. *Report of the American Society of Mechanical Engineering*; 5th Ed.,
9. Baade, P. (1970). *Lecture on Compressor Noise*. West Lafayette, IN, Purdue University.
10. Baars, E., Silveira, M., and Lampugnani, G. (1998). Compressor Noise Source Identification in Low Frequency. *Proceedings of the 1998 International Compressor Engineering Conference*, 555–560. West Lafayette, IN, Purdue University.
11. Beard, J.E., Hall, A.S., and Soedel, W. (1982). On the Classification of Compressor, Pump or Engine Designs Using Generalized Linkages. *Proceedings of the 1982 Purdue Compressor Technology Conference*, 166–172. West Lafayette, IN, Purdue University.
12. Benson, R.S. and Ucer, A.S. (1972). Some Recent Research in Gas Dynamic Modelling of Multiple Single Stage Reciprocating Compressor Systems. *Proceedings of the 1972 Compressor Technology Conference,* 491–498. West Lafayette, IN, Purdue University.
13. Biscaldi, E., Faraon, A., and Sarti, S. (1998). Numerical Prediction of the Radiated Noise of Hermetic Compressors under the Simultaneous presence of Different Noise Sources. *Proceedings of the 1998 International Compressor Engineering Conference*, 337–342. West Lafayette, IN, Purdue University.

14. Bishop, R.E.D. and Johnson, D.C. (1979). *The Mechanics of Vibration*. New York, Cambridge University Press.
15. Boswirth, L. (1980 a). Flow Forces and the Tilting of Spring Loaded Valve Plates, Parts 1 and 2. *Proceedings of the 1980 Compressor Technology Conference*, 185–197. West Lafayette, IN, Purdue University.
16. Boswirth, L. (1980 b). Hypothesis on the Failure of Spring Loaded Compressor Valve Plates. *Proceedings of the 1980 Compressor Technology Conference*, 198–206. West Lafayette, IN, Purdue University.
17. Boswirth, L. (1982). Theoretical and Experimental Study on Flow in Valve Channels, Parts 1 and 2. *Proceedings of the 1982 Compressor Technology Conference*, 38–53. West Lafayette, IN, Purdue University.
18. Boyle, R.J., Tramschek, A.B., Brown, J., and MacLaren, J.F.T. (1982). The Apportioning of Port Areas between Suction and Discharge Valves in Reciprocating Compressors. *Proceedings of the 1982 Compressor Technology Conference*, 32–37. West Lafayette, IN, Purdue University.
19. Brablik, J. (1972). Gas Pulsations as Factor Affecting Operation of Automatic Valves in Reciprocating Compressors. *Proceedings of the 1972 Compressor Technology Conference*, 188–195. West Lafayette, IN, Purdue University.
20. Brablik, J. (1974). Computer Simulation of the Working Process in the Cylinder of a Reciprocating Compressor with Piping System. *Proceedings of the 1974 Compressor Technology Conference*, 151–158. West Lafayette, IN, Purdue University.
21. Bredesen, A.M. (1974). Computer Simulation of Valve Dynamics as an Aid to Design. *Proceedings of the 1974 Compressor Technology Conference*, 171–176. West Lafayette, IN, Purdue University.
22. Brown, J., Davidson, R., and Fleming, J. (1982). Performance of Automatic Compressor Valves. *Proceedings of the 1982 Compressor Technology Conference*, 373–377. West Lafayette, IN, Purdue University.
23. Bucciarelli, M., Faraon, A., and Giusto, F.M. (1994). An Approach to Evaluate the Acoustical Characteristics of Silencers Used in Hermetic Compressors for Household Refrigeration. *Proceedings of the 1994 International Compressor Engineering Conference*, 25–30. West Lafayette, IN, Purdue University.
24. Bucciarelli, M., Giusto, F., Cossalter, V., DaLio, M., and Gardonio, P. (1992). Modal Analysis of Compressor Shell and Cavity for Emitted Noise Reduction. *Proceedings of the 1992 International Compressor Engineering Conference*, 1275–1283. West Lafayette, IN, Purdue University.
25. Bukac, H. (2002). Understanding Valve Dynamics. *Proceedings of the 2002 International Compressor Engineering Conference*, 489–498. West Lafayette, IN, Purdue University.
26. Bukac, H. (2004). Self-Excited Vibration in a Radially and Axially Compliant Scroll Compressor. *Proceedings of the 2004 International Compressor Engineering Conference*, Paper C041. West Lafayette, IN, Purdue University.
27. Buligan, G., Paone, N., Revel, G.M., and Tomasini, E.P. (2002). Valve Lift Measurements by Optical Techniques in Compressors. *Proceedings of the 2002 International Compressor Engineering Conference*, 429–436. West Lafayette, IN, Purdue University.
28. Bush, J.W., Eyo, V.A., and Housman, M.E. (1992). Design Techniques and Resulting Structural Modifications used to Reduce Compressor Noise. *Proceedings of the 1992 International Compressor Engineering Conference*, 967–975. West Lafayette, IN, Purdue University.
29. Cerrato-Jay, G. and Lowery, D. (2002). Sound Quality Evaluations of Compressors. *Proceedings of the 2002 International Compressor Engineering Conference*, 451–456. West Lafayette, IN, Purdue University.

References

30. Chen, L. and Huang, Z.S. (2004). Analysis of Acoustic Characteristics of the Muffler on a Rotary Compressor. *Proceedings of the 2004 International Compressor Engineering Conference*, Paper C015. West Lafayette, IN, Purdue University.
31. Chlumski, V. (1965). *Reciprocating and Rotary Compressors*. London, Spon.
32. Cohen, R. (1972). Valve Stress Analysis for Fatigue Problems. *Proceedings of the 1972 Compressor Technology Conference*, 129–135. West Lafayette, IN, Purdue University.
33. Collings, D.A. and Weadock, T.J. (2004). Design of Flapper Valves for a CO_2 Commercial Refrigeration Compressor. *Proceedings of the 2004 International Compressor Engineering Conference*, Paper C131. West Lafayette, IN, Purdue University.
34. Conrad, D.C. and Soedel, W. (1992). Modeling of Compressor Shell Vibrations Excited by a Rotor Imbalance. *Proceedings of the 1992 International Compressor Engineering Conference*, 759–768. West Lafayette, IN, Purdue University.
35. Cossalter, V., Doria, A., and Giusto, F. (1994). Control of Acoustic Vibrations inside Refrigerator Compressors by Means of Resonators. *Proceedings of the 1994 International Compressor Engineering Conference*, 563–568. West Lafayette, IN, Purdue University.
36. Cyklis, P. (2000). Experimental Identification of the Generalized Four-Element Transfer Matrix for the Pulsating Gas Installation Element, Part I — Theory, Part II — Experiment. *Proceedings of the 2000 International Compressor Engineering Conference*, 635–650. West Lafayette, IN, Purdue University.
37. Cyklis, P. (2002). A CFD Based Identification Method of the Transmittance for the Pulsating Gas Installation Element: Part I — Theory, Part II — Experimental Validation. *Proceedings of the 2002 International Compressor Engineering Conference*, 555–568. West Lafayette, IN, Purdue University.
38. DaSilva, A.R., Lenzi, A., and Baars, E. (2004). Controlling the Noise Radiation of Hermetic Compressors by Means of Minimization of Power Flow through Discharge Pipes Using Genetic Algorithms. *Proceedings of the 2004 International Compressor Engineering Conference*, Paper C096. West Lafayette, IN, Purdue University.
39. Dhar, M. and Soedel, W. (1978 a). *Compressor Simulation Program with Gas Pulsations*, Short Course Notes. West Lafayette, IN, Ray W. Herrick Laboratories, Purdue University.
40. Dhar, M. and Soedel, W. (1978 b). Influence of a Valve Stop and/or Suction Muffler on Suction Valve Noise of an Air Compressor. *Proceedings of the 1978 Compressor Technology Conference*, 36–44. West Lafayette, IN, Purdue University.
41. Doige, A.G. and Cohen, R. (1972). A Stress and Vibration Analysis of a Leaf-Type Compressor Valve, Parts I and II. *Refrigeration*, 47 (538 and 539): 54–70 and 57–74.
42. Dreiman, N., Collings, D., and DiFlora, M. (2000). Noise Reduction of Fractional Horse Power Hermetic Reciprocating Compressor. *Proceedings of the 2000 International Compressor Engineering Conference*, 949–956. West Lafayette, IN, Purdue University.
43. Dreiman, N. and Herrick, K. (1998). Vibration and Noise of a Rotary Compressor. *Proceedings of the 1998 International Compressor Engineering Conference*, 685–712. West Lafayette, IN, Purdue University.
44. Dusil, R. (1976). Studies of Faults in Used Valves. *Proceedings of the 1976 Compressor Technology Conference*, 99–105. West Lafayette, IN, Purdue University.
45. Dusil, R. and Appell, B. (1976). Fatigue and Fracture Mechanics Properties of Valve Steels. *Proceedings of the 1976 Compressor Technology Conference*, 82–90. West Lafayette, IN, Purdue University.
46. Dusil, R. and Johansson, B. (1978 a). Fatigue Fracture Behavior of Impact Loaded Compressor Valves. *Proceedings of the 1978 Compressor Technology Conference*, 124–128. West Lafayette, IN, Purdue University.

47. Dusil, R. and Johansson, B. (1978 b). Material Aspects of Impact Fatigue of Valve Steels. *Proceedings of the 1978 Compressor Technology Conference*, 116–123. West Lafayette, IN, Purdue University.
48. Dusil, R. and Johansson, B. (1980). Influence of Seat Positioning and Seat Design on Valve Fatigue Performance. *Proceedings of the 1980 Compressor Technology Conference*, 368–373. West Lafayette, IN, Purdue University.
49. Elson, J.P. and Soedel, W. (1972 a). A Review of Discharge and Suction Line Oscillation Research. *Proceedings of the 1972 Purdue Compressor Technology Conference*, 311–315. West Lafayette, IN, Purdue University.
50. Elson, J.P. and Soedel, W. (1972 b). Criteria for the Design of Pressure Transducer Adapter Systems. *Proceedings of the 1972 Purdue Compressor Technology Conference*, 390–394. West Lafayette, IN, Purdue University.
51. Elson, J.P. and Soedel, W. (1974). Simulation of the Interaction of Compressor Valves with Acoustic Back Pressures in Long Discharge Lines. *Journal of Sound and Vibration*, 34 (2): 211–220.
52. Elson, J.P., Soedel, W., and Cohen, R. (1976). A General Method of Simulating the Flow Dependent Nonlinear Vibrations of Compressor Reed Valves. *ASME Journal of Engineering in Industry*, 98 (3): 930–934.
53. Faulkner, L.L. (1969). Vibration Analysis of Shell Structures Using Receptances. PhD dissertation. West Lafayette, IN, Purdue University.
54. Fleming, J., Brown, J., and Davidson, R. (1982). Gas Forces on Disc Valves. *Proceedings of the 1982 Compressor Technology Conference*, 378–381. West Lafayette, IN, Purdue University.
55. Fleming, J.S. and Brown, J. (1982). An Experimental Investigation of the Aerodynamics of a Disc Valve. *Proceedings of the 1982 Compressor Technology Conference*, 21–25. West Lafayette, IN, Purdue University.
56. Forbes, D.A. and Mitchell, J. (1974). An Impact Test Rig for Annular Plate Valve Models. *Proceedings of the 1974 Compressor Technology Conference*, 312–318. West Lafayette, IN, Purdue University.
57. Frenkel, M.I. (1969). *Kolbenverdichter*. Berlin, VEB Verlag Technik.
58. Friley, J.R. and Hamilton, J.F. (1976). Characterization of Reed Type Compressor Valves by the Finite Element Method. *Proceedings of the 1976 Compressor Technology Conference*, 295–301. West Lafayette, IN, Purdue University.
59. Fu, W.C. (2004). Sound Reduction for Midsize Semi-Hermetic Compressors using Experimental Methods. *Proceedings of the 2004 International Compressor Engineering Conference*, Paper C001. West Lafayette, IN, Purdue University.
60. Funer, V. and Tauchmann, R. (1963). Moderne Messverfahren für die Untersuchung von Kompressionskältemaschinen. *Kältetechnik*, 15 (9): 282–289.
61. Futakawa, A. and Namura, K. (1980). A Fundamental Study of Valve Impact Stress in a Refrigeration Compressor. *Proceedings of the 1980 Compressor Technology Conference*, 277–285. West Lafayette, IN, Purdue University.
62. Futakawa, A., Namura, K., and Furukawa, H. (1978). Dynamic Stress of Refrigeration Compressor Reed Valve with Oval Shape. *Proceedings of the 1978 Compressor Technology Conference*, 187–194. West Lafayette, IN, Purdue University.
63. Gatecliff, G.W., Griner, G.C., and Richardson, H. (1980). A Compressor Valve Model for Use in Daily Design Work. *Proceedings of the 1980 Compressor Technology Conference*, 176–179. West Lafayette, IN, Purdue University.
64. Gatecliff, G.W. and Lady, E.R. (1972). Forced Vibration of a Cantilever Valve of Uniform Thickness and Non-Uniform Width. *Proceedings of the 1972 Purdue Compressor Technology Conference*, 316–319. West Lafayette, IN, Purdue University.

65. Gatley, W.S. and Cohen, R. (1970). Development and Evaluation of a General Method for Design of Small Acoustic Filters. *ASHRAE Transactions*, 76 (1): Paper No. 2128.
66. Gavric, L. and Badie-Cassagnet, A. (2000). Measurement of Gas Pulsations in Discharge and Suction Lines of Refrigerant Compressors. *Proceedings of the 2000 International Compressor Engineering Conference*, 627–634. West Lafayette, IN, Purdue University.
67. Gavric, L. and Darpas, M. (2002). Sound Power of Hermetic Compressors Using Vibration Measurements. *Proceedings of the 2002 International Compressor Engineering Conference*, 499–506. West Lafayette, IN, Purdue University.
68. Giacomelli, E. and Giorgetti, M. (1974). Investigation of Oil Sticktion in Ring Valves. *Proceedings of the 1974 Compressor Technology Conference*, 167–170. West Lafayette, IN, Purdue University.
69. Gilliam, D.R. and DiFlora, M.A. (1992). The Effect of the Dome Shape of a Hermetic Compressor Housing on Sound Radiation. *Proceedings of the 1992 International Compressor Engineering Conference*, 955–965. West Lafayette, IN, Purdue University.
70. Gluck, R., Ukrainetz, P., and Cohen, R. (1964). Dynamic Stress Measurement Techniques of High Speed Compressor Valves. *ASHRAE Transactions*, 70: 303–305.
71. Graff, K. (1975). *Wave Motion in Elastic Solids*. Columbus, Ohio State University Press.
72. Groth, K. (1953). Schwingungen in der Druckleitung von Kolbenverdichtern. VDI Forschungsheft 440, Bd. 19.
73. Hamilton, J.F. (1974). *Extensions of Mathematical Modeling of Positive Displacement Type Compressors*, Short Course Notes. West Lafayette, IN. Purdue University.
74. Hamilton, J.F. (1982). *Modeling and Simulation of Compressor Suspension System Vibrations*, Short Course Notes. West Lafayette, IN, Ray W. Herrick Laboratories, Purdue University.
75. Hamilton, J.F. (1988). *Measurement and Control of Compressor Noise*, Short Course Notes. West Lafayette, IN, Ray W. Herrick Laboratories, Purdue University.
76. Hatch, G. and Wollatt, D. (2002). The Dynamics of Reciprocating Compressor Valve Springs. *Proceedings of the 2002 International Compressor Engineering Conference*, 421–428. West Lafayette, IN, Purdue University.
77. Herfat, A.T. (2002). Experimental Study of Vibration Transmissibility Using Characterization of Compressor Mounting Grommets, Dynamic Stiffnesses: Part I — Frequency Response Technique Development, Analytical, Part II — Experimental Analysis and Measurements. *Proceedings of the 2002 International Compressor Engineering Conference*, 521–538. West Lafayette, IN, Purdue University.
78. Holowenko, A.R. (1955). *Dynamics of Machinery*. New York, John Wiley.
79. Hort, W. (1922). *Technische Schwingungslehre*. Berlin, Springer.
80. Hsu, M.P. and Soedel, W. (1987). Natural Modes of Irregular Shells and Interpretations, *Proceedings of the 20th Midwestern Mechanics Conference*, 268–273. West Lafayette, IN, Purdue University.
81. Huang, D.T. and Soedel, W. (1993 a). On the Vibration of Multiple Plates Welded to a Cylindrical Shell with Special Attention to Mode Pairs. *Journal of Sound and Vibration*, 166 (2): 315–339.
82. Huang, D.T. and Soedel, W. (1993 b). Study of the Forced Vibrations of Shell-Plate Combinations Using the Receptance Method. *Journal of Sound and Vibration*, 166 (2): 341–369.
83. Hwang, I., Kwon, B., and Kim, C. (2000). Low-Frequency Band Noise of Rotary Compressor. *Proceedings of the 2000 International Compressor Engineering Conference*, 1027–1032. West Lafayette, IN, Purdue University.

84. Ih, J.G. and Jang, S.H. (1998). Measurement of the Acoustic Source Characteristics of the Intake Port in the Refrigerator Compressor. *Proceedings of the 1998 International Compressor Engineering Conference*, 561–564. West Lafayette, IN, Purdue University.
85. Joergensen, S.H. (1980). Transient Valve Plate Vibrations. *Proceedings of the 1980 Compressor Technology Conference*, 286–292. West Lafayette, IN, Purdue University.
86. Johansson, R. and Persson, G. (1976). Influence of Testing and Material Factors on the Fatigue Strength of Valve Steel. *Proceedings of the 1976 Compressor Technology Conference*, 74–81. West Lafayette, IN, Purdue University.
87. Johnson, C.N. and Hamilton, J.F. (1972). Cavity Resonance in Fractional HP Refrigerant Compressors, *Proceedings of the 1972 Compressor Technology Conference*, 83–89. West Lafayette, IN, Purdue University.
88. Johnson, O., Smith, A.V., and Winslett, C.E. (1990). The Application of Advanced Analysis Methods to the Reduction of Noise from Air Compressors. *Proceedings of the 1990 International Compressor Engineering Conference*, 800–807. West Lafayette, IN, Purdue University.
89. Joo, J.M., Oh, S.K., Kim, G.K., and Kim, S.H. (2000). Optimal Valve Design for Reciprocating Compressor. *Proceedings of the 2000 International Compressor Engineering Conference*, 451–458. West Lafayette, IN, Purdue University.
90. Kawai, H., Sasano, H., Kita, I., and Ohta, T. (1988). The Compressor Noise — Shell and Steel Materials. *Proceedings of the 1988 International Compressor Engineering Conference*, 307–314. West Lafayette, IN, Purdue University.
91. Kelly, A.D. and Knight, C.E. (1992a). Dynamic Finite Element Modeling and Analysis of a Hermetic Reciprocating Compressor. *Proceedings of the 1992 International Compressor Engineering Conference*, 769–776. West Lafayette, IN, Purdue University.
92. Kelly, A.D. and Knight, C.E. (1992 b). Helical Coil Springs in Finite Element Models of Compressors. *Proceedings of the 1992 International Compressor Engineering Conference*, 779–787. West Lafayette, IN, Purdue University.
93. Khalifa, H.E. and Liu, X. (1998). Analysis of Stiction Effect on the Dynamics of Compressor Suction Valves. *Proceedings of the 1998 International Compressor Engineering Conference*, 87–92. West Lafayette, IN, Purdue University.
94. Killman, I.G. (1972 a). Aerodynamic Forces Acting on Valve Discs. *Proceedings of the 1972 Compressor Technology Conference*, 407–414. West Lafayette, IN, Purdue University.
95. Killman, I.G. (1972 b). Investigations of a Springless, Low Mass Compressor Valve. *Proceedings of the 1972 Compressor Technology Conference*, 415–422. West Lafayette, IN, Purdue University.
96. Kim, H.J. and Soedel, W. (1992 a). Remarks on the Calculation of Radiated Sound from Compressor Shell Side Walls Using Equivalent Cylinders. *Proceedings of the 1992 International Compressor Engineering Conference*, 935–946. West Lafayette, IN, Purdue University.
97. Kim, H.J. and Soedel, W. (1992 b). Transmission Loss and Back Pressure Characteristics for Compressor Mufflers. *Proceedings of the 1992 International Compressor Engineering Conference*, 1455–1464. West Lafayette, IN, Purdue University.
98. Kim H.J. and Soedel, W. (1994). Time Domain Approach to Gas Pulsation Modeling. *Proceedings of the 1994 International Compressor Engineering Conference*, 235–240. West Lafayette, IN, Purdue University.
99. Kim, J. (1992). Application of Four Pole Parameters for Gas Pulsation Analysis of Multi-Cylinder Compressors with Symmetrically Arranged Gas Cavities. *Proceedings of the 1992 International Compressor Engineering Conference*, 1487–1494. West Lafayette, IN, Purdue University.

100. Kim, J. and Soedel, W. (1987). Analytical Four Pole Parameters for Gas Filled Cylindrical Annular Cavities. *Proceedings of the 20th Midwestern Mechanics Conference*, 813–818. West Lafayette, IN, Purdue University.
101. Kim, J. and Soedel, W. (1988 a). Four Pole Parameters of Shell Cavity and Application to Gas Pulsation Modeling. *Proceedings of the 1988 International Compressor Engineering Conference*, 331–337. West Lafayette, IN Purdue University.
102. Kim, J. and Soedel, W. (1988 b). Performance and Gas Pulsations When Pumping Different Gases with the Same Compressor. *Proceedings of the 1988 International Compressor Engineering Conference*, 227–234. West Lafayette, IN Purdue University.
103. Kim, J. and Soedel, W. (1989 a). Analysis of Gas Pulsations in Multiply Connected Three-Dimensional Acoustics Cavities with Special Attention to Natural Mode or Wave Cancellation Effects. *Journal of Sound and Vibration*, 131 (1): 103–114.
104. Kim, J. and Soedel, W. (1989 b). Comments on Boundary Conditions and Source Modeling when Utilizing Acoustic Four Poles Obtained by Modal Series. *Proceedings of the 21st Midwest Mechanics Conference*, 511–512. Iowa City, Iowa State University.
105. Kim, J. and Soedel, W. (1989 c). General Formulation of Four Pole Parameters for Three-Dimensional Cavities Utilizing Modal Expansion with Special Attention to the Annular Cylinder. *Journal of Sound and Vibration*, 129 (2): 237–254.
106. Kim, J. and Soedel, W. (1990 a). Convergence of Gas Pulsations When Combining Time and Frequency Analysis Interactively. *Proceedings of the 1990 International Compressor Engineering Conference*, 641–646. West Lafayette, IN, Purdue University.
107. Kim, J. and Soedel, W. (1990 b). Development of a General Procedure to Formulate Four Pole Parameters by Modal Expansion and Its Application to Three Dimensional Cavities. *Journal of Vibrations and Acoustics*, 112 (1): 452–459.
108. Kim, J. and Soedel, W. (1990 c). Performance Study of a Prototype Reciprocating Piston Compressor with Special Attention to Valve Design and Gas Pulsations. *Proceedings of the 1990 International Compressor Engineering Conference*, 634–640. West Lafayette, IN, Purdue University.
109. Kim, J.S. and Soedel, W. (1986). Impact Stress Wave Propagation in a Compressor Valve. *Proceedings of the 1986 International Compressor Engineering Conference*, 382–391. West Lafayette, IN, Purdue University.
110. Kim, J.S. and Soedel, W. (1988). On the Response of Three-Dimensional Elastic Bodies to Distributed Dynamic Pressures, Part 1: Half-Space, Part 2: Thick Plate. *Journal of Sound and Vibration*, 126 (2), 279–308.
111. Kim, S.H. and Soedel W. (1994). Analysis of the Beating Response of Bell Type Structures. *Journal of Sound and Vibration*, 173 (4): 517–536.
112. Kim, Y.K. and Soedel, W. (1996). Theoretical Gas Pulsations in Discharge Passages of Rolling Piston Compressor, Part I: Basic Model, Part II: Representative Results. *Proceedings of the 1996 International Compressor Engineering Conference*, 611–624. West Lafayette, IN, Purdue University.
113. Kim, Y.K. and Soedel, W. (1998). Stiffening of Compressor Shell by Tension Rings, *Proceedings of the 1998 International Engineering Conference*, 565–570. West Lafayette, IN, Purdue University.
114. Koai, K. and Soedel, W. (1990 a). Contributions to the Understanding of Flow Pulsation Levels and Performance of a Twin Screw Compressor Equipped with a Slide Valve and a Stopper for Capacity Control. *Proceedings of the 1990 International Compressor Engineering Conference*, 388–397. West Lafayette, IN, Purdue University.

115. Koai, K. and Soedel, W. (1990 b). Gas Pulsations in Screw Compressors, Part I: Determination of Port Flow and Interpretation of Periodic Volume Source, Part II: Dynamics of Discharge System and Its Interaction with Port Flow. *Proceedings of the 1990 International Compressor Engineering Conference*, 369–387. West Lafayette, IN, Purdue University.

116. Kristiansen, U., Soedel, W., and Hamilton, J.F. (1972). An Investigation of Scaling Laws for Vibrating Beams and Plates, with Special Attention to the Effects of Shear and Rotatory Inertia. *Journal of Sound and Vibration*, 20 (1): 113–122.

117. Kumar, K., Lu, J., Leyderman, A., Marler, M., Nieter, J., and Peracchio, A. (1994). Reduced Noise Valve Design for a Rotary Compressor. *Proceedings of the 1994 International Compressor Engineering Conference*, 19–23. West Lafayette, IN, Purdue University.

118. Lai, P.C.C. and Soedel, W. (1996 a). Gas Pulsations in Thin, Curved or Flat Cavities due to Multiple Mass Flow Sources. *Proceedings of the 1996 International Compressor Engineering Conference*, 799–806. West Lafayette, IN, Purdue University.

119. Lai, P.C.C. and Soedel, W. (1996 b). On the Anechoic Termination Assumption when Modeling Exit Pipes. *Proceedings of the 1996 International Compressor Engineering Conference*, 815–822. West Lafayette, IN, Purdue University.

120. Lai, P.C.C. and Soedel, W. (1996c). Two-Dimensional Analysis of Thin, Shell or Plate Like Muffler Elements. *Journal of Sound and Vibration*, 194 (2): 137–171.

121. Lai, P.C.C. and Soedel, W. (1996 d). Two-Dimensional Analysis of Thin, Shell or Plate Like Muffler Elements of Non-Uniform Thickness. *Journal of Sound and Vibration*, 195 (3): 445–475.

122. Lai, P.C.C. and Soedel, W. (1996 e). Free Gas Pulsations in Acoustic Systems Composed of Two Thin, Curved or Flat, Two-Dimensional Gas Cavities Which Share a Common Open Boundary. *Journal of Sound and Vibration*, 198 (2): 225–248.

123. Lai, P.C.C., Soedel, W., Gilliam, D., and Roy, P. (1996). On the Permissibility of Approximating Irregular Cavity Geometries by Rectangular Boxes and Cylinders. *Proceedings of the 1996 International Compressor Engineering Conference*, 807–813. West Lafayette, IN, Purdue University.

124. Lai, P.C.C. and Soedel W. (1997). A General Procedure for the Analysis of Gas Pulsations in Thin Shell Type Gas Cavities with Special Attention to Compressor Manifolds. *Proceedings of the SAE Noise and Vibration Conference*, SAE Paper 971875. Traverse City, MI, SAE (Society of Automotive Engineers).

125. Lai, P.C.C. and Soedel W. (1998). Free Gas Pulsation of a Helmholtz Resonator Attached to a Thin Muffler Element. *Proceedings of the International Congress and Exposition*, SAE Paper 980281. Detroit, MI, SAE (Society of Automotive Engineers).

126. Laub, J.S. (1980). Some Considerations in Refrigerant Compressor Valve Structural Reliability and Failure Mechanisms. *Proceedings of the 1980 Compressor Technology Conference*, 386–389. West Lafayette, IN, Purdue University.

127. Laursen, M.B. (1990). Mismatching Noise Source and Resonance Spectra of Reciprocating Compressor Shells. *Proceedings of the 1990 International Compressor Engineering Conference*, 808–817. West Lafayette, IN, Purdue University.

128. Laville, F. and Soedel, W. (1978). Some New Scaling Rules for Use in Mufflers. *Journal of Sound and Vibration*, 60 (2): 273–288.

129. Lee, H., Kwon, B.H., and Park, S.O. (1998). The Reduction of the Low-Frequency Band Noise in a Hermetic Refrigerator Compressor. *Proceedings of the 1998 International Compressor Engineering Conference*, 343–348. West Lafayette, IN, Purdue University.

130. Lee, H.K., Park, J.S., and Hur, K.B. (2000). Reduction of Noise/Vibration Generated by the Discharge Valve System in a Hermetic Compressor for Refrigeration. *Proceedings of the 2000 International Compressor Engineering Conference*, 587–594. West Lafayette, IN, Purdue University.
131. Lee, J.H., Dhar, B., and Soedel W. (1985). A Mathematical Model of Low Amplitude Pulse Combustion Systems Using a Helmholtz Resonator Type Approach. *Journal of Sound and Vibration*, 98 (3): 379–401.
132. Lee, J.H. and Kim, J. (2000). Sound Transmission Through Cylindrical Shell of Hermetic Compressor. *Proceedings of the 2000 International Compressor Engineering Conference*, 933–940. West Lafayette, IN, Purdue University.
133. Lee, J.H. and Soedel, W. (1985). On the Prediction of Gas Pulsations and Exhaust Noise of Low Amplitude Pulse Combustion Systems. *Noise Control Engineering Journal*, 24 (1): 19–37.
134. Lee, J.K., Lee, S.J., Lee, D.S., Lee, B.C., and Lee, U.S. (2000). Identification and Reduction of Noise in a Scroll Compressor. *Proceedings of the 2000 International Engineering Conference*, 1041–1048. West Lafayette, IN, Purdue University.
135. Lee, J.M. and Kim, B.C. (1988). A study of the Cavity Resonance of a Compressor. *Proceedings of the 1988 International Compressor Engineering Conference*, 394–401. West Lafayette, IN, Purdue University.
136. Leemhuis, R.S. and Soedel, W. (1976). Kinematics of Wankel Compressors (or Engines) by Way of Vector Loops. *Proceedings of the 1976 Compressor Technology Conference*, 443–456. West Lafayette, IN, Purdue University.
137. Lenz, J.R. (2000). Finite Element Analysis of Dynamic Flapper Valve Stresses. *Proceedings of the 2000 International Compressor Engineering Conference*, 369–376. West Lafayette, IN, Purdue University.
138. Libralato, M. and Contarini, A. (2004). Impact Fatigue on Suction Valve Reed: New Experimental Approach. *Proceedings of the 2004 International Compressor Engineering Conference*, Paper C056. West Lafayette, IN, Purdue University.
139. Liu, Z. and Soedel, W. (1992). Modeling Temperatures in High Speed Compressors for the Purpose of Gas Pulsation and Valve Loss Modeling. *Proceedings of the 1992 International Compressor Engineering Conference*, 1375–1384. West Lafayette, IN, Purdue University.
140. Liu, Z. and Soedel, W. (1994). Discharge Gas Pulsations in a Variable Speed Compressor. *Proceedings of the 1994 International Compressor Engineering Conference*, 507–514. West Lafayette, IN, Purdue University.
141. Lowery, D.C. (1998). Visualization of Measured Acoustic Fields using a Commercial Model Analysis Code. *Proceedings of the 1998 International Compressor Engineering Conference*, 513–518. West Lafayette, IN, Purdue University.
142. Lowery, R.L. and Cohen, R. (1962). Strain Gages as Means of Analyzing High Speed Compressor Valve Vibration. *Annexe 1962-1 an Bulletin de L'Institute International du Froid*: 179–190.
143. Lowery, R.L. and Cohen, R. (1963). Experimental Determination of Natural Frequencies and Modes of Compressor Reed Valves. *ASHRAE Journal*, 5 (2): 95–98.
144. Lowery, R.L. and Cohen, R. (1971). High Speed Compressor Valve Noise and Vibration Studies, Parts I and II. *Refrigeration*, 46 (523 and 525): 27–44 and 64–76.
145. Luszczycki, M. (1978 a). Dynamic Investigation of Suction Valves in a Small Refrigeration Compressor. *Proceedings of the 1978 Compressor Technology Conference*, 382–388. West Lafayette, IN, Purdue University.

146. Luszczycki, M. (1978 b). Dynamic Investigation of Suction Valves in a Small Refrigeration Compressor. *Proceedings of the 1978 Compressor Technology Conference*, 382–388. West Lafayette, IN, Purdue University.
147. Ma, Y.C. and Bae, J.Y. (1996). Determination of Effective Force Area and Valve Behavior on the Rolling Piston Type Compressor. *Proceedings of the 1996 International Compressor Engineering Conference*, 371–376. West Lafayette, IN, Purdue University.
148. Ma, Y.C., Bolton, J.S., Jeong, H., Ahn, B., and Shin, C. (2002). Experimental Statistical Energy Analysis Applied to a Rolling Piston-Type Rotary Compressor. *Proceedings of the 2002 International Compressor Engineering Conference*, 547–554. West Lafayette, IN, Purdue University.
149. Ma, Y.C. and Min, O.K. (2000). Study of Pressure Pulsation using a Modified Helmholtz Method. *Proceedings of the 2000 International Compressor Engineering Conference*, 657–664. West Lafayette, IN, Purdue University.
150. Machu, G., Albrecht, M., Bielmeier, O., Daxner, T., and Steinrück, P. (2004). A Universal Simulation Tool for Reed Valve Dynamics. *Proceedings of the 2004 International Compressor Engineering Conference*, Paper C045. West Lafayette, IN, Purdue University.
151. MacLaren, J.F.T. (1972). A Review of Simple Mathematical Models of Valves in Reciprocating Compressors. *Proceedings of the 1972 Compressor Technology Conference*, 180–187. West Lafayette, IN, Purdue University.
152. MacLaren, J.F.T. and Tramschek, A.B. (1972). Prediction of Valve Behavior with Pulsating Flow in Reciprocating Compressors. *Proceedings of the 1972 Purdue Compressor Technology Conference*, 203–211. West Lafayette, IN, Purdue University.
153. MacLaren, J.F.T., Tramschek, A.B., Hussein, I.J., and El-Geresy, B.A. (1978). Can the Impact Velocity of Suction Valves be Calculated? *Proceedings of the 1978 Compressor Technology Conference*, 177–186. West Lafayette, IN, Purdue University.
154. MacLaren, J.F.T., Tramschek, A.B., and Kerr, S.V. (1974). A Model of a Single Stage Reciprocating Gas Compressor Accounting for Flow Pulsations. *Proceedings of the 1974 Compressor Technology Conference*, 144–150. West Lafayette, IN, Purdue University.
155. Madsen, P. (1976). Plastic Deformations of Discharge Valves in Hermetic Compressors. *Proceedings of the 1976 Compressor Technology Conference*, 302–306. West Lafayette, IN, Purdue University.
156. Marriott, L.W. (1998). Motion of the Sprung Mass of a Reciprocating Hermetic Compressor during Startup. *Proceedings of the 1998 International Compressor Engineering Conference*, 519–524. West Lafayette, IN, Purdue University.
157. Marriott, L.W. (2000). Motion of the Sprung Mass and Housing of a Reciprocating Hermetic Compressor During Standing and Stopping. *Proceedings of the 2000 International Compressor Engineering Conference*, 823–830. West Lafayette, IN, Purdue University.
158. Masters, A.R., Kim, S.J., and Jones, J.D. (1992). Active Control of Compressor Noise Radiation Using Piezoelectric Actuators. *Proceedings of the 1992 International Compressor Engineering Conference*, 325–329. West Lafayette, IN, Purdue University.
159. Matos, F.F.S., Prata, A.T., and Deschamps, C.J. (2002). Numerical Simulation of the Dynamics of Reed Type Valves. *Proceedings of the 2002 International Compressor Engineering Conference*, 481–488. West Lafayette, IN, Purdue University.
160. McLaren, R.J.L., Papastergiou, S., Brown, J., and MacLaren, J.F.T. (1982). Analysis of Bending Stresses in Cantilever Type Suction Valve Reeds. *Proceedings of the 1982 Compressor Technology Conference*, 89–97. West Lafayette, IN, Purdue University.

161. Medira, V., Parkovic, B., and Smoljan, B. (2004). The Analysis of Shaft Breaks on Electric Motors Coupled with Reciprocating Compressors. *Proceedings of the 2004 International Compressor Engineering Conference*, Paper C134. West Lafayette, IN, Purdue University.
162. Moaveni, M., Cohen, R., and Hamilton, J.F. (1972). The Prediction of Dynamic Strain in Leaf-Type Compressor Valves with Variable Mass and Stiffness. *Proceedings of the 1972 Compressor Technology Conference*, 156–163. West Lafayette, IN, Purdue University.
163. Mochizuki, T., Ishijima, K., and Asami, K., (1988). Research on Electromagnetic Noise of Rotary Compressor for Household Refrigerator. *Proceedings of the 1988 International Compressor Engineering Conference*, 315–320. West Lafayette, IN, Purdue University.
164. Morimoto, K., Kataoka, Y., Uekawa, T., and Kamiishida, H. (2004). Noise Reduction of Swing Compressors with Concentrated Winding Motors. *Proceedings of the 2004 International Compressor Engineering Conference*, Paper C051. West Lafayette, IN, Purdue University.
165. Morimoto, T., Yamamoto, S., Hase, S., and Yamada, S. (1996). Development of a High SEER Scroll Compressor. *Proceedings of the 1996 International Compressor Engineering Conference*, 317–322. West Lafayette, IN, Purdue University.
166. Motegi, S. and Nahashima, S. (1996). A Study of Noise Reduction in a Scroll Compressor. *Proceedings of the 1996 International Compressor Engineering Conference*, 605–610. West Lafayette, IN, Purdue University.
167. Mutyala, B.R.C. and Soedel, W. (1976). A Mathematical Model of Helmholtz Resonator Type Gas Oscillation Discharges of Two-Stroke Engines. *Journal of Sound and Vibration*, 44 (4): 479–491.
168. Nieter, J.J. and Kim, H.J. (1998). Internal Acoustics Modeling of a Rotary Compressor Discharge Manifold. *Proceedings of the 1998 International Compressor Engineering Conference*, 531–536. West Lafayette, IN, Purdue University.
169. Paczuski, A.W. (2004). Defining Impact. *Proceedings of the 2004 International Compressor Engineering Conference*, Paper C018. West Lafayette, IN, Purdue University.
170. Pandeya, P. and Soedel, W. (1978). Analysis of the Influence of Seat-Plating or Cushioning on Valve Impact Stresses in High Speed Compressors. *Proceedings of the 1978 Compressor Technology Conference*, 169–176. West Lafayette, IN, Purdue University.
171. Papastergiou, S., Brown, J., and MacLaren, J.F.T. (1980). The Dynamic Behavior of Valve Reeds in Reciprocating Gas Compressors — Analytical and Experimental Studies. *Proceedings of the 1980 Compressor Technology Conference*, 263–276. West Lafayette, IN, Purdue University.
172. Papastergiou, S., Brown, J., and MacLaren, J.F.T. (1982 a). Impact Velocities of Valve Reeds. *Proceedings of the 1982 Compressor Technology Conference*, 249–256. West Lafayette, IN, Purdue University.
173. Papastergiou, S., Brown, J., and MacLaren, J.F.T. (1982 b). The Dynamic Behavior of Half-Annular Valve Reeds in Reciprocating Compressors. *Proceedings of the 1982 Compressor Technology Conference*, 240–248. West Lafayette, IN, Purdue University.
174. Park, J.I. and Adams, D.E. (2004). Modeling and Simulation of the Suction Process in a Multi-Cylinder Automotive Compressor. *Proceedings of the 2004 International Compressor Engineering Conference*, Paper C110. West Lafayette, IN, Purdue University.

175. Park, S.K., Kim, H., Lee, C., Youn, H., Cho, S., and Im, G. (1994). The Design of Compressor Mufflers. *Proceedings of the 1994 International Compressor Engineering Conference*, 247–252. West Lafayette, IN, Purdue University.
176. Payne, J.G. and Cohen, R. (1971). Vibration of Flexible Ring Valves Used in Refrigeration Compressors. *Proceedings of the XII International Congress of Refrigeration*, Paper No. 3.30. Washington, DC.
177. Plastinin, P.N. (1984). *Compressor Valves*. Moscow, Baumann University.
178. Pollak, E., Soedel, W., Friedlaender, F.J., and Cohen, R. (1978). Mathematical Model of an Electrodynamic Oscillating Refrigeration Compressor. *Proceedings of the 1978 Purdue Compressor Technology Conference*, 246–259. West Lafayette, IN, Purdue University.
179. Pollak, E., Soedel, W., Cohen, R., and Friedlaender, F.J., (1979). On the Resonance and Operational Behavior of an Oscillating Electrodynamic Compressor. *Journal of Sound and Vibration,* 67 (1): 121–133.
180. Qvale, E.V., Soedel, W., Stevenson, M.J., Elson, J.P., and Coates, D.A. (1972). Problem Areas in Mathematical Modeling and Simulation of Refrigerating Compressors. *ASHRAE Transactions,* 78 (1): 75–84.
181. Ramani, A., Rose, J., Knight, C.E., and Mitchell, L.D. (1994 a). Finite Element Modeling of a Refrigeration Compressor for Sound Prediction Purposes. *Proceedings of the 1994 International Compressor Engineering Conference*, 7–12. West Lafayette, IN, Purdue University.
182. Ramani, A., Rose, J., Knight, C.E., and Mitchiner, R.G. (1994 b) Investigation of Crankcase Resonances in Compressor Models for Sound Prediction. *Proceedings of the 1994 International Compressor Engineering Conference*, 13–17. West Lafayette, IN, Purdue University.
183. Rauen, D.G. and Soedel, W. (1980 a). Compressor Originated Noise in a Diving System. *Proceedings of the 1980 Purdue Compressor Technology Conference*, 249–258. West Lafayette, IN, Purdue University.
184. Rauen, D.G. and Soedel, W. (1980 b). Simulation of Pressure Pulsations in the Breathing Circuit of a Deep-Sea Diving System by Using the Four-Pole Method. *Journal of Sound and Vibration,* 71 (2): 283–297.
185. Reddy, H.K. and Hamilton, J.F. (1976). Accurate Experimental Determination of Frequencies, Mode Shapes and Dynamic Strains in Plate Valves of Reciprocating Compressors. *Proceedings of the 1976 Compressor Technology Conference*, 90–294. West Lafayette, IN, Purdue University.
186. Reynolds, D.D. (1981). *Engineering Principles of Acoustics — Noise and Vibration Control*. Boston, MA, Allyn and Bacon.
187. Richardson, H., Gatecliff, G.W., and Griner, G.C. (1980). Verification of Flapper Suction Valve Simulation Program. *Proceedings of the 1980 Compressor Technology Conference*, 180–184. West Lafayette, IN, Purdue University.
188. Roy, P.K. and Hix, S.G. (1998). Higher Efficiency, Lower Sound, and Lower Cost Air Conditioning Compressor. *Proceedings of the 1998 International Compressor Engineering Conference*, 713–718. West Lafayette, IN, Purdue University.
189. Roys, B. and Soedel, W. (1989). On the Acoustics of Small High-Speed Compressors: A Review and Discussion. *Noise Control Engineering Journal,* 32 (1): 25–34.
190. Schwerzler, D.D. and Hamilton, J.F. (1972). An Analytical Method for Determining Effective Flow and Force Areas for Refrigeration Compressor Valving Systems. *Proceedings of the 1972 Compressor Technology Conference*, 30–36. West Lafayette, IN, Purdue University.

191. Seve, F., Berlioz, A., Dufour, R., Charreyron, F., and Audouy, L. (2000). On the Unbalance Response of a Rotary Compressor. *Proceedings of the 2000 International Compressor Engineering Conference*, 831–838. West Lafayette, IN, Purdue University.
192. Seve, F., Berlioz, A., Dufour, R., Charreyron, M., Peyaud, F., and Audouy, L. (2000). Balancing of a Variable Speed Rotary Compressor: Experimental and Numerical Investigations. *Proceedings of the 2000 International Compressor Engineering Conference*, 839–846. West Lafayette, IN, Purdue University.
193. Shapiro, U. (1992). The Role of Estimating the Stiffness of Rolling Element Bearings in the Analysis of Semi-Hermetic, Twin Screw Compressors. *Proceedings of the 1992 International Compressor Engineering Conference*, 1295–1306. West Lafayette, IN, Purdue University.
194. Shiva-Prasad, B.G. and Wollatt, D. (2000). Valve Dynamic Measurements in a VIP Compressor. *Proceedings of the 2000 International Compressor Engineering Conference*, 361–368. West Lafayette, IN, Purdue University.
195. Simmons, R.A. and Soedel, W. (1996). Surging in Coil Springs. *Proceedings of the 1996 International Compressor Engineering Conference*, 721–727. West Lafayette, IN, Purdue University.
196. Simonich, J. (1978). Mechanical Stresses of Valve Plates on Impact against Valve Seat and Guard. *Proceedings of the 1978 Compressor Technology Conference*, 162–168. West Lafayette, IN, Purdue University.
197. Singh, R. and Soedel, W. (1974). A Review of Compressor Lines Pulsation Analysis and Muffler Design Research, Part I: Pulsation Effects and Muffler Criteria, Part II: Analysis of Pulsating Flows. *Proceedings of the 1974 Compressor Technology Conference*, 102–123. West Lafayette, IN, Purdue University.
198. Singh, R. and Soedel, W. (1976). Fluid Dynamic Effects in Multicylinder Compressor Suction and Discharge Cavities. *Proceedings of the 1976 Purdue Compressor Technology Conference*, 271–281. West Lafayette, IN, Purdue University.
199. Singh, R. and Soedel, W. (1975). On Discretized Modeling of Flow Pulsations in Multicylinder Gas Machinery Manifolds. *Proceedings of the XIII International Congress of Refrigeration*, Paper B.2.15. Moscow, USSR, International Institute of Refrigeration.
200. Singh, R. and Soedel, W. (1978 a). An Efficient Method of Measuring Impedances of Fluid Machinery Manifolds. *Journal of Sound and Vibration*, 56 (1): 105–125.
201. Singh, R. and Soedel, W. (1978 b). Assessment of Fluid-Induced Damping in Refrigeration Machinery Manifolds. *Journal of Sound and Vibration*, 57 (3) 449–452.
202. Singh, R. and Soedel, W. (1978 c). Simulation of Compressor Gas Pulsations. *Progress in Refrigeration Science and Technology*, Vol. II, 740–772. Moscow, International Institute of Refrigeration.
203. Singh, R. and Soedel, W. (1979). Mathematical Modeling of Multicylinder Compressor Discharge System Interactions. *Journal of Sound and Vibration*, 63 (1): 125–143.
204. Singh, R. and Soedel, W. (1985). Forced Vibration and Acoustic Power Radiation Response of a Pipe Transporting Oscillating Compressible Fluids. *Proceedings of the 1985 Pressure Vessels and Piping Conference*, 229–236. Chicago, IL, ASME (American Society of Mechanical Engineering).
205. Smith, I.O. (1978). The Metallography of Impact Fatigue. *Proceedings of the 1978 Compressor Technology Conference*, 111–115. West Lafayette, IN, Purdue University.
206. Smith, J.P., Kiel, D.H., and Hurst, C.J. (1992). Intensity Measurements and Radiated Noise Reduction for a Freon Compressor. *Proceedings of the 1992 International Compressor Engineering Conference*, 947–954. West Lafayette, IN, Purdue University.

207. Soedel, D.T. and Soedel, W. (1992). Development of a Simplified Design Formula for the Low Frequency Cut-Off of a Small Two Volume Silencer. *Proceedings of the 1992 International Compressor Engineering Conference*, 465–1476. West Lafayette, IN, Purdue University.
208. Soedel, W. (1971). Similitude Approximations for Vibrating Thin Shells. *Journal of the Acoustical Society of America*, 49 (5): 1535–1541.
209. Soedel, W. (1972). *Introduction to Computer Simulation of Positive Displacement Compressors*, Short Course Notes. West Lafayette, IN, Purdue University.
210. Soedel, W. (1973). A natural frequency analogy between spherically curved panels and flat plates. *Journal of Sound and Vibration*, 29(4): 457–461.
211. Soedel, W. (1974 a). On the Simulation of Anechoic Pipes in Helmholtz Resonator Models of Compressor Discharge Systems. *Proceedings of the 1974 Purdue Compressor Technology Conference*, 136–140. West Lafayette, IN, Purdue University.
212. Soedel, W. (1974 b). On Dynamic Stresses in Compressor Valve Reeds or Plates During Colinear Impact on Valve Seats. *Proceedings of the 1974 Compressor Technology Conference*, 319–328. West Lafayette, IN, Purdue University.
213. Soedel, W. (1976 a). A Formula for Estimating Dynamic Pressure Changes in Compressor Suction and Discharge Plenums. *Proceedings of the 1976 Compressor Technology Conference*, 322–325. West Lafayette, IN, Purdue University.
214. Soedel, W. (1976 b). The Time Response of a Continuous Gas Column to a Nonharmonic Forcing Flow at its Entrance. *Proceedings of the 1976 Purdue Compressor Technology Conference*, 245–248. West Lafayette, IN, Purdue University.
215. Soedel, W. and Baum, J.M. (1976). Natural Frequencies and Modes of Gases in Multi-Cylinder Compressor Manifolds and Their Use in Design. *Proceedings of the 1976 Purdue Compressor Technology Conference*, 257–270. West Lafayette, IN, Purdue University.
216. Soedel, W. (1978). *Gas Pulsations in Compressors*, Short Course Notes. West Lafayette, IN, Purdue University.
217. Soedel, W. (1980 a). Simple Mathematical Models of Mode Splitting of Hermetic Compressor Shells that Deviate from Axisymmetry. *Proceedings of the 1980 Purdue Compressor Technology Conference*, 259–262. West Lafayette, IN, Purdue University.
218. Soedel, W. (1980 b). Simple Mathematical Models of the Vibration and Force Transmission of Discharge or Suction Tubes as Function of Discharge and Suction Pressures. *Proceedings of the 1980 Purdue Compressor Technology Conference*, 69–73. West Lafayette, IN, Purdue University.
219. Soedel, W. (1982). Analysis of the Motion of Plate Type Suction or Discharge Valves Mounted on Reciprocating Pistons. *Proceedings of the 1982 Compressor Technology Conference*, 257–261. West Lafayette, IN, Purdue University.
220. Soedel, W. (1984). *Mechanics and Design of Compressor Valves*, Short Course Notes. West Lafayette, IN, Purdue University.
221. Soedel, W. (1992). *Mechanics, Simulation and Design of Compressor Valves, Gas Pulsations and Pulsation Mufflers*, Short Course Notes. West Lafayette, IN, Purdue University.
222. Soedel, W. (2004). *Vibrations of Shells and Plates*, 3rd Edition. New York, Marcel Dekker.
223. Soedel, W., Padilla-Navas, E, and Kotalik, B.D. (1973). On Helmholtz Resonator Effects in the Discharge System of a Two-Cylinder Compressor. *Journal of Sound and Vibration*, 30 (3): 263–277.
224. Soedel, W., Strader, D.L., and Mutyala, B.R. (1976). Redesign of Diving Compressor Manifolds to Avoid Excessive Power Usage: A Case Study. *Proceedings of the 1976 Compressor Technology Conference*, 326–340. West Lafayette, IN, Purdue University.

225. Song, H.J. and Soedel, W. (1998). Pulsations in Liquid-Gas Mixtures. *Proceedings of the 1998 International Compressor Engineering Conference*, 367–372. West Lafayette, IN, Purdue University.
226. Sprang, J.O., Nilsson, J.O., and Persson, G. (1980). A Method of Measuring the Impact Velocity of Flapper Valves. *Proceedings of the 1980 Compressor Technology Conference*, 381–385. West Lafayette, IN, Purdue University.
227. Suh, K., Lee, H., Oh, W.S., and Jung, W.H. (1998). An Analysis of the Hermetic Reciprocating Compressor Acoustic System. *Proceedings of the 1998 International Compressor Engineering Conference*, 361–366. West Lafayette, IN, Purdue University.
228. Suh, K.H., Kim, J.D., Lee, B.C., and Kim, Y.H. (2000). The Analysis of a Discharge Muffler in a Rotary Compressor. *Proceedings of the 2000 International Compressor Engineering Conference*, 651–658. West Lafayette, IN, Purdue University.
229. Svendsen, C. (2004). Acoustics of Suction Mufflers in Reciprocating Hermetic Compressors. *Proceedings of the 2004 International Compressor Engineering Conference*, Paper C029. West Lafayette, IN, Purdue University.
230. Svenzon, M. (1976). Impact Fatigue of Valve Steel. *Proceedings of the 1976 Compressor Technology Conference*, 65–73. West Lafayette, IN, Purdue University.
231. Tavakoli, M.S. and Singh, R. (1990). Modal Analysis of a Hermetic Can. *Journal of Sound and Vibration*, 136 (1): 141–145.
232. Tojo, K., Saegusa, S., Sudo, M., and Tagawa, S. (1980). Noise Reduction of Refrigerator Compressor. *Proceedings of the 1980 Compressor Technology Conference*, 235–242. West Lafayette, IN, Purdue University.
233. Tramschek, A.B. and MacLaren, J.F.T. (1980). Simulation of a Reciprocating Compressor Accounting for Interaction between Valve Movement and Plenum Chamber Pressure. *Proceedings of the 1980 Compressor Technology Conference*, 354–360. West Lafayette, IN, Purdue University.
234. Trella, T.J. and Soedel, W. (1971). Sound Radiation from a Reciprocating Compressor Orifice without Valve. *Journal of the Acoustical Society of America*, 49 (6): 1722–1728.
235. Trella, T.J. and Soedel, W. (1972). On Noise Generation of Air Compressor Automatic Reed Valves. *Proceedings of the 1972 Compressor Technology Conference*, 64–68. West Lafayette, IN, Purdue University.
236. Trella, T.J. and Soedel, W. (1974). Effect of Valve Port Inertia on Valve Dynamics, Parts 1 and 2. *Proceedings of the 1974 Compressor Technology Conference*, 190–207. West Lafayette, IN, Purdue University.
237. Tsui, C.Y., Oliver, C.C., and Cohen, R. (1972). Discharge Phenomena of a Two-Dimensional Poppet-Type Valve. *Proceedings of the 1972 Purdue Compressor Technology Conference*, 212–220. West Lafayette, IN, Purdue University.
238. Upfold, R.W. (1972). Designing Compressor Valves to Avoid Flutter. *Proceedings of the 1972 Purdue Technology Conference*, 400–406. West Lafayette, IN, Purdue University.
239. Ventimiglia, J., Cerrato-Jay, G., and Lowery, D. (2002). Hybrid Experimental and Analytical Approach to Reduce Low Frequency Noise and Vibration of a Large Reciprocating Compressor. *Proceedings of the 2002 International Compressor Engineering Conference*, 445–450. West Lafayette, IN, Purdue University.
240. Waltz, J. and Soedel, W. (1980). On the Development of a Reed Valve Impact Fatigue Tester. *Proceedings of the 1980 Compressor Technology Conference*, 96–400. West Lafayette, IN, Purdue University.
241. Wambsganss, M.W., Jr. (1964). Time Characteristics of the Nozzle-Flapper Relay. *ISA Transactions*, 3 (1): 13–19.

242. Wambsganss, M.W. Jr., Coates, D.A, and Cohen, R. (1967). Simulation of Reciprocating Gas Compressor with Automatic Reed Valves. *Simulation Magazine*, 8 (4): 209–214.
243. Wambsganss, M.W. Jr. and Cohen, R. (1967). Dynamics of Reciprocating Compressors with Automatic Reed Valves: Part I — Theory and Simulation and Part II — Experiments and Evaluation. *Proceedings of the XII International Congress of Refrigeration*, 779–799.
244. Wang, S.M., Kang, J.H., Park, J.C., and Kim, C. (2004). Design Optimization of a Compressor Loop Pipe Using the Response Surface Method. *Proceedings of the 2004 International Compressor Engineering Conference*, Paper C088. West Lafayette, IN, Purdue University.
245. Wankel, F. (1963). *Rotary Piston Machines*. London, Iliffe Books.
246. Weiss, H.H. and Boswirth, L. (1982). A Simple but Efficient Equipment for Experimental Determination of Valve Loss Coefficients under Compressible and Steady Flow Conditions. *Proceedings of the 1982 Compressor Technology Conference*, 69–76. West Lafayette, IN, Purdue University.
247. Wollatt, D. (1972). Some Practical Applications of Modern Compressor Valve Technology. *Proceedings of the 1972 Compressor Technology Conference*, 170–179. West Lafayette, IN, Purdue University.
248. Wollatt, D. (1974). A Simple Numerical Solution for Compressor Valves with One Degree of Freedom. *Proceedings of the 1974 Compressor Technology Conference*, 159–165. West Lafayette, IN, Purdue University.
249. Wollatt, D. (1980). Increased Lift for Feather Valves by Elimination of Failures Caused by Impact. *Proceedings of the 1980 Compressor Technology Conference*, 293–299. West Lafayette, IN, Purdue University.
250. Wollatt, D. (1982). Estimating Valve Losses when Dynamic Effects are Important. *Proceedings of the 1982 Compressor Technology Conference*, 13–20. West Lafayette, IN., Purdue University.
251. Yang, Q., Engel, P., Shiva-Prasad, B.G., and Wollatt, D. (1996). Dynamic Response of Compressor Valve Springs to Impact Loading. *Proceedings of the 1996 International Compressor Engineering Conference*, 353–358. West Lafayette, IN, Purdue University.
252. Yee, V. and Soedel, W. (1980). Comments on Blade Excited Rigid Body Vibrations of Rotary Vane Compressors. *Proceedings of the 1980 Purdue Compressor Technology Conference*, 243–248. West Lafayette, IN, Purdue University.
253. Yee, V. and Soedel, W. (1983). Pressure Oscillations During Re-Expansion of Gases in Rotary Vane Compressors by a Modified Helmholtz Resonator Approach. *Journal of Sound and Vibration*, 91 (1): 27–36.
254. Yee, V. and Soedel, W. (1988). On Transfer Slot Design in Rotary Sliding Vane Compressors with Special Attention to Gas Pulsation Losses. *Proceedings of the 1988 International Compressor Engineering Conference*, 539–545. West Lafayette, IN, Purdue University.
255. Yoshimura, T., Akashi, H., Yagi, A., and Tsuboi, K. (2002). The Estimation of Compressor Performance Using a Theoretical Analysis of the Gas Flow Through the Muffler Combined with Valve Motion. *Proceedings of the 2002 International Compressor Engineering Conference*, 507–514. West Lafayette, IN, Purdue University.
256. Yoshimura, T., Akashi, K., Inagaki, K., Kita, I., and Yabiki, J. (1994). Noise Reduction of Hermetic Reciprocating Compressor. *Proceedings of the 1994 International Compressor Engineering Conference*, 253–258. West Lafayette, IN, Purdue University.

257. Yoshimura, T., Koyama, T., Morita, I., and Kobayashi, M. (1992). A Study of the Vibration Reduction of Rolling Piston Type Rotary Compressors. *Proceedings of the 1992 International Compressor Engineering Conference*, 1257–1266. West Lafayette, IN, Purdue University.
258. Yun, K.W. (1996). Changes in Sound Characteristics of Rotary Compressor with Run Time. *Proceedings of the 1996 International Compressor Engineering Conference*, 715–720. West Lafayette, IN, Purdue University.
259. Zhou, W. and Kim, J. (1998). Formulation of Four Poles of Three-Dimensional Acoustics Cavities Using Pressure Response Functions with Special Attention to Source Modeling. *Proceedings of the 1998 International Compressor Engineering Conference*, 549–554. West Lafayette, IN, Purdue University.
260. Zhou, W. and Kim, J. (2001). New Iterative Scheme in Computer Simulation of Positive Displacement Compressors Considering the Effect of Gas Pulsations. *ASME Journal of Mechanical Design*, 123 (2): 282–288.
261. Zukas, J.A. (1982). *Impact Dynamics*. New York, John Wiley.

Author Index

A

Adams, D.E., 195, 327
Adams, G.P., 18, 317
Adams, J.A., 174, 317
Ahn, B., 326
Akashi, H., 195, 317, 332
Akella, S., 195, 317
Albrecht, M., 326
Anantapantula, V.S., 317
Appell, B., 137, 319
Asami, K., 16, 317, 327
Audouy, L., 329

B

Baade, P., 315, 317
Baars, E., 305, 317, 319
Badie-Cassagnet, A., 305, 320
Bae, J.Y. 174, 326
Baum, J.M., 215, 330
Beard, J.E., 11, 317
Benson, R.S., 181, 317
Berlioz, A., 329
Bielmeier, O., 326
Biscaldi, E., 294, 317
Bishop, R.E.D., 113, 318
Bolton, J.S., 326
Boswirth, L., 148, 150, 187, 318, 332
Boyle, R.J., 174, 318
Brablik, J., 181, 195, 318
Bredesen, A.M., 174, 318
Brown, J., 150, 174, 305, 318, 320, 326, 327
Bucciarelli, M., 32, 305, 318
Bukac, H., 17, 174, 318
Buligan, G., 305, 318
Bush, J.W., 19, 318

C

Cerrato-Jay, G., 305, 318, 331
Charreyron, F., 329
Charreyron, M., 329
Chen, L., 195, 319
Chlumski, V., 137, 185, 319
Cho, S., 328
Coates, D.A., 143, 174, 181, 328

Cohen, R., 137, 143, 148, 149, 150, 168, 174, 180, 181, 185, 195, 305, 310, 319, 320, 321, 325, 327, 328, 331, 332
Collings, D.A., 174, 319
Conrad, D.C., 19, 319
Contarini, A., 174, 325
Cossalter, V., 195, 318, 319
Cyklis, P., 305, 319

D

DaLio, M., 318
Darpas, M., 305, 321
DaSilva, A.R., 294, 319
Davidson, R., 318, 320
Daxner, T., 326
Deschamps, C.J., 326
Dhar, B., 325
Dhar, M., 179, 195, 211, 319
DiFlora, M.A., 19, 319, 321
Doige, A.G., 174, 319
Doria, A., 319
Dreiman, N., 16, 305, 319
Dufour, R., 329
Dusil, R., 137, 193, 319, 320

E

El-Geresy, B.A., 326
Elson, J.P., 149, 174, 178, 181, 195, 215, 305, 320, 328
Engel, P., 332

F

Faraon, A., 317, 318
Faulkner, L.L., 44, 320
Fleming, J., 148, 150, 305, 318, 320
Forbes, D.A., 137, 305, 320
Frenkel, M.I., 137, 320
Friedlaender, F.J., 328
Friley, J.R., 174, 320
Fu, W.C., 305, 320
Funer, V., 305, 320
Furukawa, H., 320
Futakawa, A., 174, 193, 320

G

Gardonio, P., 318
Gatecliff, G.W., 150, 174, 320, 328
Gatley, W.S., 195, 305, 321
Gavric, L., 305, 321
Giacomelli, E., 148, 321
Gilliam, D.R., 19, 321, 324
Giorgetti, M., 148, 321
Giusto, F.M., 318, 319
Gluck, R., 305, 321
Graff, K., 193, 321
Griner, G.C., 150, 320, 328
Groth, K., 195, 321

H

Hall, A.S., 11, 317
Hamilton, J.F., 15, 67, 95, 101, 135, 139, 143, 168, 174, 181, 296, 305, 307, 313, 314, 317, 320, 321, 322, 324, 327, 328
Hase, S., 327
Hatch, G., 113, 321
Herfat, A.T., 305, 321
Herrick, K., 16, 319
Hix, S.G., 195, 328
Holowenko, A.R., 109, 321
Hort, W., 151, 321
Housman, M.E., 318
Hsu, M.P., 32, 44, 51, 321
Huang, D.T., 44, 45, 321
Huang, Z.S., 195, 319
Hur, K.B., 325
Hurst, C.J., 329
Hussein, I.J., 326
Hwang, I., 305, 321

I

Ih, J.G., 305, 322
Im, G., 328
Inagaki, K., 332
Ishijima, K., 317, 327

J

Jang, S.H., 305, 322
Jeong, H., 326
Joergensen, S.H., 174, 322
Johansson, R., 138, 193, 319, 320, 322
Johnson, C.N., 15, 95, 296, 307, 322
Johnson, D.C., 113, 318
Johnson, O., 195, 322
Jones, J.D., 326
Joo, J.M., 174, 322
Jung, W.H., 331

K

Kamiishida, H., 327
Kang, J.H., 332
Kataoka, Y., 327
Kawai, H., 294, 322
Kelly, A.D., 19, 113, 322
Kerr, S.V., 181, 326
Khalifa, H.E., 174, 322
Kiel, D.H., 329
Killman, I.G., 133, 148, 150, 151, 160, 162, 322
Kim, B.C., 305, 325
Kim, C., 321, 332
Kim, G.K., 322
Kim, H., 328
Kim, H.J., 195, 262, 264, 270, 283, 290, 322, 327
Kim, J., 32, 178, 195, 322, 323, 325, 333
Kim, J.D., 331
Kim, J.S., 186, 187, 323
Kim, S.H., 322, 323
Kim, S.J., 326
Kim, Y.H., 331
Kim, Y.K., 32, 57, 58, 323
Kita, I., 322, 332
Knight, C.E., 19, 113, 322, 328
Koai, K., 18, 123, 195, 323, 324
Kobayashi, M., 333
Kotalik, B.D., 181, 330
Koyama, T., 333
Kristiansen, U., 135, 324
Kumar, K., 16, 324
Kwon, B., 321
Kwon, B.H., 324

L

Lady, E.R., 174, 320
Lai, P.C.C., 195, 208, 296, 298, 299, 303, 324
Lampugnani, G., 317
Laub, J.S., 138, 324
Laursen, M.B., 70, 305, 324
Laville, F., 264, 267, 324
Lee, B.C., 325, 331
Lee, C., 327
Lee, D.S., 325
Lee, H., 305, 324, 331
Lee, H.K., 195, 325
Lee, J.H., 32, 195, 325
Lee, J.K., 305, 325
Lee, J.M., 305, 325
Lee, S.J., 325
Lee, U.S., 325

Leemhuis, R.S., 11, 325
Lenz, J.R., 174, 325
Lenzi, A., 319
Leyderman, A., 324
Libralato, M., 174, 325
Liu, X., 174, 322
Liu, Z., 195, 325
Lowery, D., 318, 331
Lowery, R. L., 185, 305, 310, 325
Lu, J., 324
Luszczycki, M., 174, 179, 325, 326

M

Ma, Y.C., 174, 195, 305, 326
Machu, G., 174, 326
MacLaren, J.F.T., 151, 174, 181, 195, 318, 326, 327, 331
Madsen, P., 174, 326
Marler, M., 324
Marriott, L.W., 67, 326
Masters, A.R., 294, 326
Matos, F.F.S., 174, 326
McLaren, R.J.L., 174, 326
Medira, V., 314, 327
Min, O.K., 195, 326
Mitchell, J., 305
Mitchell, L.D., 137, 320, 328
Mitchiner, R.G., 328
Moaveni, M., 174, 327
Mochizuki, T., 314, 327
Morimoto, K., 314, 327
Morimoto, T., 17, 327
Morita, I., 333
Motegi, S., 17, 305, 327
Mutyala, B.R.C., 195, 327, 330

N

Nahashima, S., 17, 305, 327
Namura, K., 193, 320
Nieter, J.J., 195, 324, 327
Nilsson, J.O., 331

O

Oh, S.K., 322
Oh, W.S., 331
Ohta, T., 322
Oliver, C.C., 151, 331

P

Paczuski, A.W., 187, 327
Padilla-Navas, E., 181, 330

Pandeya, P., 187, 193, 327
Paone, N., 318
Papastergiou, S., 174, 326, 327
Park, J.C., 332
Park, J.I., 195, 327
Park, J.S., 325
Park, S.K., 195, 328
Park, S.O., 324
Parkovic, B., 327
Payne, J.G., 174, 181, 328
Peracchio, A., 324
Persson, G., 138, 322, 331
Peyaud, F., 329
Plastinin, P.N., 137, 151, 328
Pollak, E., 13, 328
Prata, A.T., 326

Q

Qvale, E.V., 195, 328

R

Ramani, A., 81, 87, 328
Rauen, D.G., 195, 328
Reddy, H.K., 174, 328
Revel, G.M., 318
Reynolds, D.D., 101, 288, 290, 328
Richardson, H., 150, 174, 320, 328
Rose, J., 328
Roy, P.K., 195, 324, 328
Roys, B., 1, 314, 328

S

Saegusa, S., 331
Sarti, S., 317
Sasano, H., 322
Schwerzler, D.D., 139, 168, 174, 181, 328
Seve, F., 16, 329
Shapiro, U., 18, 329
Shin, C., 326
Shiva-Prasad, B.G., 305, 329, 332
Silveira, M., 317
Simmons, R.A., 112, 329
Simonich, J., 174, 329
Singh, R., 19, 44, 195, 215, 230, 249, 252, 294, 305, 329, 331
Smith, A.V., 322
Smith, I.O., 138, 329
Smith, J.P., 294, 329
Smoljan, B., 327
Soedel, D.T., 195, 222, 223, 270, 279, 330

Soedel, W., 1, 11, 15, 18, 19, 28, 30, 31, 32, 39, 41, 43, 44, 45, 49, 50, 51, 55, 57, 58, 59, 62, 65, 68, 69, 71, 73, 75, 76, 77, 81, 82, 83, 87, 90, 93, 97, 100, 113, 116, 123, 135, 137, 138, 139, 149, 150, 151, 162, 166, 174, 178, 179, 181, 184, 186, 187, 192, 193, 195, 198, 208, 211, 215, 222, 223, 226, 230, 249, 252, 262, 264, 267, 270, 278, 279, 283, 290, 294, 296, 298, 299, 303, 305, 314, 317, 319, 320, 321, 322, 323, 324, 325, 327, 328, 329, 330, 331, 332
Song, H.J., 195, 331
Sprang, J.O., 138, 331
Steinrück, P., 326
Stevenson, M.J., 328
Strader, D.L., 330
Sudo, M., 331
Sugimoto, S., 317
Suh, K., 195, 331
Suh, K.H., 195, 331
Svendsen, C., 195, 331
Svenzon, M., 137, 193, 331

T

Tagawa, S., 331
Tanaka, H., 317
Tauchmann, R., 305, 320
Tavakoli, M.S., 19, 44, 331
Tojo, K., 178, 331
Tomasini, E.P., 318
Tramschek, A.B., 174, 181, 318, 326, 331
Trella, T.J., 150, 151, 178, 223, 294, 331
Tsui, C.Y., 151, 331

U

Uekawa, T., 327
Ucer, A.S., 181, 317

Ukrainetz, P., 185, 305, 321
Upfold, R.W., 174, 331

V

Venkateswarlu, K., 317
Ventimiglia, J., 305, 331

W

Waltz, J., 137, 331
Wambsganss, Jr., M.W., 143, 148, 149, 150, 168, 174, 180, 181, 331, 332
Wang, S.M., 96, 332
Wankel, F., 11, 332
Weadock, T.J., 174, 319
Weiss, H.H., 150, 332
Winslett, C.E., 322
Wollatt, D., 113, 151, 174, 305, 321, 329, 332

Y

Yabiki, J., 332
Yagi, A., 317
Yamada, S., 327
Yamamoto, S., 327
Yang, Q., 113, 332
Yee, V., 15, 224, 332
Yoshimura, T., 195, 305, 317, 332, 333
Youn, H., 327
Yun, K.W., 305, 308, 333

Z

Zhou, W., 195, 333
Zukas, J.A., 193, 333

Subject Index

A

Acoustics,
  one dimensional, 227–284
  three dimensional, 285–293
  two dimensional, 296–304
Anechoic termination, 208–219, 246
Attenuator, 271–276, 282

B

Bernoulli effect, 147, 148
Back pressure, 177, 182, 263
Barreling, 41, 42
Bending moment resultants, 26, 27
Bessel,
  equation, 299
  functions, 291, 300, 301
Boundary conditions,
  casing, 84
  cylindrical shell, 32
  housing, 30

C

Casing, 5, 81–94, 101–119
  isolation, 9, 101–119
  material, 2
  modes, 87–94
Chladni figures, 173
Complex receptances, 116
Compression process, 1, 125–128
  isothermal, 2
  isentropic, 2
Compressor,
  dynamic, 1
  high-side, 6
  low-side, 6
  multicylinder, 211–218, 250, 251
  multiple discharges, 18
  multistage, 138, 139
  positive displacement, 1
  reciprocating, 11–13, 107–109, 125–128
  rolling piston, 15
  rotary vane, 14, 15, 224, 225
  scotch-yoke, 13
  screw, 17, 123, 124
  scroll, 16, 123
  slider valve, 123
  swing, 13
Convergence, 100
Coordinates,
  Cartesian, 20, 84
  curvilinear, 81
  cylindrical, 20
  polar, 20
  spherical, 20, 21
Curvature, 24, 25, 41–44

D

D'Alembert's solution, 231
Damping,
  dissipative, 70, 71, 178, 195
  friction, 76–80
  hysteretic, 71
  structural, 62
  valve, 178, 179
Decible scale, 294–296
Decoupling volume, 4
Diagnostic modifications, 310–312
Discharge and suction
  manifolds, 195–284
  mufflers, 195–284
  tubes, 95–100
Displacement, 24, 25
Dynamic absorber, 70, 72–75

E

Electromotor, 313
End caps, 43–46
Energy expressions, 28, 83
Equations of motion
  casing, 81–86
  elastic solid, 81–86
  housing, 19–30
  pipe, 95, 97, 100
  shell, 28, 29
  valve reed, 162

F

Finite gas column, 224–226, 234–238, 256–261
Flexible tubes, 5, 311
Forced response,

339

casing, 81–94
housing, 61–80
impact, 90, 91
modal expansion, 61–63
pipe, 98, 99
steady state,
    harmonic, 63, 64, 67–69
    periodic, 67–69, 92–94
valve, 165–167
Four pole concept, 236–238
Friction damping, 76–80
Fundamental form, 21

G

Gas column,
    modes, 224, 258
    natural frequencies, 258
    response, 256–261
Gas natural frequencies and modes,
    annular cylinder, 296–298
    circular disk, 299–302
    cylinder-disk combination, 303, 304
Gas pulsation,
    back pressure, 177
    cavity, 296–304
    manifolds, 198–224

H

Hamilton's principle, 28
Hankel function, 292
Helmholtz,
    resonator, 198–201
    simplification, 198–223
Housing, 6–8
    asymmetric, 47–54
    residual stresses, 58, 59
    stiffening, 55–57

I

Intercoolers, 252
Images,
    method of, 306, 307
Impact, 90, 91
    colinear, 187–193
    failure, 137
Isolation springs, 9, 101–121
    excitation, 104–109
    idealized, 101–111
    reciprocating compressor, 107–111
    rotating unbalance, 104–106
    standard approach, 101–103

L

Lamé parameters, 21, 22
Life expectancy, 138
Low pass filter, 4, 222, 223
Lubrication oil, 314, 315

M

Material, 2, 137, 138
Measurements, 305–312
    exterior, 309
    identification, 310–312
    interior, 309
    repeatability, 8, 307, 308
    sound, 294, 305, 306
Membrane resultants, 26, 27
Modal
    damping, 65, 66, 70, 71, 76–80
    expansion, 61–63, 88–94, 259–262
    forcing, 65, 66
    mass, 65, 66
    series, 61–63, 88–94, 259–262
    stiffness, 65, 66
Mode splitting, 47–54
    experimental modes, 51–54
    mass variation, 47–49
    ovalness, 50
    stiffness variation, 47–49
Momentum, 91
Monopole source, 288, 289, 306
Mufflers, 195–283
    anechoic pipe, 209–221, 246–248, 252–255
    attenuators, 271–275, 282
    branched tubes, 242–245
    cut-off frequency, 222
    dissipative, 195
    four poles, 236–248
    Helmholtz resonator, 198–223, 280–283
    impact noise, 197
    impedance, 249
    low pass filter, 196
    multicylinder, 211–214, 250, 251
    reaction, 195
    resonators,
        series, 201–222
        side branch, 196
    synthesis, 280, 281
    turbulence, 198
    two cylinder, 211–214
    wave equation,
        damped, 231
        one dimensional, 227–233
        three dimensional, 285–287
        undamped, 232, 233

Subject Index

N

Natural frequencies and modes,
  casing, 87, 88
  cylindrical shell, 32–40
  housing, 19–46
  modification, 55–60
  pipe, 96–100
  ring, 57
  stiffened shell, 55–58
  valve, 151–164, 172
Node
  lines, 31, 38, 52, 53, 172, 173
  points, 31

O

Oil sump, 10
Orifice,
  equation, 149
  flow meter, 148

P

Pipe excitation 98, 99
Plate, 20, 162
Poppet valve, 151–161

R

Radiation of sound, 288–293
  circular cylindrical housing, 290–293
Receptances, 44, 55, 56, 73–75,
  113–116
Reference surface, 20, 21
Repeatability,
  measurements, 8, 307, 308
  sound, 8, 58
Residual stresses, 58–60
Response, 61–80, 89–94
  harmonic, 63, 64, 67–69
  impact, 90, 91
  periodic, 67–69, 92–94
Resonances, 63–75, 95
  damping, 70–80, 96, 118
  detuning, 55, 119–121
Resultants,
  bending moment, 27
  Kirchhoff shear, 30
  membrane force, 26, 27
  transverse shear, 27
Rocking vibration, 304

S

Shear modulus, 23
Shell,
  homogeneous, 19
  isotropic, 19
  resonances, 63–80
Simulation, 180–182
Sound
  cancellation, 306
  rooms, 305, 306
  sources, 1–18
Stiffening, 55–57
Strain, 22–25
Stress, 22, 23
  residual, 58, 59
Suction
  manifolds, 196–226
  mufflers, 196–284
  tubes, 95–100
Surging in coil springs, 112–122
  housing resonances, 119–122
  vibration isolation, 101–122

T

Thermodynamic process, 125–128, 132
Transmission loss, 262–284
  definition of, 262
Transmission mechanism, 98, 99
Tubes, 95–100
  curvature, 97
  force transmission, 98–100
  inertia, 269
  mass flow effect, 100, 101
  pressure changes, 96
  straight, 95
Turbulence, 230

U

Unbalance,
  harmonics, 109
  primary, 109
  reciprocating, 107–110
  rotating, 104–107
  secondary, 109

V

Valves, 2, 3, 123–194
  automatic, 2
  Bernoulli effect, 3, 147
  closing, 158–160

damping, 154, 178, 179
design, 125–138
dynamics, 151–181
    plate, 162–173
    poppet, 151–161
    reed, 162–173
effective spring rate, 132
fatigue, 136–138
floating, 2, 133
flow area, 129–131, 139–142, 148, 149
flow force, 132, 168
flow velocity, 174–177
flutter, 2, 3, 132
force area, 143, 150
gate, 2, 3
impact, 132, 134, 137, 185–192
    colinear, 187–193
    failure, 137
    seat, 158–162
lift, 131
material, 137, 138
measurements, 148–150
multicylinder, 138
natural frequencies and modes, 163, 164
noise, 195–198
orifices, 139–141
    in series, 140
    parallel, 141
passage mass, 223, 224
poppet, 151–162
port areas, 129, 130
pumping oscillations, 174–177
reed, 134, 135, 162–173, 185
reliability, 136–138, 182–193
response to forcing, 165–167
ring, 125, 142–143, 151
setting velocity, 136, 137
simulation, 180, 181
sticktion, 148
stop, 131, 178–180
stresses,
    bending, 182–184
    impact, 185–194
    measurement, 185
thermodynamic consideration, 125–128
timing, 126–128
types, 123, 124
Vibration,
    isolation, 101–121
    localization, 44–46
Viscosity, 230
Volume penetration, 276–278

W

Wave equation,
    acoustics, 285–287
    one dimensional, 86, 227–284
    solids, 81–86
    three dimensional, 81–85
    two dimensional, 296–304
Waves,
    acoustic, 227–293, 307
    compression, 186–192
    shear, 187
    tension, 186–192

Related Titles

Vibration and Shock Handbook, 1580
Clarence De Silva, University of British Columbia, Vancouver, Canada
ISBN: 0849315808

Human Response to Vibrations, TF1483
Neil J. Mansfield, Loughborough, Leicestershire, U.K.
ISBN: 041528239X

Vibration Simulation Using MATLAB and ANSYS, C2050
Michael R. Hatch, Hatch Consulting, Mountain View, CA
ISBN: 1584882050

Ultrasonic Data, DK8307
Dale Ensminger, Consultant, Columbus, OH
ISBN: 0824758307